Fundamentals of
Plant Breeding and
Hybrid Seed Production

Rattan Lal Agrawal BSc, (Ag) MSc (Ag)
Former Professor
Department of Genetics and Plant Breeding
GB Pant University of Agriculture amd Technology
Pantnagar, Uttarakhand

Oxford & IBH Publishing Co. Pvt. Ltd.
New Delhi
(*A Unit of* CBS Publishers & Distributors Pvt Ltd)

CBSPD

CBS Publishers & Distributors Pvt Ltd

New Delhi • Bengaluru • Chennai • Kochi • Kolkata • Lucknow • Mumbai
Hyderabad • Jharkhand • Nagpur • Patna • Pune • Uttarakhand

Fundamentals of
Plant Breeding and
Hybrid Seed Production

ISBN: 978-81-204-1217-0

1998, Rattan Lal Agrawal

First Edition 1998
  Reprint: 2017, **2024**

OXFORD & IBH

New Delhi
(*A Unit of* CBS Publishers & Distributors Pvt Ltd)

Published by **Satish Kumar Jain** and produced by **Varun Jain** for

CBS Publishers & Distributors Pvt Ltd

4819/XI Prahlad Street, 24 Ansari Road, Daryaganj, New Delhi 110 002, India.
Ph: 011-23289259, 23266861   Website: www.cbspd.com
                           e-mail: delhi@cbspd.com

Corporate Office: 204 FIE, Industrial Area, Patparganj, Delhi 110 092
Ph: 011-4934 4934   Fax: 011-4934 4935
                        e-mail: publishing@cbspd.com; publicity@cbspd.com

Branches

- **Bengaluru:** Seema House 2975, 17th Cross, KR Road, Banasankari 2nd Stage, Bengaluru 560 070, Karnataka, India
  Ph: +91-80-26771678/79   Fax: +91-80-26771680   e-mail: bangalore@cbspd.com
- **Chennai:** 7, Subbaraya Street, Shenoy Nagar, Chennai 600 030, Tamil Nadu, India
  Ph: +91-44-26680620, 26681266   Fax: +91-44-42032115   e-mail: chennai@cbspd.com
- **Kochi:** 42/1325, 1326, Power House Road, Opp KSEB, Power House, Ernakulum Kochi 682 018, Kerala, India
  Ph: +91-484-4059061-65,67   Fax: +91-484-4059065   e-mail: kochi@cbspd.com
- **Kolkata:** 147, Hind Ceramics Compound, 1st Floor, Nilgunj Road, Belghoria, Kolkata-700056, West Bengal, India
  Ph: +033-25633055, 033-25633056   e-mail: kolkata@cbspd.com
- **Lucknow:** Basement, Khushnuma Complex, 7 Meerabai Marg (Behind Jawahar Bhawan), Lucknow-226001, UP, India
  Ph: +0522-4000032   e-mail: tiwari.lucknow@cbspd.com
- **Mumbai:** PWD Shed, Gala no 25/26, Ramchandra Bhatt Marg, Next to JJ Hospital Gate no. 2, Opp. Union Bank of India,
  Noorbaug, Mumbai-400009, Maharashtra, India
  Ph: 022-66661880/89   e-mail: mumbai@cbspd.com

Representatives

• Hyderabad	0-9885175004	• Jharkhand	0-9811541605	• Nagpur	0-8692091830
• Patna	0-9334159340	• Pune	0-9664372571	• Uttarakhand	0-9716462459

Printed at SRK Graphics, Delhi (India)

And he gave it for his opinion that whoever could make two ears of corn or two blades of grass to grow upon a spot where only one grew before, would deserve better of mankind, and do more essential service to his country.

Gulliver's Travels
Jonathan Swift (1726)

Preface

Man's intervention in the evolution of plants, to adapt them more closely to his needs, constitutes the practice of plant breeding. Historically regarded as an art and later both as an art and a science, plant breeding is now considered an integrating technology (Riley, 1978). Be what it may, it certainly is an interdisciplinary science which involves co-operation between various scientists in the related disciplines, between plant breeding research centres, genetic resources centre(s) and other agencies involved in the testing and evaluation of new varieties. The success of such an approach is amply evident from the achievements of co-operative crop improvement research programmes at both the national and global level. The large number of improved varieties, promising breeding material evolved under the auspices of co-operative programmes has helped several nations to raise their agricultural production and ensure better food security. This has also served as a vehicle for transformation of agriculture and adoption of new emerging farming technologies by farmers. Organized seed production has further helped to realize the dividends from this endeavour. This is what plant breeding and seed production are all about, and known about.

With several new developments, such as hybrids of vegetable crops, mass micropropagation through tissue culture of some horticultural and plantation crops, advancements in the field of genetic engineering, one may hope that plant breeding will touch new heights. Also, the enactment of plant variety protection legislation in several countries has provided the impetus necessary to private investment on a significant scale in plant breeding research and organized seed production.

Plant-breeding literature has become too voluminous over the period of years to readily assimilate. Yet one needs to keep himself abreast with the latest developments. This book is an earnest attempt to explain the fundamental principles and to provide an up-to-date exposure to the subject. Besides discussion on conventional breeding procedures, considerable attention is given to current breeding approaches and possible solutions suggested. The chapters on tissue culture and genetic engineering cover all the essential aspects related to plant breeding. Students are thereby updated on recent developments. Chapters pertaining to variety release, variety maintenance, plant variety protection and hybrid seed production have been included from the viewpoint of the seed industry. Also, a plant breeder is often required to shoulder the responsibility of maintaining nucleus and breeder seed stocks, and to impart training in seed production, especially hybrid seed production. I earnestly hope that the aggregate information compiled in this book will be of immense practical value to students and all others engaged in plant breeding research and seed production.

Thanks are due to my daughter Radhika, herself a student of Botany, who assisted me in various ways during the preparation of the manuscript. Thanks are also due to Dr. G.K. Garg for vetting chapters on Tissue Culture and Genetic Engineering, Dr. Arvind Shukla for vetting the chapter on Crop Germplasm, Dr. Anil Kumar for reading proofs and Mr. V.M. Lal for preparing line drawings. I am also thankful to G.B. Pant University of Agriculture and Technology, Pantnagar for necessary encouragement to complete this work. I also wish to record my sincere gratitude to (Ms) Margaret Majithia for thoroughly editing the manuscript and making valuable suggestions for its further improvement.

A-731 Indira Nagar   RATTAN LAL AGRAWAL
Lucknow-226016, INDIA
May, 1998

Contents

PART THREE : SEED PRODUCTION

PART I

INTRODUCTORY TOPICS

1

Introduction

Goals of Plant Breeding

The quotation from *Gulliver's Travels*, Jonathan Swift (1726) with which this book begins sets the ultimate goal for a plant breeder. A plant breeder aims at developing varieties which will perform better than existing ones in one or more ways in relation to their economic use. The various goals of plant breeding in order of priority thus are :

1. Raising the yield ceiling, that is, increasing yield of grain, fodder, fibre, oil, sugar, vegetables, fruits, timber, etc.
2. Development of varieties suited to cropping systems/ecosystems and integrated farming systems. For example, improved varieties for dryland areas, flood areas, drought areas, high-altitude areas, sodic lands, water-logged conditions, deserts, etc. Development of early maturing varieties for increased cropping intensity and very high disease- and pest-tolerant varieties which are more ecofriendly is also an important goal.
3. Improving the nutritional quality.

What is Plant Breeding ?

Plant breeding is the applied science of botany that deals with the hereditary (genetic) improvement of crop plants of economic importance. It is essentially an interdisciplinary science in that it uses the knowledge and techniques from many basic science areas. Its contribution to agricultural progress is measured not only by information, but by material products as well, such as improved varieties, hybrids, clones, etc.

According to Vavilov 'in effect breeding is man's interference in the morphological formation of animals and plants'. In other words, 'it is evolution directed by the will of man'. He maintained that the Russian word *selektsia*, that is, the study of selection in the wider sense, defines with sufficient exactitude and embraces in its entirety the general content of that system which we have in mind. This word perhaps has a truer claim than the French description *améliorationdes plantés cultivées et du bétail* or the German word *züchtung* which literally means breeding, rearing, growing, cultivation. The English expression plant or animal breeding is no better. The production of new varieties and 'selection' involves, more than variability and heredity.

Selection necessitates the consideration of original material and of the physiological, biochemical and other varietal differences as well as the complex process of segregation and creation of desired forms.

Smith (1966) defined plant breeding 'as the art and science of improving the genetic pattern of plants in relation to their economic use. Usually and ideally it involves effective cooperation with and help from workers in somewhat remote disciplines.'

Poehlman and Borthakur (1969) in their book on the subject defined plant breeding 'as the art and science of changing and improving the heredity of plants.'

Art and Science of Plant Breeding

The plant breeder does not have to produce something out of nothing. He, however needs to provide right direction. The art of plant breeding implies the ability of a plant breeder to discern by observation the differences in the plant materials he handles and the selections he makes for further increase. As an art its origin goes back to the beginning of agriculture. A large number of most valuable varieties of fruit trees of the present time are the creation of the art of the plant breeder. Many of the best varieties of pears, apples, grapes originated in remote ages. The great variety of new ornamental plants as exemplified in new varieties of roses, dahlias, chrysanthemums, tulips, cannas, and gladioli, bears eloquent testimony to breeding as an art.

The science of plant breeding, on the other hand, implies the application of principles of genetics embodied in the study of heredity and variation, cytology and the techniques (screening techniques) and knowledge gained in related disciplines, such as plant pathology, entomology, plant physiology, soil science, etc. (described under the relationship of plant breeding to other sciences). The science of plant breeding is the study of developing varieties for human needs. As a science, the scope of breeding is defined by the need for obtaining practical results.

The definition of plant breeding as a science rests chiefly on the fact that it not merely selects a certain part of other sciences, but transforms and differentiates them to the extent necessary for achievement of its objective, that is, a new variety. On the basis of discipline it works out its own methods and establishes a regular procedure to be followed in the development of a new variety.

Relationship of Plant Breeding to other Sciences

Botany: Plant breeding is the applied branch of botany. A plant breeder must be an accomplished botanist, very well versed with the taxonomy, morphology, anatomy and reproductive behaviour (floral biology) of the given crop. This knowledge is required to his day-to-day work, in germplasm classification and indexing, maintenance of germplasm and the use of germplasm in the selection of suitable sources as donor parents for achieving specific breeding objectives, as well as in interspecific and intergeneric hybridization, etc.

Genetics and Cytology: A precise knowledge of genetic and cytological behaviour of important economic characters is necessary for manipulating the heredity of plants which a plant breeder attempts in breeding improved crop varieties possessing the desired combination of characters. The basic principles of genetics embodied in the study of heredity and variation, and cytological behaviour forms the very basis of the methods employed for improving the heredity of crop varieties.

Biotechnology: Advances in the field of genetic engineering and plant tissue culture techniques have provided new means to the plant breeder for introducing single gene into plant without changing any other constituent of the plants genome. Also, it has made it possible to transfer gene(s) between plant species that can not interbreed.

Plant Physiology: A knowledge of plant physiology is important for manipulating the plant type according to maximum physiological efficiency and also for determining plant type response to diverse situations, such as drought, cold, salinity, etc. The application of principles of plant physiology is necessary in breeding varieties for specific adaptation, namely, high plant population, drought resistance, cold resistance, salinity tolerance, heat tolerance, etc.

Agronomy: A knowledge of agronomy is necessary for raising good crops, *per se*. the first requirement and in breeding for specific agronomic practices, namely, minimum tillage, low fertility, dryland conditions, multiple-cropping systems, mechanized harvesting and intensive farming systems.

Plant Pathology: A knowledge of host-pathogen relationships and techniques for screening and evaluation of plant materials in respect of important crop diseases is necessary in breeding disease-resistant varieties. A plant breeder must have adequate training in plant pathology.

Entomology: A knowledge of host-insect relationships and techniques for screening and evaluation of plant materials in respect of important insect pests is necessary in breeding insect-resistant varieties.

Biochemistry: A knowledge of biochemistry, more particularly the techniques for evaluation of nutritive quality of plant materials is necessary in breeding for quality characteristics, and in some cases development of varieties devoid of toxic substances.

Microbiology: A knowledge of microbiology is important in breeding varieties for high symbiotic associations of micro-organisms, such as *Rhizobium* and *Azospirillum*, in crop plants.

Biometry and Statistics: A knowledge of biometry is necessary for estimation of genetic parameters of populations, effects of inbreeding and cross-breeding, genotype-environment interaction and selection response in respect to quantitative characters. This information enables a plant breeder to predict the relative potential and limitations of alternate breeding methods.

A knowledge of statistics helps in proper planning of experiments, collection and evaluation of data and its interpretation.

Contribution of Plant Breeding

Most of the cultivated species of crop plants have been modified by plant breeding to increase their usefulness as food, feed, and industrial products. Older varieties are replaced by newer combinations on continued basis. The net result is the development of higher yielding types better adapted to the environment of the area of their culture.

The improvements effected by the plant breeder are gradual and progressive in that any single new variety represents but a small advance over its predecessor. Improved crop varieties give the grower an opportunity for greater profits and consequently lead to innovation and the adoption of new agricultural technology. Sometimes the advance is spectacular and can transform the importance of a crop in an area. Plant breeding thus has contributed immensely towards greater food security, productivity and improved quality of the economic product. The major achievement has been in developing varieties resistant to various diseases, insects and abiotic stresses.

2

Historical Résumé

Remote Past

The selection of useful plants from among wild plants began ages ago. The selected materials served as a base and were maintained through further selection. The selection of a particular type of plant was largely a matter of experience and an art. This is evident from the instructions that can be found in the works of Theophrastus (372-28 B.C.), Virgil (70-19 B.C.) and Columella (1st Century A.D.).

Sexuality in plants: The knowledge of sexuality in plants in Egypt and Assyria dates back to 860 B.C., as is evident from an Assyrian relief on which the pollination of female date palms from the male date palm trees is symbolically depicted. Caeselpinus wrote about sexuality in plants in 1583. Camerarius (1694) rediscovered sexuality in plants and published his essay *De Sexuplantarum* in 1694. Thereafter, the knowledge of sex in plants expanded and the crossing of new types through artificial hybridization began to take shape. Thomas Fairchild (1719) crossed two species within the genus *Dianthus*.

Early History

Linnaeus published *Species Plantarium* in 1753, which provided the basis for classification of plants. Much work by botanists, cytologists, geneticists and statisticians during the early period has helped plant breeding directly or indirectly. The most significant developments in the various disciplines having bearing on plant breeding are described below.

(a) Heredity

Lamarckism : Jean Baptiste Lamarck (1744-1829) propounded the *theory of acquired characters*, which states that,

i) variation in an individual is brought about by conscious effort, reaction to environmental effects and effects of use and disuse; and

ii) heredity carries forward the change that is acquired during the lifetime of an individual.

This theory was later disproved by Weismann.

Darwinism : Charles Darwin (1809-1882) propounded the *theory of natural selection* in his famous book *Origin of Species*. He also wrongly believed in the heritability of acquired characters. His theory propounded that

i) variation is constant in nature,

ii) overproduction of offspring brings about a struggle for existence,

iii) natural selection operates by elimination of the unfit and survival of the fit, and

iv) heredity continues the line of survivors.

Darwin tried to explain the transmission of acquired characters by supposing that hereditary materials were drawn from all parts of an organism, which he termed *pan-genes*. These formed the germ cells which gave rise to a new individual. Darwin also recognized and described spontaneous changes or sports. From the present point of view, his greatest service consisted in that he opened the limitless influence of human reason and will on the variability of plants and animals. His book *Cross and Self-fertilization in the Vegetable Kingdom* appeared in 1876 wherein he concluded that cross-fertilization is generally beneficial while self-fertilization is injurious.

Weismannism: August Weismann (1834-1914), a follower of Darwin, conducted controlled experiments on mice to verify the heritability of acquired characters. He cut off the tails of mice when they were young, generation after generation for 22 generations. The tails still persisted. Weismann thus disproved earlier theories related to heredity of acquired characters and propounded the *germplasm theory*. The germplasm theory states that

i) the hereditary material is separated into the germplasm at the very early stage of development of an individual, while the rest of the body (somatoplasm) is only a house for the germplasm;

ii) any change or mutilations affecting the somatoplasm and not reaching the germplasm is not heritable; and

iii) while the somatoplasm dies at the death of each individual, the germplasm lives on. It is immortal.

The *germplasm theory* laid the foundation of modern genetical thought.

Mendelism: Gregor John Mendel (1822-1884). His now famous work, *Mendel's Laws of Inheritance* was published by the Natural History Society of Brünn (now Brno) in 1866. He crossed different varieties of peas and studied their progenies. Instead of taking the parents or the progenies as units for study, he chose individual characters and thus introduced a new concept, that an organism is a composite of a large number of independently behaving unit characters. Mendel's laws of inheritance are discussed in Chapter 8.

Unfortunately, the great importance of Mendel's work was not recognized until after a gap of 35 years when in 1900 de Vries, Correns and Tschermak independently discovered his work. The rediscovery of *Mendel's Laws of Inheritance* led to a new era of rapid development of the science of heredity.

Genetics: Bateson (1906) coined the term genetics to cover all matters concerning heredity and variation. Bateson and Punnet (1906) showed that sometimes traits tend to be inherited together and do not segregate easily.

Linkage and chromosome theory of heredity: Morgan (1912) enunciated the theory of linkage and the chromosome theory of heredity.

Cytology: The development of the *cell theory* by Schleiden (1838) and Schwann (1839) was an important event. The next major advance was Virchow's theory of *cell lineage*, put forward in 1858. Virchow's observation gave the *cell theory* its impact in terms of heredity, development and evolution, for if present cells have come from pre-existing cells, then all the cells trace their ancestry back to the first formed cell in an unbroken line of descent.

The importance of the nucleus and its remarkable process of division were described by Strasburger, Van Beneden and Flemming and the latter gave it the name of *mitosis*. Hertwig (1884), Strasburger (1884) and Weismann (1892) concluded that the meticulous accuracy in division and distribution of chromosomes confirmed that they were concerned in the transmission of hereditary material. The term *gametes* and *chromosomes* were suggested by Strasburger in 1877 and Waldeyer in 1888. In 1902, Sutton and Boveri drew attention to the parallelism between the behaviour of the Mendelian factors and that of chromosomes during *meiosis*. This established the physical basis of heredity.

Cytogenetics: With the establishment of a physical basis for heredity the study of genetics and cytology became so intimate and interpenetrating that in many instances it became impossible to delimit the one from the other. The term *cytogenetics* came to be applied for such studies. Winkler (1916) introduced the term *genome* for a set of chromosomes. Winge (1917) proposed the *theory of polyploid origin of plant species* by multiplication of whole genomes. Sakamura and Kihara (1918-21) classified wheat species on the basis of chromosome number and groups. Kihara (1919) concluded that two of the genomes of hexaploid wheat were equivalent to those of tetraploid wheat. Sax (1922) suggested that a genome of wheat might be related to a diploid species. Percival (1926) explained the polyploid origin of 28 and 42 chromosome wheats. Sears (1939) published his cytogenetic studies with wheat and its relatives and described chromosomal aberrations including monosomics. Nishiyama (1929) published extensive studies on the cytogenetics of *Avena*.

Mutation: de Vries (1902) discovered another important genetic phenomenon, the mutation, and its important role in the evolution of new species. Stadler (1928) described the mutagenic effects of X-rays in barley. His evidence of the differential response of corn genes to radiation treatments in 1930 was introductory to the vast amount of mutation breeding research that followed.

Biometry: Galton (1889) and his students studied continuous variation. Using statistical methods, they were able to demonstrate that it is at least partly heritable. Galton recognized the *blending* type of inheritance.

Yule (1906) published his paper, a preliminary note on the theory of inheritance of quantitative characters, based on Mendel's laws. Nilsson Ehle (1908) proposed the *multiple factor hypothesis*, which assumes that a given quantitative character is controlled by a series of independent genes which are cumulative in effect. This hypothesis was amply confirmed in corn by East in 1910.

Hardy (1908) and Weinberg (1909) independently developed a fundamental law of population genetics known as the *Hardy-Weinberg Law*. Fisher (1918) published papers on quantitative inheritance and correlation between relatives. Mathematical derivations of genotypic variance and its division into additive, dominance and epistatic portions are provided in his work. Wright (1921) published the biometrical relations between parents and offspring. Fisher, Immer and Tedin (1932) proposed an experimental approach for separating and measuring additive and dominance effects using second and third degree statistics. Sewall Wright (1935) in one of his many contributions gave a detailed division and designation of genotypic variance into additive genetic variance, dominance variance and epistatic variance. Smith (1936) described the use of 'discriminant function' to determine the genetic value of a plant or line.

Evolution : The most significant contribution made during this period was that of Vavilov (1935) on species origin, variation and plant breeding, which was published in a 2500-page symposium report entitled *Scientific Basis of Plant Breeding*. Dobzhansky (1937) published his book *Genetics and the Origin of Species* two years later. Huzlay (1940) edited *New Systematics*, which summarized current viewpoints on broad problems of plant differentiation and evolution.

Polyploids : Dustin (1934) discovered colchicine. Blackeslee and Avery, and Nebel and Ruttle (1937) showed how colchicine could be used to double chromosome numbers.

Cytoplasmic male sterility : Rhodes (1933) discovered cytoplasmic male sterility in corn. It provided the principle needed for the subsequent utilization for hybrids in sorghum and other plants. East's extensive review of the distribution of self-sterility in flowering plants was published posthumously in 1940.

(b) Plant Breeding

EARLY WORK ON ARTIFICIAL HYBRIDIZATION

Kölreuter (1761-65) fully realized the potential of artificial hybridization and his work marked the importance of crossing in crop plants. He did extensive crosses in many species of the genus *Nicotiana* and made the following very significant observations on crossing and pollination behaviour.

1. Only crosses between related species would generally be successful, and even then not always.
2. The F_1 of some interspecific crosses was sterile.
3. In most cases reciprocal crosses were similar.
4. Continued self-fertilization of successive generations of hybrids includes types which closely resemble parents.
5. There is a possibility that certain characters of one of the parents are

dominant in F_1 plants, and the others are intermediate between the characters of two parents.

6. F_1 plants sometimes exceeded the best parents in growing power.

Thomas Andrew Knight (1759-1838) used artificial hybridization to develop several new fruit varieties. He became best known for his work on peas in 1823 from which he drew the following conclusions:

1. male and female parents make an equal contribution to an F_1 offspring, and

2. segregation occurred in the F_2 generation.

He also made a passing reference to the great growing power of F_1.

Vön Gartner (1849) is credited with making 10,000 crosses in 700 species and 80 genera from which he obtained 250 hybrids. Many of his F_1s were vigorous. He also noted the relationship between F_1, F_2 and parents.

Beal (1878-81) observed increased yields from hybrids between corn varieties and suggested the use of varietal cross in corn production.

East and Shull (1904, 1905) began inbreeding in corn and the study of inbred lines. Their work, along with that of Jones, collectively provided extensive information on the behaviour of corn under self- and cross-pollination. Shamel (1905) reported yields of two lines of corn inbred for three generations and their hybrids. This was probably the first report of yields of hybrids between inbred lines. Bruce (1910), and Keeble and Pellew (1910) suggested that hybrid vigour in characters was due to the operation of favourable dominance in two or more of its components. Systematic inbreeding in corn was started by Wallace, Hayes, Richey, Kyle, and Holbert (1913-1916). Shull (1916) suggested the term *heterosis* for hybrid vigour. McFadden (1917) described wheat-rye hybrids. Jones (1917) made the first commercial crossed corn, *the Burr Leaming Hybrid*. In 1920 he proposed double-cross commercial hybrids.

PURE-LINE SELECTION/SELECTION FOLLOWED AFTER ARTIFICIAL HYBRIDIZATION

a) Cereal crops

Patrick Shirreff (1819-73) utilized pure-line selection to develop new lines of oats and wheat and later began planting in plots for progeny testing of wheat. Subsequently, he suggested that for hybridization (followed by pure line selection in segregating generations), the parents used in crosses must possess desirable characters if superior segregates are to be expected. Lécouteur (1837) was another to apply 'pure-line selection' in wheat. Lécoq (1845) recommended the performance of crosses to serve breeding purposes. Hallet (1857) practised single-plant selections in wheat, oats and barley. Henry Lévéque de Vilmorin (1884) started his combination crosses in wheat on a large scale. He was most successful. The well-known plant breeding institute at Svalöf (Sweden) was founded in 1886. Bröekema (1886) crossed wheat varieties in order to combine productivity with quality. In 1889 he practised backcross. Otto Pitsch (1886) crossed and selected winter barley and wheat through line selection. Mansholt (1886) applied line selection in the local varieties and later to the hybrid populations.

Johannsen (1903) expounded the concept of *pure line*. Harlan and Pope (1922) described the use of backcross in the breeding of small grains.

Harlan et al. (1940) compared methods of barley breeding and showed the relative values of the pedigree and bulk systems of breeding using progenies of a large number of diallel crosses. This work led to the suggestion of composite crosses which included many varieties crossed in complex pairing schemes.

Disease resistance : Biffen (1902) reported that resistance to stripe rust in wheat was due to a single recessive gene. Freeman (1904) made efforts to improve the stem rust resistance in wheat. Hayes et al. (1920) successfully transferred durum stem rust resistance to bread wheats in his famous *Marquis* x *lumillo* cross. Jones and Gillman (1916) released a yellows-resistant cabbage. Hayes and Stakman (1921) stressed the importance of considering individual races in testing wheat varieties for stem rust resistance. Craigie (1927) provided valuable insight into how new rust races are formed through hybridization and mutation in rust fungi. McFadden (1930) reported the successful transfer of stem rust resistance from tetraploid to hexaploid wheats and the development of Hope and H44 wheats from the cross of Marquis and Yaroslavemmer.

b) Other crops

Louis Lévéque de Vilmorin (1856) expounded the importance of *selection génealogique* in sugar-beet breeding, which is equivalent to family selection in terms of cross-fertilized crops. Goodrich (1857) practised selection in potato which widely led to the practice of selection by others subsequently, and a good number of potato varieties were developed. Burbank's first potato variety *Early Rose* was selected from a single seed ball in 1872. Hopkins (1899) described the ear-to-row method of corn breeding. Jones (1920) proposed the double-cross pattern for commercial hybrids. Richey (1927) proposed the method of convergent improvement; it was further described by Richey and Sprague in 1931.

Davis (1931) suggested the use of inbred-variety cross (topcross) for testing corn inbreds. Jenkins and Brunson (1932) used the topcross method to give comparative tests of combining ability. Jenkins (1934) evaluated methods for predicting double-cross performance from the yields of component single crosses.

c) Use of statistics

Plant breeders became cognizant of sampling and experimental errors. Wood and Stratton (1910) suggested the use of check plots to correct for soil variability in yield trials. Mercer and Hall (1911) introduced the questions of number and arrangement of replications and size and shape of plots. Harris (1912) suggested the use of χ^2 (Chi-square) for goodness of fit test for segregation ratios and in 1915 he contributed evidence on the degree of substrate heterogeneity in field experiments. By 1935 the variance analysis methodology was in wide use. Experimental designs, variance analysis methods and proper field plot technique were explained comprehensively by Snedecor in the mid-1930s and Lush in 1936.

New Developments

By the early twentieth century, plant breeding had developed well and was remarkably successful. There has been a flurry of genetic, cytogenetic and plant breeding research since then. Many people and events have helped plant breeding to advance and become more precise and beneficial. The International Agricultural Research Centre(s) (IARCs) and the International Plant Genetic Resources Institute (IPGRI) established during past few decades have immensely benefited plant breeding research throughout the world. These centres have greatly helped the developing nations to achieve better food supply and food security.

The scientific basis of modern plant breeding is much broader and uses, as conceptual and technical tools, cytology, systematics, physiology, pathology, entomology, chemistry and statistics. There have been many significant advances. These include demonstration of the auto- and allopolyploid nature of many crop plants and their origin. This opened the way to the production of artificial polyploid forms and their direct or indirect exploitation. The discovery of the possibility of increasing the frequency of mutations by means of radiations or chemicals was hailed as a means of increasing the variability available to the breeder. Also there have been rapid advances in quantitative genetics, resistance breeding and conservation and characterisation of genetic resources. More recent advances in plant tissue culture and genetic engineering, namely, gene cloning and genetic transformation, have allowed breeders to design new methods. These methods hold great promise for the future. There has been increased interest in the development of hybrid varieties. These new enabling technologies, all adding to the overall science of plant breeding, are considered in detail in this book.

3

Crop Germplasm

Definitions

The term *crop germplasm*, broadly speaking, can be defined as an array of plant materials, a reservoir of genes, be it land races, improved varieties and wild relatives (wild races, species, etc.) that serves as a basis for crop improvement or related research. It is in fact the assemblage of genetic diversity accumulated in crop plants through years of evolution under domestication and natural selection. Crop germplasm also implies the documented information available on germplasm, in relation to a crop species in the gene banks worldover for use by plant breeders.

The term *genetic stocks* of crop plants is restricted to stocks of specific genes and gene combinations that have direct usefulness in genetic analysis. Genetic stocks are usually held and maintained by individual geneticists.

A new term encompassing germplasm, *genetic resources* (Genetic Resources Centre(s) was coined by the FAO Panel of Experts on Plant Exploration and Introduction in 1970. The principal objective of these Genetic Resources Centre(s) is long-term conservation of germplasm, its propagation and distribution.

Importance of germplasm collection

The assemblage of genetic diversity, that is, collection of germplasm from crop plants, offers enormous variability in plant-breeding programmes for the development of new superior crop varieties with desired traits. It can be readily realized that the greater the range of initial material available to the plant breeder, the greater his chance of producing the desired types. This simple principle was first realized by plant breeders in the former USSR headed by Vavilov during the early years of twentieth century. They sent out expeditions to make worldwide collections of all the plants grown or likely to be grown in the USSR. This work of survey and collection of economic plant material, including allied species of plants, and their classification with regard to useful characters is now recognized to be of paramount importance. Indeed, it is so important as to have long been a primary objective of Bureaus of Plant Introduction and Exploration (or Board/Bureaus of Plant Genetic Resources) in most countries, including the International Plant

Genetic Resources Institute (IPGRI) and International Agricultural Research Centres (IARCs).

Categories of Germplasm Collection

There are two major categories of germplasm collection.

1. Working Collections, World Collections (Active Collections) : The collections of specific crops that are held in adequate storage (controlled storage conditions), documented and available for immediate use by all concerned are referred to as *Working Collections* or *Active Collections*. When the number and geographical origin of accessions has reached such a magnitude as to result in world-wide request for seeds the working collections acquire the title of *World Collections*. Sometimes *World Collections* implies a broadly represented collection even though there is no attempt to distribute samples.
2. Conserved Stocks (Base Collections) : These are broad segments of germplasm held in national and international seed storages for long-term conservation (conservation centres). These stocks duplicate working collections but are released only when the latter have been depleted.

Sources of Germplasm

There are three main sources of crop germplasm diversity.

1. Wild species and primitive forms of crops nurtured in primary centres of diversity. These are collected through well-organized expeditions to regions of crop diversity.
2. Plant migrants nurtured in secondary centres of culture where their diversity may be augmented. These are also collected through expeditions to appropriate regions.
3. Products of plant breeding, including induced polyploidy, mutations, and the combining of multiple traits into useful breeding lines.

CENTRES OF CROP PLANT DIVERSITY
(Centres of Origin of Crop Plants)

Alphonse de Candolle (1886) is credited with postulating the idea of *Centres of Origin* of cultivated plants. He suggested that the centre of origin of a crop plant, that is, the area of crop plant diversity could be determined by identifying the area where a plant was first domesticated (brought under human management) and then linking it with the area(s) where its wild progenitors still continue to occur. This should be supported by historical and linguistic evidence.

N.I. Vavilov opened a whole new approach to the problem of locating pools of germplasm and the practical application of this knowledge. The tremendous mass of material assembled at the All-Union Institute of Plant Industry, Leningrad under his able guidance (1920-1936) made possible a geographic survey of crop plants which has never been duplicated. The most useful and important generalization of the Vavilovian work was the concept of geographic centres of variability or gene centres (Smith, 1969). The fact that these are not necessarily the centres of origin does not make this concept less useful. Geographic concentration of variability is a real phenomenon, reflecting certain geographic and agronomic connotations. Vavilov also developed the concept of *homologous series* in genetics which implies that similar variation in two or more related crops could be observed in a geographic area.

Vavilov's Dicta for Establishing Centres of Origin

1. Establishment of area with greatest varietal diversity of a species.
2. Elucidation of the system of varietal diversity of a species.
3. Delimitation of the differential geography of wild relatives of a cultivated plant.
4. Recognition of varietal endemism in a crop species.
5. Absence of interspecific hybridization.
6. Location of centres of genetically dominant characters.
7. Information from archaeological, historical and linguistic data.

It is apparent from the first two dicta that Vavilov was evolving an idea of areas of genetic diversity. He summarized his information on the centres of origin of cultivated plants and proposed eight principal world centres of crop origin (Table 3.1)

It is obvious from Table 3.1 that Vavilov realized that many plants were brought into domestication in several areas (see *Cicer arietinum* L.).

Primary and Secondary Centres

There have been remarkable shifts of many crop species from their primary centres to distant lands for cultivation purposes; this led to the concept of secondary centres of diversity.

Primary centres: In primary centres of diversity the species is in competition with other natural elements of its environments.

Secondary centres: In secondary centres new biological and physical stresses become selective factors. Primitive cultivars in these secondary centres are important segments of germplasm because of man's involvement in assuring their success although the age element is very brief in comparison to natural evolvement of crop species in their primary centres of origin. In many respects these secondary centres have become significant sources of germplasm, especially in respect to different biological ecosystems.

Since the publication of Vavilov's work on the centres of origin of cultivated plants much new information has accumulated on this subject. Zhukovsky (1968), a close associate of Vavilov, rearranged the world map of gene centres and distribution of plant resources. Large areas had to be divided into microcentres[1] specific for certain species. Several taxa were regrouped. A new Vavilov-Zhukovsky world map of gene centres and plant resources was constructed. (Fig. 3.1).

---

[1] Smaller areas/pockets of varietal/racial diversity within a Vavilovian centre.

Table 3.1. Principal world centres of crop origin of cultivated plants (centres of plant diversity)

Centre(s) of origin	Selected plants*
I. Chinese centre of origin	*Glycine hispida* Maxim. *Eleocharis tuberosa* Schult. *Brassica chinensis* L. *Prunus persica* L.
II. Indian centre of origin	*Oryza sativa* L. *Cicer arietinum* L. *Solanum melongena* L. *Saccharum officinarum* L.
IIa. Indo-Malayan centre of origin	*Musa paradisiaca* L. *Artocarpus communis* Forst. *Cocos nucifera* L. *Saccharum officinarum* L.
III. Central Asiatic centre of origin	*Triticum vulgare* Vill. *(T. aestivum)* *T. compactum* Host. *Cicer arietinum* L. *Linum usitattissimum* L.
IV. Near Eastern centre of origin	*Triticum monococcum* L. *T. vulgare* Vill. *(T. aestivum)* *Cicer arietinum* ssp. *pisiforme* Pop. *Linum usitattissimum* L. *Cucurbita pepo* L.
V. Mediterranean centre of origin	*Triticum diococcum* Schrank. *Triticum spelta* L. *Cicer arietinum* L. *Linum usitattissimum* L.

VI. Abyssinian centre of origin	*Triticum durum* ssp.
	  abyssinicum Vav.
	T. diococcum ssp.
	  abyssinicum Stol.
	Eragrostis abyssinicum L.
	Cicer arietinum L.

VII. South Mexican and central American centre of origin	*Zea mays* L. *Phaseolus vulgaris* L. *Cucurbita moschta* Duch. *Cucurbita mixta* Pang.
VIII. South American centre of origin	*Zea mays* L. *gr. amylacea* *Phaseolus lunatus* L. *gr. macrospermus* *Phaseolus vulgaris* L. *Lycopersicon esculentum* M. *Cucurbita maxima* Duch.

*Some of the principal crop plants and the centres to which Vavilov assigned them.

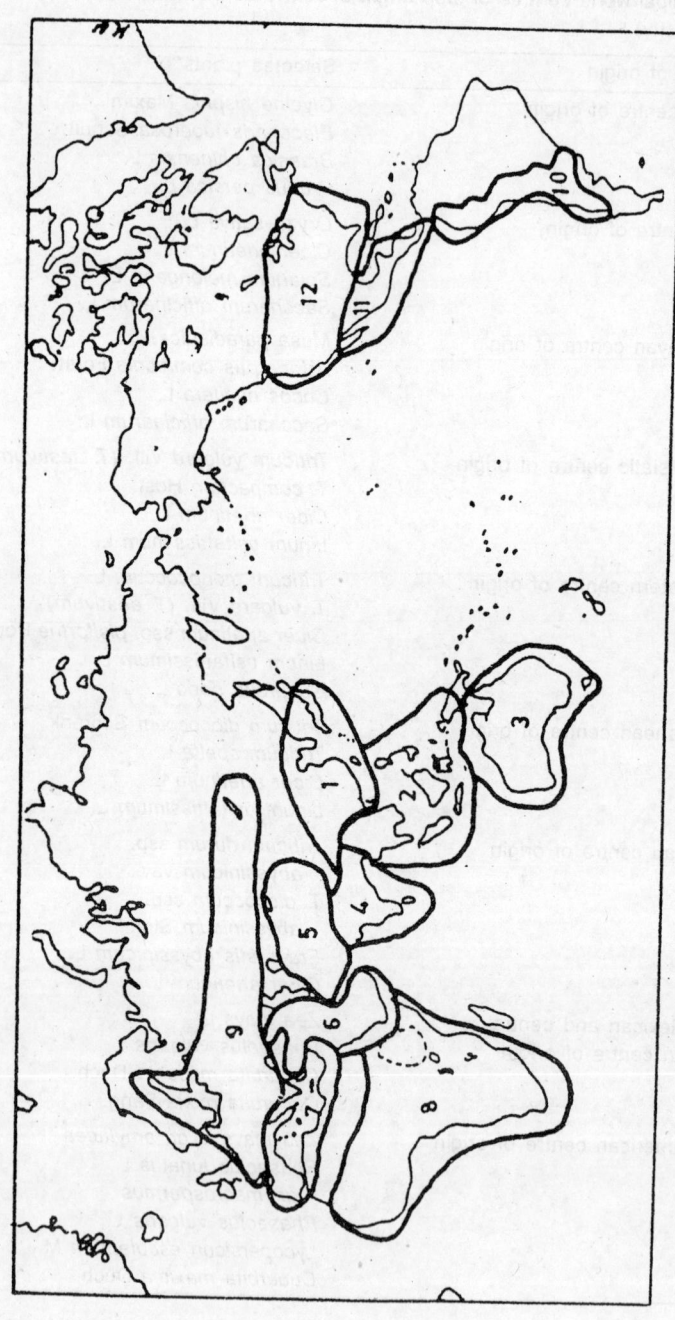

Fig. 3.1:   Macrocentres of cultivated plants according to Vavilov and Zhukovsky. (Redrawn from Zhukovsky's original map.)
1. Chinese-Japanese 2. Indonesian-Indo Chinese 3. Australian 4. Indostanian 5. Middle-Asian 6. Near Eastern 7. Mediterranean 8. African,
including Ethiopia 9. Euro-Siberian 10. Central American 11; South. American 12. North American.
(Adopted from Leppik, E.E., 1969).

NUCLEAR CENTRES AND REGIONS OF DIVERSITY

Hawkes (1983) postulated the idea of *nuclear centres* and *regions of diversity*. He clearly envisaged centres of agricultural origin from where farming spread into one or more regions for which he proposed the rare nuclear centres or regions of diversity. He linked up his nuclear centres with archaeological evidence to provide strong proofs of agricultural origin. He recognized the following nuclear and regional centres.

I. Nuclear Centres
    A. Northern China
    B. Near East (Fertile Crescent)
    C. Southern Mexico to Guatemala (from Tehuacan southwards)
    D. Peru (Andes and Coastal belt as well as the Andean slopes).

II. Regions of Diversity (Table 3.2)

There are several regions in which it would seem that perhaps crops actually did not originate. This argument is primarily based on the absence of wild progenitors and absence of archaeological remains to suggest antiquity of a crop species. These are the regions into which crops perhaps spread from nuclear centres in the dim past and in which spatial isolation in time and intensive human selection played a dominant role in the development of several additional cultigens and land races.

Table 3.2. Regions of diversity

Region	Remarks
I. Chinese Region	
II. Indian Region (excluding Pakistan)	Related to Chinese Centre (A)
IIa. Indo-Malayan Region	
III. Central Asiatic Region	
IV. Near East Region	Related to Near East
V. Mediterranean Region	Nuclear Centre (B)
VI. Ethiopian Region	
VII. Meso/American Region	Related to Nuclear Centre (C)
VIII. Northern Andean Region (Colombia, Ecuador, Peru, Bolivia and adjacent lowland)	Related to Nuclear Centre (D)
IX. West African Region (not mentioned by Vavilov)	Related to Nuclear Centre (B) through Ethiopian region.

III. MICROCENTRES

Hawkes (1983) also identified small microcentres for several crops. These are shown in Table 3.3.

Table 3.3. Microcentres of crop diversity

Centre	Selected plants
New Guinea	Sugar-cane (*Saccharum officinarum* L.)
Solomon Islands and Fiji	*Musa* spp.
Europe	*Avena strigosa*
	Secale cereale
Chile	*Bromus mango*
Brazil	*Manihot esculenta*
	Ananas cosmosus
USA	*Helianthus annus*
	H. tuberosus

GENETIC RESOURCES CENTRES

Global Network

A. International and Regional Centres

1) INTERNATIONAL PLANT GENETIC RESOURCES INSTITUTE (IPGRI)

IPGRI (formerly IBPGR) was established in 1974 by the Consultative Group on International Agricultural Research (CGIAR) with its headquarters at FAO, Rome (Italy) to further the study, collection, preservation, documentation, evaluation and utilization of the genetic diversity of useful plants for the benefit of people throughout the world. It acts as a catalyst both within and outside the CGIAR system in stimulating the action needed to sustain a viable network of institutions for the conservation of genetic resources of plants. The research component was added in 1984 to ensure better functioning of genetic resources activity.

The IPGRI's network includes more than 600 institutes throughout 100 countries. For each country included in the IPGRI programme, the IPGRI provides a package of assistance to various genetic resources activities tailored to the needs of the country's genetic resources programme.

The various activities of the IPGRI are as follows:
 i) research on conservation and diversity,
 ii) germplasm collection, conservation, characterization and its evaluation,
 iii) documentation, and
 iv) training

2) INTERNATIONAL AGRICULTURAL RESEARCH CENTRES (IARCs) AND REGIONAL CENTRES

The following IARCs and regional centres are actively involved at the crop level in various genetic resources activities.

International Rice Research Institute (IRRI), Los Banos, Philippines

International Maize and Wheat Improvement Centre (CIMMYT), El Batan, Mexico

International Institute of Tropical Agriculture (IITA), Ibadan, Nigeria

International Centre for Tropical Agriculture (CIAT), Cali, Colombia

Tropical Agricultural Research and Training Centre (CATIE), Turrialba, Costa Rica

Dutch/German Potato Gene Bank, Braunschweig, Germany

International Crops Research Institute for the Semi-Arid Tropics (ICRISAT), Patencheru (Hyderabad), India

International Potato Centre (IPC), Lima, Peru

Nordic Gene Bank, Lund, Sweden

International Centre for Agricultural Research in Dry Areas (ICARDA), Aleppo, Syria

Asian Vegetable Research and Development Centre (AVRDC), Shangwa, Taiwan

Algean Regional Agricultural Research Institute (ARARI), Izmir, Turkey.

B. National Organizations

At the national level the following organizations are involved in various countries in genetic resources activities.

Australia	: Commonwealth Scientific and Industrial Research Organisation (CSIRO), New South Wales Dept. of Agriculture.
Brazil	: National Genetic Resources Centre (CENARGEN)
France	: Institut of Tropical Research and Food Crops (IRAT), National Institute of Agricultural Research (INRA), Overseas office for Scientific and Techniqueal Research (ORSTOM).
Germany	: Society for Techniqueal Cooperation (GTZ), Central Institute for Genetics and Plant Breeding.
India	: National Bureau of Plant Genetic Resources (NBPGR), New Delhi.
Indonesia	: National Biological Institute, Bogor
Italy	: Laboratorio del Germoplasma, University of Bari
Japan	: National Institute of Agricultural Sciences, Kyoto University
Mexico	: National Institute of Agricultural Research
Netherlands	: Institute for Horticultural Plant Breeding, Institute for Plant Breeding, Government Institute for Research on Varieties of Cultivated Plants, Seed Testing Station
Sweden	: Scandinavian Gene Bank
UK	: Royal Botanic Gardens, John Innes Institute, Scottish Plant Breeding Station, Plant Breeding Institute
USA	: Agricultural Research Service, USDA
USSR	: Vavilov All-Union Institute of Plant Industry (VIR)

Functions of genetic resources centre :

1) Exploration and collection of material and collaboration with national centres.

2) Identification and preliminary evaluation of materials.

3) Initial planting of introduced material according to quarantine laws of the country.
4) Exchange and distribution of seed and vegetative stocks including appropriate introduction of breeding lines and advanced cultivars.
5) Maintenance and storage of seed and vegetative stocks for medium- and long-term preservation.
6) Documentation and exchange of information with other centres in the network in an internationally accepted form.
7) Organization of genetic stock rejuvenation by national centres wherever possible or otherwise by regional centres.
8) Organizing training programmes for personnel in collaboration with national or international training schemes.

PLANT EXPLORATION

The exploration for, and introduction of, seed-reproduced crops involves direct collection in the primary and secondary centres of diversity. Present-day explorations are conducted in a systematic fashion by crop specialists. The team generally includes broadly trained agronomists, geneticists, entomologists and pathologists, besides local persons who are familiar with the terrain. Most explorations are of short duration, coinciding with the time of seed ripening for the species of primary interest as governed by altitudinal and latitudinal range.

Adequate attention is now paid to sampling techniques as well as population distribution. Both random and directed sampling methods of collection are employed and accurate documentation of useful data has lately become of increasing significance as a means of communication between the collector and evaluator. Collection of vegetatively propagated plants poses additional difficulties associated with transit survival, danger of insect and disease transportation, quarantine regulations, etc.

PLANT INTRODUCTION

What is Plant Introduction?

According to Frankel (1957) 'plant introduction is the transposition of a genetic entity from an environment to which it is attuned to one in which it is untried'. Plant introduction consists of taking a genotype or a group of genotypes of plants into new environments in which they have not been grown before.

Primary introduction: When the introduced plant material is so well suited to the new environment, that it is eventually released as a variety for cultivation with no alteration it is referred to as primary introduction.

Secondary introduction: When the introduced plant material is subjected to selection, and or is used in hybridization for incorporation of a gene(s), this is referred to as secondary introduction.

Goals of Plant Introduction

There are three major goals of plant introduction in a plant-breeding programme:
1) To secure for possible general use any outstanding varieties grown elsewhere in any part of the world.
2) To introduce new crops and to initiate a plant-breeding programme in respect of such crops.
3) To collect germplasm for use in the genetic improvement of crops in a plant-breeding programme.

Procedure of Plant Introduction

The new germplasm from other countries, IPGRI, IARCs etc. is usually obtained by the official agency (ies) in a country under the overall policy framework formulated from time to time. The new introductions may also be made by personal exploratory search, and through correspondence with the fellow scientists, including those engaged at IPGRI and IARCs. The important thing is to necessarily follow the specified procedure and plant quarantine regulations.

Plant Quarantine

The term 'plant quarantine' refers to those quarantines and regulations concerned with material(s) entering from foreign countries. Quarantine precautions are directed against international spread of pests and diseases in order to prevent their introduction to uninfested areas. There are two kinds of true quarantine pathogens.

(1) Those whose spread is explosive, even when the amount of inoculum is in traces. Sampling and testing for such pathogens is unrealistic as a safeguard and no guarantee for freedom from diseases can be made.

(2) Those who spread at such a slow rate that pocket infections can be checked by appropriate barrage precautions, and in some cases eradicated. Careful testing of representative samples in respect of such pathogens is acceptable.

This distinction of the two epidemic types is the criterion applied for the categorization of quanrantine pathogens and insect pests into *prohibited* or *restricted*. Each country has specific quarantine provisions to regulate the exchange of plant materials.

General requirements: These include demands such as, Phytosanitary Certificate, Official Permit, and inspection by Plant Quarantine Inspector.

Specific requirements: These include specification of pathogens and insect pests against which an importing country wishes to be protected; and specification of inspection procedures to ensure that the exporting country/ agency be precisely informed how to test a seed lot/planting material before a certificate is issued which asserts freedom from infection.

Import Permits

Request for Official Permit to import plant material(s) should be made by the importer to the Official Agency in advance of shipment. The application should include information, namely, the kind of seeds/fruits (botanical name), vegetables, plants or plant products to be imported, the country and locality where the restricted products were produced, the manner of importation (by baggage, cargo or mail); port or ports of arrival and whether other shipments are contemplated; and the name and address of the importer.

Quarantine Inspection

1. Port Inspection: Inspection at sea ports, airports and border crossing points is a major safeguard provided under plant quarantine laws and regulations against the entry and spread of foreign plant pests.
2. Pre-entry Inspection: In a number of special cases where it has been practical to work out necessary arrangements for safeguarding against pest entry inspection and clearance of regulated plant materials or products are effected in countries where plants are grown.

Post-entry Quarantine Surveillance

Imported germplasm is grown in isolation for a reasonable period of time to ensure that imported plant materials are free from exotic pathogens and insect pests. The post-entry quarantine surveillance is very necessary for seed-transmitted pathogens, especially for virus diseases. Fumigation may be necessary to control storage insects in many instances.

Introductions not conforming to quarantine requirements, or suspected to be contaminated, are either destroyed or returned to the sender.

EVALUATION AND CHARACTERIZATION OF GERMPLASM

The germplasm is evaluated and multiplied at the selected sites, usually for 2-3 seasons to determine its likely potential and adaptability. The sites are selected on the basis of origin of the material so as to provide nearly similar growing conditions. Generally, the germplasm material meant for evaluation is planted in an augmented design (incomplete block design) with checks interspersed at regular intervals, and/or at random for comparison. Data on various aspects and characters is recorded as per 'descriptors' list (list of characters prepared for a crop) uniformly for all the entries. Evaluation for resistance to diseases and pests, stress conditions and biochemical traits, etc. is also done. The recorded data along with part of the seed is sent to the long-term storage (conservation) section and the remaining part stored as a 'working collection'.

DOCUMENTATION OF GERMPLASM

An efficient system of documentation is necessary for effective use of germplasm sources. Ready retrieval of information is crucial for supply of needed information and material besides being necessary for good housekeeping, namely, periodic germination tests, regeneration and locating accession in storage.

Passport data: All collections are given an *Accession No.* A prefix, namely, *EC, IC* or *IW* may be added to this no. for exotic, indigenous and indigenous wild collections respectively. Information such as date of collection, acquisition data (provided by the collector of the sample) including principal attributes and distribution status are recorded in the passport data.

Sample data: This includes Accession No., initial date of storage, location in storage, date of initial germination test, initial germination percentage, seed moisture percentage, number of seeds/or weight of seeds and information related to regeneration.

Cataloguing: The documented information is published in the form of catalogues. These catalogues are primarily meant for all end-users of the germplasm. Catalogues should have all relevant information, such as Accession No., species, variety, country of origin, adaptation and other attributes (recorded during evaluation and characterization) besides any other information that may be available in respect of an accession.

CONSERVATION OF GERMPLASM

Germplasm may be conserved either *in situ*, that is, in *gene sanctuaries* or in long-term storage under low to very low temperature conditions, i.e. in gene banks.

Gene Sanctuaries: These are the areas in which the whole ecosystem is preserved under minimal human interference. Here the threatened species are preserved in their natural environment (habitat). Continuous evolution is the salient feature of this type of conservation.

Gene Banks: These are the repositories of seeds of germplasm, seed stores, where the seeds are preserved for medium to long terms under controlled storage conditions. The seeds are dried to low moisture content (5-7%) and packed in hermetically sealed containers (glass bottles, aluminium metal cans, etc.):

a) medium-term storage (5 to 10 years): storage temperature range varies from 0 to −10°C.

b) long-term storage (over 10 years): storage temperature (−20°C).

Plant tissue culture repository: The germplasm (seeds, pollen, meristems/shoot apices, embryos, etc.) is conserved *in vitro* in this repository. This is gaining significance in respect of vegetatively propagated crops. Minimal growth and cryopreservation (storage in liquid nitrogen, −196°C) are the two main features of *in-vitro* conservation technology.

LOCAL GERMPLASM

Indigenous or local varieties, acclimatized varieties[2], domesticated plant

---

[2] Adapted varieties are those accustomed and/or attuned to the climate of an area (acclimatized); the term is specifically used in relation to varieties introduced through plant introduction.

materials[3], wild species or races, etc. growing in wilderness in a given area constitute local germplasm.

Salient Features of Local Germplasm

1. *Good agronomic base* : Long-standing indigenous varieties possess wonderful adaptability to the environmental complex of a given geographic region and provide an excellent agronomic base to a plant breeder for further improvement.

2. *Genetic variability* : Indigenous varieties are usually heterogeneous populations consisting of many forms differing from one another morphologically and physiologically. The genetic variability in indigenous varieties is far more due to natural outcrossing, spontaneous mutations, etc. than what is generally imagined by plant breeders. This variability should be thoroughly explored by systematically combing the actual growing centres and collecting a sufficiently large number of samples. This variability provides ample scope to a plant breeder for selection of desirable traits.

3. *Wild species:* Wild relatives of a crop may also be found growing in wilderness in a given area along with cultivated varieties. The need for collecting them and wherever feasible using them as a donor parent, more particularly for transfer of disease/insect resistance genes is well recognized.

4. *Local taste:* Indigenous varieties suit the palate of local people and usually cater to the industrial needs of the region.

---

[3] Refers to the process of bringing wild species, subspecies, races, etc. under the management of man.

4

Reproductive System

A good insight into reproductive systems of crop species is necessary for a plant breeder for the following reasons.

1. Plant breeding methodology differs for the most part according to the system of reproduction of a crop species. To determine the most appropriate methodology for a crop species precise knowledge of its reproductive system is therefore essential.

2. A plant breeder is required to make artificial hybridization for genetic studies to elicit information regarding inheritance of specific traits and interspecies relationships and for transferring desired characters from one breeding material to another. Precise knowledge of floral structure, floral biology, sex expression, mode of natural pollination and extent of natural outcrossing, etc. for the crop species with which a plant breeder is working is thus essential.

3. Precise knowledge of reproductive behaviour is also needed to control pollination to avoid unwanted hybrids (outcrosses) both in breeding nurseries and in pure seed production for commercial purposes, and in the development of hybrid varieties and hybrid seed production.

The possible range of reproductive systems in crop species is rather large. The reproductive systems of interest to a plant breeder are discussed below.

SEXUAL REPRODUCTION

The normal sexual life cycle of a plant is shown in Fig. 4.1. Sexual reproduction involves the union (amphimixis) of female gamete (egg cell) with a male gamete (generative nucleus) as a result of which fertilization takes place and a zygote is formed which develops into a seed with embryo.

From the viewpoint of plant breeding sexually reproducing species are

$$\underset{(2n)}{\text{Sporophyte}} \xrightarrow{\text{meiosis}} \underset{(1n)}{\text{Spore}} \rightarrow \underset{(1n)}{\text{Gametophyte}} \xrightarrow{\text{fertilization}} \underset{(2n)}{\text{Zygote}} \rightarrow \underset{(2n)}{\text{Seed with embryo}}$$

Fig. 4.1. Normal sexual life cycle of a plant. (after Grant, 1971)

grouped into self-pollinated crop species (autogamous crop species), cross-pollinated crop species (allogamous crop species), and often cross-pollinated crop species. This grouping is based on the extent of natural cross-pollination (outcrossing)[1].

1. Autogamy (Self-pollinated, Self-fertilized Crop Species)

Crop species in which the pollen of a flower usually pollinates the stigma of the same flower are grouped as self-pollinated crops. The extent of self-pollination varies in different crop species. The following sex expressions favour self-pollination.

a) Cleistogamy

Refers to the condition wherein pollination takes place in unopened flowers, thus ensuring total self-pollination. The extent of natural cross-pollination in such species, if any, is negligible.

b) Chasmogamy

Refers to the condition wherein the flower opens only after pollination has taken place. Since the flower opens some outcrossing may take place, though the extent is usually very low.

c) Floral Structure

In many species, although pollination usually takes place after the flower opens, self-pollination is ensured by the floral structure. The following floral structures favour self-pollination:

1) Stigma may be closely surrounded by anthers. The position of anthers in relation to stigma ensures self-pollination (e.g. tomato, egg-plant).
2) Stamens and stigmas are hidden by other floral organs (e.g. legumes).
3) Stigma becomes receptive and elongates through the staminal column, which ensures predominant self-pollination.

In some crop species the extent of natural cross-pollination is higher than that seen in cleistogamous crop species. As a rule, when the extent of natural cross-pollination in a crop species is less than 10%, it is grouped as self-pollinated. At least 90% of the progeny is self-pollinated (Mayo, 1980)[2].

2. Allogamy (Cross-pollinated, Cross-fertilized Crop Species)

The crop species in which the stigma of a flower on a plant is usually pollinated by a pollen from flowers on other plants are grouped as cross-pollinated crops. The extent of cross-pollination varies in different crop species. At least 50% of the progeny is outcrossed (Mayo, 1980). The following sex expressions favour cross-pollination.

---

[1] The extent of natural outcrossing may vary, however, in different varieties of a crop species. Environmental conditions and insect populations (in insect-pollinated crops) also influence the extent of natural outcrossing.

[2] According to Hayes et al. (1955) in naturally self-pollinated crops as a rule the extent of cross-pollination is less than 4 percent.

a) *Dichogamy*

Protandry refers to the condition when stamens ripen before stigmas.

Protogyny refers to the condition when stigmas ripen before stamens.

b) *Dicliny (Unisexuality)*

Monoecism refers to the condition wherein male and female flowers are physically separate on the same plant.

Dioecism refers to the condition wherein male and female flowers are borne on different plants.

Modification of Sex Expression

Andromonoecism refers to the condition wherein a plant bears some male flowers and some hermaphrodite flowers.

Gynomonoecism refers to the condition wherein a plant bears some female flowers and some hermaphrodite flowers.

Androdioecism, refers to the condition wherein some plants are male and some are hermaphrodite.

Gynodioecism refers to the condition wherein some plants are female and some are hermaphrodite.

Androecism, the term is used in reference to an inbred line that predominantly produces male flowers (>90%).

Gynoecism, the term is used in reference to an inbred line that predominantly produces female flowers (>90%).

c) *Floral Characteristics*

In some species (e.g. alfalfa) the stigma is covered by a waxy cuticle and becomes receptive only when this is broken. This is done by pollinating insects which also effect cross-pollination. (see alfalfa in Chapter 7).

Male sterility refers to the condition wherein the anthers are either absent or non-functional. (see Chapter 5)

Self-incompatibility refers to the condition wherein the pollen from a flower is unable to fertilize its own stigma, or the stigmas of other flowers on the same plant, or any other plant having a particular phenotype, although both the stigmas and pollen are functional (see Chapter 6).

3. Often Cross-pollinated Crops

In this group the extent of natural self-pollination is more than 50% while natural cross-pollination (out-crossing) may vary from 10-49%.

Classification of Crop Species

Crop species are classified as self-, cross- or often cross-pollinated on the basis of natural outcrossing percentage. Table 4.1 lists the important crop species belonging to the three groups described above.

Determination of Mode of Sexual Reproduction[3]

The mode of sexual reproduction, i.e., whether a crop is self-pollinated or cross-pollinated is reasonably well known for agricultural species. However,

---

[3] Mode of reproduction is determined on the basis of extent of natural cross-pollination.

Table 4.1. Classification of crop species on the basis of pollination behaviour

Self-pollinated crops
Rice, wheat, barley, oats, coarse millets (common millet, finger millet, Italian millet, little millet, barnyard millet), lentil, chick-pea, cowpea, French bean, *Dolichos* bean, groundnut, flax, sesame, soybean, mountain brome grass, tomato, egg-plant, lettuce, apricot, citrus, peach, etc.

Cross-pollinated crops
Pearl millet, corn, *Brassica campestris*, sunflower, niger, sunnhemp, Egyptian clover, alfalfa, fenugreek, red clover, white clover, crimson clover, sweet clover, rye-grass, timothy, smooth brome grass, sugar-beet, sugar-cane, cucurbitaceous vegetables, cauliflower, cabbage, spinach beet, amaranthus, garden beet, radish, turnip, carrot, onion, etc.

Often Cross-pollinated crops
Sorghum, *Brassica juncea*, castor, safflower, cotton, chili, okra, etc.

if it is not known in respect of a new crop species it can be easily determined. The various steps involved in determination are :

1) *Examination of floral structure:* The presence of unisexual flowers, dichogamy or other mechanisms (conditions) would indicate the likelihood of cross-pollination.

2) *Seedset in isolated single plants:* Grow single plants in isolation[4] and observe whether seeds are produced or not.

   i) If no seeds are produced, this is a certain indication that the crop species in question is cross-pollinated.

   ii) If seeds are produced, then it is necessary to grow the selfed seed. Self-pollinated crops do not show inbreeding depression or adverse effects. On the contrary, in cross-pollinated crops the adverse effect of inbreeding is noticeable to varying degrees.

3) Extent of natural outcrossing: This is determined by interplanting with strains carrying a recessive marker gene. Seeds are harvested from the recessive type and the amount of natural outcrossing is calculated from the proportion of recessive and dominant progeny.

Importance of Determination of Prevalence of Natural Outcrossing
Determination of prevalence of natural outcrossing is not only important from the viewpoint of developing appropriate breeding methodology for a crop species, but more significantly from the viewpoint of pure seed production of improved varieties and their maintenance. Determination of adequate isolation requirements for the maintenance of inbred lines, synthetics and other open pollinated varieties, pure lines, etc. is an essential requirement for pure seed production. In the absence of adequate isolation during the seed production cycle varieties/lines will deteriorate genetically due to contamination from pollen from other sources (varieties, breeding materials, etc.). This is particularly evident in cross-pollinated crops and often cross-pollinated crops.

---

[4] Preferably the plant should be grown in isolation in a greenhouse or in the open rather than bagged or caged.

Factors Affecting Extent of Natural Outcrossing

The prevalence of natural outcrossing in a crop species is greatly affected by the mode of natural pollination and whether pollen dispersal occurs through wind or insects. Quite obviously, wind direction and speed, or size of insect population will affect the extent of natural outcrossing. Besides this, varietal mass, time of flowering of other varieties grown in the vicinity, temperature, rainfall and humidity conditions, floral structure, soil moisture etc. during the flowering period of the crop inevitably affect in one way or the other the extent of natural outcrossing.

Thus it is imperative that experiments involving interplanting strains carrying a recessive marker gene with strains carrying the dominant alternative allele be conducted over a number of environments, including seasons and locations for both plant breeding and seed production purposes (to determine isolation requirements). Also, in a crop species the different genotypes may exhibit differences in the prevalence of natural outcrossing. A plant breeder should be mindful of such differences.

ASEXUAL REPRODUCTION

Asexual reproduction means reproduction through vegetative parts of the plant, such as, buds (roses), bulbs (tulips), tubers (potato), stems (sugar-cane) or other parts, that are utilized to produce new individuals. In addition to these, there are types of asexual reproduction included under the term apomixis which we shall detail below.

Apomixis

Apomixis is the substitution of a sexual process with an asexual process. All forms of asexual reproduction in higher plants that replace or act as substitutes for sexual reproduction are known as apomixis.

Types of Apomixis

1. AGAMOSPERMY

In this type of apomixis sexual organs or related parts take part but the seeds are formed asexually without the union of male and female gametes. This is due to modification of the normal meiotic and fertilization processes (apomeiosis).[5] Agamospermy is divided into three main groups:

a) Gametophytic apomixis

This refers to the type of apomixis in which the daughter sporophyte develops from a diploid gametophyte in a manner suggesting alternation of generation (Fig. 4.2). The gametophyte (2n) develops from a somatic cell of the ovule (2n) as a result of a series of purely mitotic divisions (apospory); or from archesporial cell (2n) through omission or restitution of meiosis. Meiosis may progress to the formation of dyads and semi-heterotypic division occur

---

[5] This includes replacement processes which result in diploid gametophytes functioning in meiosis (apospory).

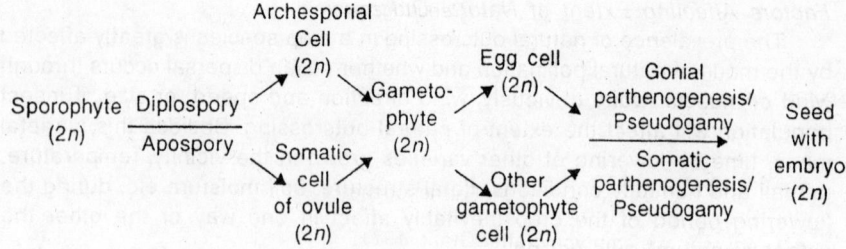

Fig. 4.2, Gametophytic apomixis (after Grant, 1971)

(diplospory). In essence, the meiosis is modified to prevent pairing and reduction of chromosomes. The term diplospory, as used here, covers the phenomenon of diplospory, semi-apospory and generative apospory as defined by Fagerlind (1944), and aneuspory and gonial apospory defined by Battaglia (1963). The resultant gatemophyte has the somatic number of chromosomes (2n). Unreduced gametophytes formed by apospory or diplospory may give rise to embryos by parthenogenesis[6] or pseudogamy[7].

Regardless of the mechanisms involved, gametophytic apomixis may be *obligate* or *facultative.* In obligate apomicts the sexual process is completely replaced by asexual methods of reproduction; while in facultative apomicts the sexual and asexual methods of reproduction coexist in one plant. The proportion of apomictic progeny produced by facultative agamosperms has been shown to vary with several factors, namely the chromosome numbers and the environmental conditions under which seed is produced.

(b) Adventitious embryony (Adventive embryony, sporophytic embryony)
This refers to the condition wherein the embryo develops directly from the nucleus or adjacent tissues with no intervening formation of embryo sacs and egg cell (Fig. 4.3).

c) Non-recurrent apomixis
Non-recurrent apomixis refers to the condition in which normal meiosis yield a haploid sporophyte via haploid gametophyte.

---

[6] Refers to the development of a sporophyte from a female gamete in the absence of pollination or indeed any stimulus, without, syngamy. Such an individual may be either a diploid or haploid (diploid parthenogenesis or haploid parthenogenesis). When the embryo is derived from the egg cell it is known as *gonial parthenogenesis* and when derived from some other cell in the gametophyte as *somatic parthenogenesis* (apogamety).

[7] *Pseudogamy* signifies that pollination is necessary to initiate or complete embryo formation although the male gametophyte makes no contribution to the embryo. *Gonial pseudogamy* refers to the development of the embryo from the egg cell. *Somatic pseudogamy* refers to its development from some other cell of the gametophyte.
Semigamy refers to maternally and/or paternally derived offspring after induction by pollination but no gametic fusion.
Polyembryony refers to the occurrence of more than one embryo in the seed.

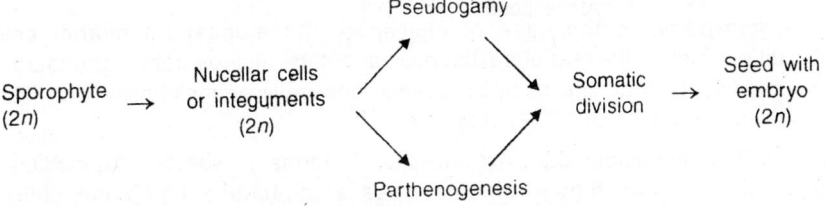

Fig. 4.3. Adventitious embryony (after Grant, 1971)

2. VEGETATIVE APOMIXIS

Vegetative apomixis refers to reproduction through vegetative parts, such as bulbils or other propagules which replace flowers or inflorescences. However, the term also covers reproduction by stolons, rhizomes, runners and the like where the normal sexual processes are not functioning or are markedly reduced.

Determination of Apomixis[8]

The breeding behaviour of agamospermous plants is like that of an interspecific hybrid.

1. Apomictic reproduction is mostly observable in an offspring that is more uniform than expected, the uniform plants being similar to the mother plant. The percentage of uniform plants gives the degree of apomixis in the female parent. When apomixis is the only form of seed formation (obligate apomicts) this holds true for all offsprings. The only deviating types are rare triploids due to fertilization of an unreduced egg cell. On the other hand, facultative apomicts combine apomictic and sexual reproduction, resulting in a progeny of maternal types and aberrant types originating either from fertilization, and also due to cytological irregularities both at the male and female side. The uniformity of offsprings therefore depends on the degree of apomictic reproduction in the parent plant, which is a genotypic character, somewhat influenced by the environment.

2. When the pollen parents have no or little influence on the progeny, apomixis is very likely

3. Apomictic plants are highly heterozygous. When self-pollination is possible, the lack of inbreeding depression in most of the progeny plants is another indication of apomixis.

4. An enhanced number of multiple seedlings too refers to apomictic behaviour.

5. The choromosome number is irregular. In many species rather a wide range is found, for example in *Poa pratensis* the chromosome number may vary from $2n = 28 - 154$. An uneven chromosome number combined with an undisturbed seed setting is typical of apomictic reproduction.

---

[8] Confirmation of apomixis may be done through embryological studies. The occurrence of several embryo sacs in the young ovule is an important criterion.

Types of Apomixis Useful in Plant Breeding

1. *Diplospory*: In the case of diplospory the megaspore mother cell differentiates like that of sexual ovules, but its nucleus does not undergo meiosis. Instead the nucleus divides mitotically or meiotically with first division restitution (FDR) (Fig. 4.4).

   In FDR the nucleus does not undergo the normally expected disjunctional separation of homologous chromosomes at anaphase I and the entire diploid complement divides mitotically to give rise to a dyad with two unreduced spores. The embryo sac originates from such spores having $2n$ chromosomes.

2. *Apospory*: In this case the types of division are similar to that described for diplospory. The embryo sac, however, originates from any cell of the nucleus.

3. *Second division restitution (SDR) or haploid fusion*: Restitution in the second division of fusion of two haploid nuclei also yield diploid reproductive cells (Fig. 4.4).

4. *Semigamy*: In semigamy, although the male nucleus penetrates the egg cell, it does not fuse with the egg nucleus. The embryo shows the coexistence of both male and female nuclei.

Genetics of Apomixis

   For genetic studies apomictic genotypes are crossed with related sexual forms. It is generally believed that apomixis is genetically controlled. The

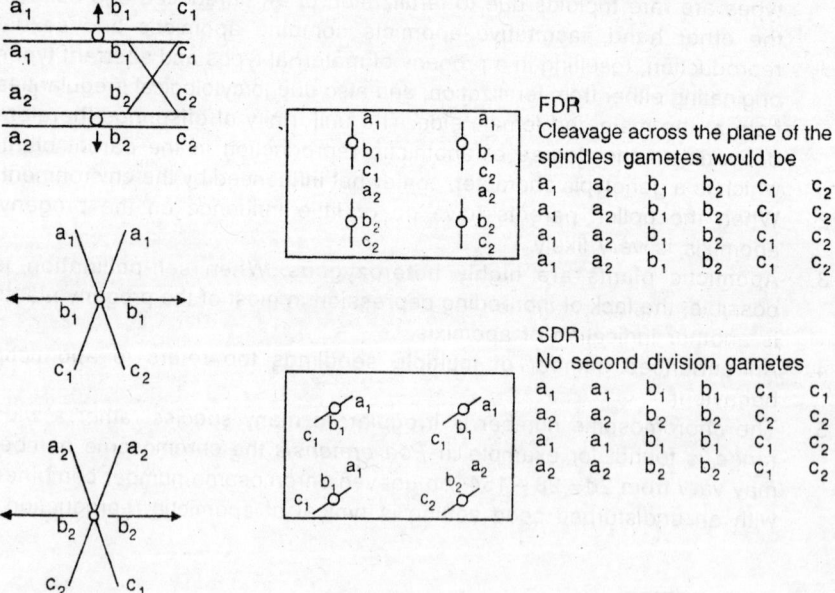

Fig. 4.4. Consequences of FDR and SDR in unreduced gametes. (from Thomas, 1993)

formation of unreduced gametes seems to be under complex genetic control, and the system varies in different species from simple monogenic control to inexplicable genetic situations. Also, the various genetic mechanisms may exist even within the same species.

The inheritance of adventitious embryony, on the other hand, appears to be simple and dominant over sexual embryony and very few genes may be involved.

Environmental factors may have some influence on the proportion of sexual to apomictic seeds, which has these two implications for a plant breeder:

1) introducing greater sexuality through environmental manipulations (for example photoperiod, controlled greenhouse conditions) to facilitate hybridization, and

2) changes in method of reproduction by the environment may pose a danger when the varieties are multiplied outside the region where they have been developed.

5

Male Sterility

Male sterility, that is, the incapacity of a plant or plants to produce or to release functional pollen may be true sterility, functional male sterility or sterility induced by a chemical gametocide.

a) True male sterility: This type of male sterility is either due to unisexual flowers, that is, absence of male sex organs in female flowers, as in *dioecious* species and *monoecious* species, or due to pollen sterility because of irregularities in the development of microspores, causing abortion of the pollen grain in bisexual flowers (*hermaphrodite* species).

b) Functional male sterility: This type of male sterility results from a defect in the dehiscence mechanism of the anther due to which pollen are not released and thus fail to fertilize, even though they are fertile.

c) Induced male sterility: This type of male sterility results when certain chemicals (gametocides) are used at a particular stage of crop growth in appropriate doses which interfere with the development of male gametophyte leading to aborted anthers.

TRUE MALE STERILITY

1. *Dioecious Species*

In dioecious species (for example asparagus, hemp, spinach, papaya, date palms) male and female flowers or inflorescences are borne on different plants. Sex determination is under the control of one major gene pair, *mm* female and *Mm* male. In all such cases, in which specific sex chromosomes have been identified, the female and male plants are designated as *XX* (female) and *XY* (male).

2. *Monoecious Species*

In *monoecious* species, for example corn, male and female flowers or inflorescences are borne on the same plant but at different places. In some crops, for example cucurbits, the male and female flowers develop successively.

3. *Pollen Sterility (Pollen Abortion)*

Pollen sterility, that is, the failure of the stamens to produce viable pollen may be due to chromosomal factors, extrachromosomal hereditary particles (cytoplasm), or could be due to both the cytoplasmic and chromosomal factors (genetic factors). Non-functional or abortive pollens may be formed in varying degrees. Three types are generally recognized (Fig. 5.1).

a) GENETIC MALE STERILITY

Genetic male sterility is usually governed by one recessive gene (*ms*). In some crops more than one loci are involved. Examples are corn, barley, sorghum and tomatoes. The genetic male sterile line is maintained by crossing it with a heterozygous counterpart fertile line. Half of the resultant progeny is male sterile and the half is male fertile. In crossing blocks (hybrid seed production blocks) the fertile plants in the female parent lines (genetic male sterile lines) are rogued out as soon as they can be recognized. In some cases in which male sterile plants can be identified through closely linked marker genes at an early stage, for example in sunflower there is close linkage between the allele for male fertility *MS* and the allele for red anthocyanin coloration of the entire plant *T* (Le Clercq, 1966) or by some pleiotropic effect of the male sterile gene, the task of producing hybrid seed is straightforward. If, on the other hand, one has to identify male sterile plants through careful examination of flowers, it may not always be possible to remove the fertile siblings in female parent lines before they shed pollen. This is a serious drawback to practical exploitation. Genetic male sterility is thus of limited value as a tool for hybrid seed production.

Balanced tertiary trisomic system (BTT) in barley (Ramage and Tullen, 1964; Ramage, 1965) : The essential feature of this system is the presence of an extra chromosome composed of parts of two non-homologous chromosomes. The extra chromosome contains the *Ms* allele and the normal homologous chromosomes contain the *ms* gene. This supernumerary chromosome is not transmitted by the pollen grains. Consequently, diploid progeny of a trisomic are always sterile (*ms ms*), while the trisomics descendants are fertile (*Ms ms ms*).

The two homologues of the second chromosome included in the extra chromosome carry the recessive allele (*r*, green stem) of a marker gene, whereas the extrachromosome possesses the matching allele (*R*, red stem). Thus fertile and sterile segregates can be easily distinguished. The major limitation of this system, however is that BTT plants are weak, have reduced female fertility and low pollen production, further aggravated by the problem of poor pollen dispersal in the barley. Also, the genetic purity of seeds produced is lowered due to the appearance of some fertile plants (trisomic plants) in the male sterile population which fertilizes a higher proportion of the neighbouring male sterile sister plants than will the intended pollen parent.

b) CYTOPLASMIC MALE STERILITY

In this type the dysfunction of the anther or abortion of the developing microspores is the result of the action of cytoplasm. The sterilizing cytoplasm is usually designated as *S* and the normal cytoplasm as *F* or *N* cytoplasm. The entire progeny of a cross between a cytoplasmic male sterile plant and

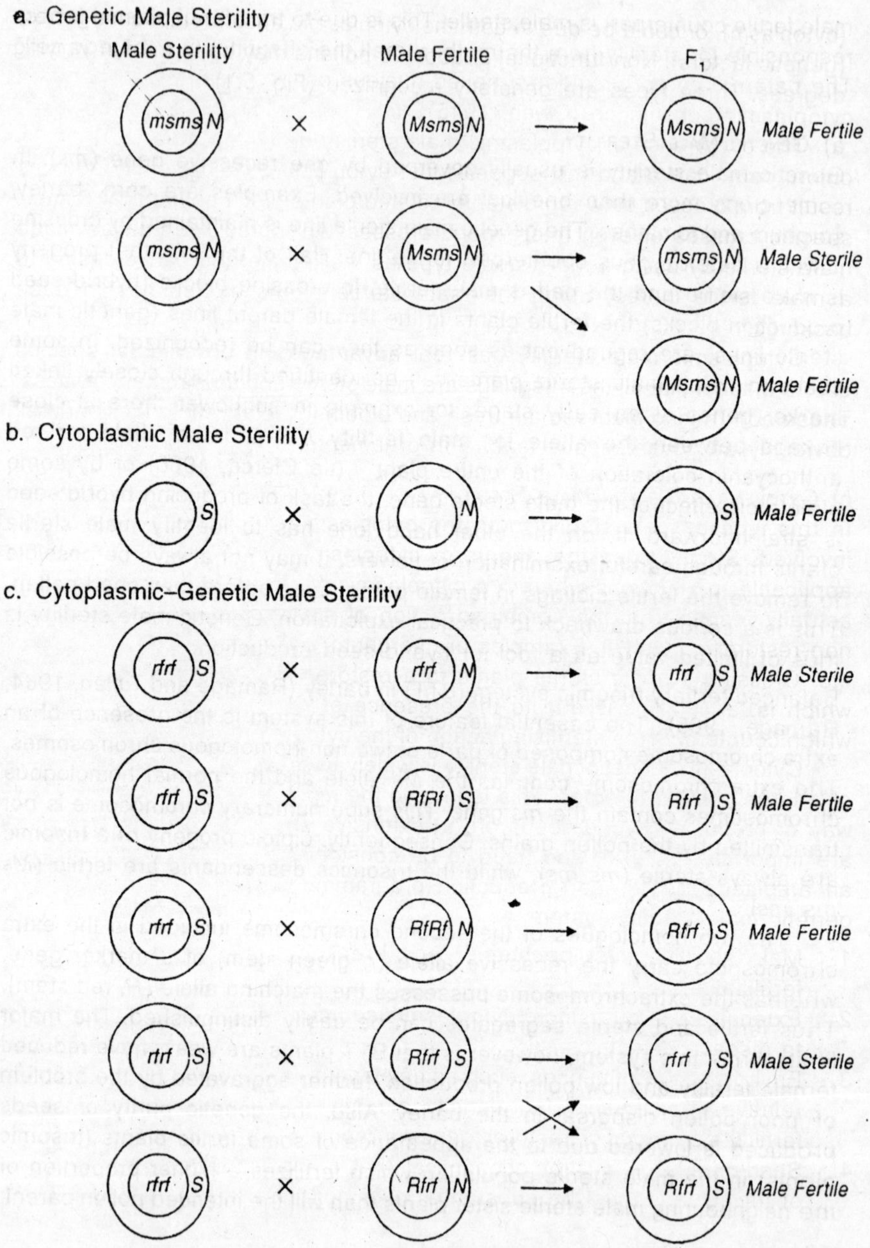

38

Fig. 5.1. Methods of inheritance of male sterility. The letters in inner circles represent genetic factors, and in the outer circle cytoplasmic factors. The gene *Ms* is dominant over small *ms*. The gene *Rf* is dominant over *rf* and has the ability to restore fertility. The cytoplasmic factors are transmitted only by a female parent. *N*–Normal (fertile), and *S*–sterile cytoplasm.

male fertile counterpart is male sterile. This is due to transfer of plasmagenes, responsible for sterility from the mother to all the offsprings via the egg cell. The paternal parent provides a naked nucleus only and does not affect the cytoplasmic constitution (Owen, 1945).

The occurrence of cytoplasmic male sterility has been reported in beet, onion, corn, carrot and red pepper. Most cytoplasmic male sterility sources result from crosses between distantly related plants, such as intra- and inter-specific crosses or even intergeneric crosses. An inbred line can be made male sterile by using a male sterile type as the female and backcrossing to a male fertile inbred, using the male sterile plants as females in each backcross.

Cytoplasmic male sterility has real advantages in ornamental species due to the fact that all the offsprings are male sterile and thus remain fruitless. These non-fruiting plant remain fresh and bloom for relatively longer periods, obviously an advantageous feature in ornamentals.

c) Cytoplasmic-Genetic Male Sterility

In this type of male sterility both the cytoplasmic and genetic factors are involved. Strictly speaking, the term cytoplasmic-genetic male sterility is applicable to situations wherein a cytoplasmic-genetic male sterility system actually results from the combined action of sterilizing cytoplasm (S) and non-restorer genes (rf). Examples are sorghum, pearl millet, sunflower, rice, etc. Crossing of male sterile plants to a restorer line gives rise to progeny which is fertile. This is due to the presence of restorer genes (Rf) genes, which counteracts the sterilizing action of the cytoplasm and restores fertility.

Cytoplasmic-genetic male sterility is widely used in hybrid seed production of a large number of crops. This is indeed the most convenient and practical way of producing hybrid seeds on an extensive scale. The following factors are important however, and need to be considered before embarking upon an ambitious hybrid seed production programme based upon cytoplasmic-genetic male sterility system.

1. Male sterility must be stable. Any breakdown leads to partial self-fertilization.
2. Extensive use of a single cytoplasmic source may increase vulnerability to disease epidemics.
3. Environmental influences, such as day length, light intensity, temperature, relative humidity and plant density, etc. have been reported to affect the fertility behaviour of restorer genes in some crops.
4. Seed production must be economical, that is, yield and quality advantages of the hybrids must be substantial over non-hybrids.

FUNCTIONAL MALE STERILITY

In some instances although the plant produces normal amount of viable pollen, the pollen is not released because of a defective dehiscence mechanism. The anthers do not open and shed the pollen. The various aspects of use of functional male sterility have been more widely investigated in tomato, wherein it has been reported that non-dehiscence is highly sensitive to environment. For this reason functional male sterility has seldom been used in commercial F_1 hybrid seed production of tomato.

Chemical Induction of Male Sterility

A chemical, i.e., a chemical gametocide, having the ability to induce pollen sterility or check pollen shed may be suitably employed in hybrid seed production. The following chemical hybridizing agents (gametocides), for example in rice have reportedly been used in China to induce male sterility for hybrid seed production.

MG1 (monosodium methane arsenate)
MG2 (sodium methyl arsenate)

Criteria for a good gametocide

1. Once standardized, it should give consistent, reproducible results. It should cause pollen abortion only. Ovule fertility should not be impaired in any way.
2. It should be free from any mutagenic effect.
3. The method of treatment should be simple and easy to adopt in the field.
4. It should be economical.

Limitations of chemical hybridizing agents (CHAs)

The available gametocides for hybrid seed production have not been put to extensive practical use, primarily for three reasons:
1) pollen abortion is incomplete and erratic, treatment is effective for short periods and at a particular stage in plant development and hence needs to be repeated;
2) female fertility is somewhat adversely affected, and
3) cost is relatively high.

6

Self-incompatibility

Self-incompatibility denotes the inability of a plant to set seeds upon self-pollination, despite the fact that both ovule and pollen development are normal and viable. This is due to some physiological hindrance to fertilization, for example slow pollen tube growth. Self-incompatibility differs from male sterility. The important differences are given in Table 6.1.

Table 6.1. Differences between male sterility and self-incompatibility

Male sterility	Self-incompatibility
1. Due to failure of processes such as pollen development	Due to some physiological hindrance to fertilization. Failure to obtain seeds is due to slow pollen tube growth.
2. Pollen are non-functional (abortive)	Pollen and ovule are normal and functional.
3. Genetic, cytoplasmic or both cytoplasmic and genetic factors control the expression of male sterility.	Genetic and physiological factors control the expression of self-incompatibility

SYSTEMS OF SELF-INCOMPATIBILITY

Factors such as genetics, physiology and floral morphology determine the system of self-incompatibility. The salient features of self-incompatibility systems are summarized in Table 6.2.

Heteromorphic System

The characteristic features of heteromorphic systems are:

Heterostyly: Each mating type (flower) is morphologically distinct, characterized by relative positions of its stigma and anthers. Examples are *Primula* species in which two types of flowers, namely, thrum and pins are borne on different plants of a population (distyly); and *Lythrum* in which three different types of flowers are borne on different plants (tristyly). The relative length of style determines the types of flower.

Table 6.2. Self-incompatibility systems in flowering plants (after Lewis, 1954)

Floral morphology	Genetic control				Physiology of action
	No. of loci	Alleles per locus	S allele action		
			in pollen tube	in style	
Heteromorphic					
distyly	1	2	Sporophytic dominance	Dominance	Complementary stimulant or oppositional inhibitor
tristyly	2	2			
Homomorphic	1	many	Sporophytic dominance or individual action	Dominance or individual action	Oppositional inhibitor
	1 or 2	many	Gametophytic individual action	Individual action	Oppositional inhibitor

Genetic control of heterostyly

a) Distyly

The difference between the two flower types, for example thrums and pins, is governed by a complex of six differentiating characters which are genetically controlled. The characters are: style length (*Gg*), stigmatic surface (*Ss*), stylar incompatibility (I_1 i_1), pollen incompatibility (I_2 i_2), pollen size (*Pp*) and anther size (*Aa*). This entire system, however. is controlled by a single gene '*S*' having two alleles. Lewis (1954) proposed supergene structure for the *S*-locus as follows:

Phenotype	Genotype
Thrum (short style, long anthers)	Ss $GS I_1 I_2 PA/g si_1 i_2 pa$
Pins (long style, short anthers)	ss $g s i_1 i_2 pa/g si_1 i_2 pa$

Although self-incompatibility is primarily a reaction between haploid pollen tubes and diploid style, thrum pollen. despite segregation of *S* and *s* behaves as if it were all *S*-type. This is because of sporophytic determination of the pollen reaction and *S* being dominant over *s*.

Mating occurs only between anthers and stigmas of the same length, that is,

Pin × pin	ss × ss	incompatible mating
Pin × thrum	ss × Ss	1 Pin SS : 1 thrum Ss
Thrum × pin	Ss × ss	1 Thrum Ss : 1 pin ss
Thrum × thrum	Ss × Ss	Incompatible mating

The occurrence of homomorphic plants in this group, although rare, may be explained by assuming intralocus crossing over.

b) Tristyly

The difference between the flower types is governed by two genes, both with two alleles (*Ss* and *Mm*); double dominants have the shortest style while double recessives develop the longest styles.

Homomorphic Systems

All mating types are morphologically alike in this group and can be recognized only by breeding tests. The incompatibility locus is characterized by a large multiple alleles series. The incompatibility systems among homomorphic flowering plants are broadly divided into two groups, namely, the gametophytic system and the sporophytic system (Table 6.3).

Gametophytic Determination of Pollen Reaction

Examples: Species belonging to Solanaceae, Onagraceae, Rosaceae, Gramineae, Leguminosae, Scrophulariaceae

a) Single Gene S with Multiple Allele Forms

The gametophytic incompatibility system originally called the *oppositional factor system*, is controlled by a single gene S, which is usually characterized by a large number of allelic forms, such as, S_1, S_2, S_3, S_x, etc. It gives rise to three main types of pollination:

1) fully incompatible: for example, $S_1S_2 \times S_1S_2$ in which both alleles are common;
2) 50% incompatible and 50% compatible: for example, $S_1S_2 \times S_1S_3$ in which one allele is different and one is common;
3) fully compatible: all the pollen is compatible: for example, $S_1S_2 \times S_3S_4$ in which the alleles involved are different.

Table 6.3. Salient features of gametophytic and sporophytic systems

Gametophytic system	Sporophytic system
Mating type of pollen is determined by the S-allele pollen	Mating type of pollen is determined sporophytically (i.e. by the parental S-genotype)
Controlled by alleles at one locus, or alleles at two loci	Controlled by a single locus
Alleles of the incompatibility gene(s) act individually in the style	Alleles of the incompatibility gene can either express dominance or individual action, both in the pollen and style
The incompatible pollen inhibited in the style	The incompatible pollen tubes can be inhibited on the stylar surface or incompatibility may be expressed between gametes after fertilization

The number of multiple alleles involved is very large, for example in *Trifolium* (212), *Oenothera* (37), and *Nicotiana* (17). This is necessary as well since the gametophytic determination of pollen reaction at a single *S* locus cannot perpetuate itself. Consequently, plants are virtually always heterozygous at this locus. The possible ways in which the unusually high number of *S*-alleles may have arisen are:

i) evolution of the new alleles through direct mutation,

ii) an allele may give rise to a new one in several steps due to some balanced changes in its internal structure, and

iii) all the *S*-alleles may have evolved independently, from the original *S*-allele during the course of evolution of the incompatibility system, excluding the possibility of direct mutation of one allele into another.

Self-compatibility allele

1) East and Yarnell (1929) found plants having a self-compatible allele at *S*-locus. Plants with a genotype $S_f S_x$, where S_x is any other allele in the series, are self-compatible and on selfing give self-compatible plants of the constitution $S_f S_f$ and $S_f S_x$.

2) Anderson and de Winton (1931) found an allele having a special effect. This allele S_f is able to inhibit the growth of pollen tubes carrying the S_f allele.

3) Lewis and Crowe (1954, 1958) studied spontaneous and X-ray-induced mutations of the *S*-locus in *Prunus avium* and *Oenothera organensis*. The results suggested that the *S* locus was composed of two cistrons, one governing pollen reaction, the other controlling stylar reaction. Mutation could affect one or both of these units, changing *S*-alleles into S_f alleles, which led to self-fertility. Such S_f alleles of the *S* series are also known to occur naturally in several species. Later reports confirmed this interpretation, that the *S*-locus is a complex of at least two parts and showed further that mutations of the two components of the *S*-complex may be revertible or permanent.

4) In addition to the multiple alleles of the *S*-locus, certain other loci are also reported to affect the growth of pollen tubes in compatible matings, for example *P*-locus in *Nicotiana*, *F*-locus in *Antitrrhinum*, *R*-locus in *Solanum* and *T*-locus in *Oenothera*.

b) Two Genes and Two Separate Loci

1. Pollen control may also be exerted gametophytically by the operation of two separate incompatibility loci requiring joint identity in the style for both the factors (*S* and *Z*) of the pollen for full incompatibility reaction to be produced, for example in *Secale cereale*, *Festuca pratensis*, *Phalaris coerulescens*, *Hordeum bulbosum* and *Physalis ixocarpa*. In these cases a weakening of self-incompatibility results since absolute identity of all the four alleles would be very rare.

2. Mather (1944) observed that while *S*-alleles control the working mechanism of self-incompatibility, the essential basis of the system rests

upon a delicate balanced polygenic complex, thereby suggesting a polygenic control. This hypothesis is supported by Thore (1963) from his work on *Trifolium pratense* wherein he observed that the self-compatibility system may be strengthened leading to complete outbreeding or weakened to the extent of complete self-fertility by way of genetic segregation and selection of modifying genes depending upon the direction of selection pressure.

Sporophytic Determination of Pollen Reaction

Examples: *Cosmos bipinnatus, Cardamine pratensis, Iberia amara and Brassica* species.

Genetic control of this system is by a single locus having multiple alleles which may show either dominance or may have individual action both in the pollen and style. Because of this variation there is a great complexity in incompatibility relationships. Thus, instead of having two simple pollination patterns, as in the gametophytic system, there are several patterns in this system. In general, the greater individual action in operation, whether in pollen or in the style, the more the cross incompatibility.

Further details are given in Chapter 21.

Distinguishing the Sporophytic System from the Gametophytic System

1. *Variety of mating patterns:* As mentioned earlier, there are just two simple mating patterns in the gametophytic system, versus a far greater variety of mating patterns in the sporophytic system.
2. *Reciprocal differences:* In the sporophytic system there are frequent reciprocal differences in matings.
3. *Incompatibility of female parent:* Incompatibility can occur with the female parent in sporophytic system.
4. *Number of incompatibility groups:* A family can consist of three incompatible groups in the sporophytic system.
5. *Occurrence of homozygotes:* Homozygotes are a normal part of the sporophytic system.
6. *Number of genotypes:* In a sporophytic system an incompatible group may comprise two genotypes.
7. *Effect of polyploidy upon the action of self-incompatibility mechanism:* The breakdown of self-incompatibility in induced autotetraploids is of wide occurrence only in the gametophytic system (where self-incompatibility is controlled by a single locus). It is believed that the presence of two alleles in the diploid pollen creates an altogether new situation for the operation of monogenic gametophytic control where a competitive interaction may occur between alleles contained in the same pollen, resulting in the breakdown of the whole mechanism.

In the sporophytic system, on the other hand, two alleles are normally present at the time of *S*-allele action and hence there is obviously no effect of autotetraploidy.

8. *Time of S-allele action:* In a sporophytic system the time of S-allele action is premeiotic, that is, after meiotic division II but prior to separation of the microspores from common cytoplasm of the pollen mother cell; in the gametophytic system the S-allele action is post-meiotic, that is, after separation of the four microspores (Pandey, 1958).

PHYSIOLOGY OF SELF-INCOMPATIBILITY

Incompatibility alleles are present in both the female and male tissue, that is, ovule, ovary, pollen grain and pollen tube. It is important to note that self-incompatibility reaction takes place only when identical genetic information is present in both the male and female tissues, the same allele acting in different ways in pollen and megasporangial tissue. Several theories have been advanced on the physiological and chemical nature of self-incompatibility.

East's Immunological Theory (East, 1929): This theory assumes that incompatibility is due to a positive inhibitory reaction rather than absence of a specific stimulant since pollen grains are capable of germinating and growing in a laboratory medium and also in styles of different species. The inhibitory reaction is highly specific in the sense that only pollen and styles with like alleles react together. The secretions of the pollen tube act as antigens against the stylar tissue which forms antibodies against such a pollen tube and in this way inhibits growth.

Straub's Theory (Straub, 1946, 1947): This theory envisages a substance is formed in the pollen grain in a definite amount and with a specific quality determined by the S-allele present. This substance (PGF) is regarded as necessary for the growth of the pollen tube in the style and probably because of its enzyme-like property, is able to decompose the stylar tissue. When all the substance is consumed in growth, the tubes do not grow further.

The East and Straub theories differ in that Straub envisages that the pollen component of the reacting system has a vital function in tube growth as well as in its specific incompatibility reaction with a stylar component, whereas in East's theory no such growth function is involved. From an experimental viewpoint, there are two types of self-incompatible plants:

1) Plants in which incompatibility reaction takes place during growth of the pollen tubes through the styles, for example in Solanaceae, Unagraceae and Scrophulariaceae.
2) Plants in which incompatibility reaction takes place on the surface of the stigma, for example, Cruciferae, Compositeae.

For these two types of incompatibility systems, Brewbaker (1957) observed a correlation between pollen cytology and the place of incompatibility inhibition in *homomorphic* plants.

a) Gametophytic: Observed primarily in species having binucleate pollen with inhibition occurring during pollen tube growth.

b) Sporophytic : Linked with trinucleate pollen grains and inhibition occurring on the stigmatic surface or during very early tube growth.

On the basis of Linskens (1955) work on the germination and growth of pollèn *in vitro* showing the possibility of sucrose as a key to successful germination and growth *in vivo*, Brewbaker (1959) revived Straub's consumption theory and proposed that physiological action of incompatibility alleles may be mediated simply through the control of sucrose uptake or metabolism by the pollen grain or pollen tube. It appears possible that trinucleate pollen grains might be relatively sucrose- or metabolite-deficient at maturity due to the occurrence of a second mitotic division in microsporogenesis since germination of most binucleate pollen may readily be accomplished in 3-10% sugar or sugar-agar solutions, especially upon addition of 10-100 ppm boric acid, whereas trinucleate pollen generally requires higher sugar concentrations.

Recent studies have shown, however, that in the first place the cuticle of the stigma is the incompatibility barrier in the case of incompatibility reaction which takes place on the stigmatic surface. This cuticular material can be broken down enzymatically. It may be concluded that this type of incompatibility barrier is linked with the cutinase enzyme system.

In the case of an incompatibility barrier which acts by inhibiting pollen tube growth within the style before its reaching the embryo sac, numerous reports have indicated that inhibition of pollen tube growth in an incompatible style may be caused by an antigen-antibody type of reaction between a specific protein in the pollen and a protein with homologous specificity in the style (Linskens, 1961).

SELF-INCOMPATIBILITY IN PLANT BREEDING

Genetically controlled self-incompatibility in flowering plants offers unique economic opportunities as described below. it is a potential tool in breeding F_1 hybrids, synthetics, triploids, etc.

1. Production of hybrids (triploids) : Self-incompatibility can facilitate production of triploids by planting alternate lines of self-incompatible diploids and tetraploids. However, this can only be done in those cases in which self-incompatibility does not break down at the tetraploid level.

2. Production of F_1 hybrids : In plants in which S-homozygotes can occur (i.e., in the sporophytic system), reversible mutants of the S-locus are of particular interest for this purpose since S-homozygotes will constitute half of their progeny. If two inbred lines could be made homozygous for their S-alleles, say S_1 and S_2, only hybrid seed would be produced when these two lines S_1S_1 and S_2S_2 are interplanted.

Double-cross hybrids could also be produced, by making four inbred lines homozygous, say S_1S_1, S_2S_2, S_3S_3 and S_4S_4. Alternate planting of two single crosses, say, $S_1S_2 \times S_3S_4$ would give double-cross hybrid seeds. The enormous economic advantages from such a system are obvious. Further details are given in Chapter 21.

3. Ornamental plants: The blooming period of ornamentals and the vegetative phase in vegetables (say leafy and root vegetables) can be prolonged by preventing seed production.

4. Horticultural crops: In many fruit crops, for example apple, the presence of self-incompatibility necessitates mixed plantings which leads to uneven quality. Development of self-fertile strains from self-incompatible ones by inducing stable type mutants of S-alleles conferring self-fertility would be highly desirable. Selfing in these fruit trees would release potential variability and would permit segregation of rare recessive phenotypes in this highly heterozygous material.

Limitations of Self-incompatibility

1) To breed F_1 hybrids development of homozygous lines is necessary. For this repeated selfing of lines must be possible.

2) Production of hybrid seed involving the homozygous lines produced in the manner outlined earlier is possible, provided the possibilities of self- and sib-fertilization are completely excluded.

7

Techniques of Artificial Hybridization

A knowledge of the principles of artificial hybridization (crossing) and self-pollination (selfing) and the techniques involved, is a basic requirement for a plant breeder, because he is required to make hundreds of crosses or self-pollinations in his attempts to breed improved varieties. This is so because, whether it is the incorporation of desirable genes from one line to another, or genetic studies, or development of hybrid varieties, he is required to artificially hybridize. Similarly, a plant breeder has to make a large number of self-pollinations or sib-pollinations to breed inbred lines or to maintain inbred lines (Fehr, 1980).

ARTIFICIAL HYBRIDIZATION[1]

General Principles

Fehr (1980) outlined the general principles for successful artificial hybridization.

1. Selection of parents : The careful selection of parents consistent with the breeding objective is the first prerequisite for achieving success through artificial hybridization. Parents should be carefully selected from the available germplasm on the basis of documented information and personal acquaintance with the breeding materials. Prior knowledge regarding polyploidy, chromosome numbers and feasibility of hybridization is especially important in making wide-crosses. Once the choice of parents has been made, usually the more vigorous of the two is chosen as female. In wide-crosses, it depends on which way hybridization would be successful.

2. Floral biology of the crop : A knowledge of the floral biology of a crop is necessary for success in artificial hybridization or self-pollination. Success in artificial hybridization surely depends upon the good availability of pollen when the female is in flowering and the stigmas are receptive. A study of the floral biology of the crop concerned to gather information, such as flowering behaviour, time and period of anthesis (dehiscence of anthers), pollen viability, storability and stigma receptivity

---

[1] Methods employed for hybrid seed production are discussed in Chapter 32.

is thus essential. When male sterility or self-incompatibility is involved, the breeder should also know the system involved.

The floral biology of a crop may be somewhat influenced by the environment and hence needs to be studied at several different locations and in different seasons to ascertain the most appropriate time of anthesis and stigma receptivity.

3. Floral Structure : A breeder should be thoroughly familiar with the various flower parts, more particularly with the sexual organs, their morphology, number and arrangement in the flower, besides the mechanism of natural pollen dehiscence and fertilization in respect of the crop(s) he is working with. This knowledge is needed to manipulate sexual organs of the flower to carry out artificial hybridization or self-pollination work.

Procedures of Artificial Hybridization

For the sake of clarity we shall discuss the procedures of artificial hybridization under the headings: (1) Emasculation and (2) Pollination.

1. Emasculation

Emasculation refers to removal of anthers, or making the anthers non-functional (through the use of natural or induced male sterility or even self-incompatibility). It is the first step in artificial hybridization. Successful emasculation depends upon several factors (Fehr, 1980).

FACTORS AFFECTING SUCCESSFUL EMASCULATION

1) Identification of flowers for emasculation : Flowers that are too immature are not only difficult to emasculate, but the percentage of successful pollinations is also low. On the other hand, flowers that are relatively old may have already been selfed or naturally pollinated. It is therefore imperative to select flowers for emasculation at the right stage. One way to develop correct judgment is to open a few flowers and examine them, if necessary under magnification, to determine pollen shed/fertilization. Any presence of pollen on the stigmatic surface or rupturing of anthers or withering of stigma would indicate that pollen has already been shed or the flower is already pollinated. There exists a relationship between floral appearance and development of sex organs. The secret is to know the right flower stage for emasculation.

2) Preparation of flowers : In crops in which the flowers are unisexual, preparation of the female involves simple bagging of the unopened flowers so as to prevent natural cross pollination. The usual practice is to cover at least one day before the flower is likely to open. In crops in which the flowers are bisexual and it is necessary to prevent undesirable pollinations, female flowers must be covered with bags, such as glassine bags, cloth bags, soda straws or paper bags. The need for covering female flowers prior to emasculation is not universal and varies among crop species and environments, for example greenhouse conditions or field conditions.

3) Time of emasculation : The right time for emasculation of a flower is before the anthers are mature and have a chance to shed pollen and the stigmas are receptive. It should also not coincide with the time of anther dehiscence. In crops in which the usual anthesis occurs in the morning hours emasculations should invariably be done in the afternoons and vice versa. The actual timings are based on the floral biology of the crops. In some crops it is desirable to emasculate one or more days before pollination is done, while in others it may be desirable to emasculate and simultaneously pollinate the flowers (the period of stigma receptivity also influences the time lag between emasculation and pollination).

EMASCULATION METHODS

Flowers are emasculated in a number of ways depending on the floral characteristics and mode of natural pollination of the crop. (Fehr, 1980).

1) No emasculation is needed: Genotypes with male sterility, self-incompatibility, or protogyny; monoecious and dioecious species do not require emasculation. When self-pollinated species with fertile male flowers are not emasculated, the stigma must be receptive at least a day before anthers of the flower can shed pollen. The stigma is pollinated immediately and then covered with a bag. The usual practice is to cover the unopened flower with a bag to avoid accidental pollen, and subsequently pollinate it when the flower open.

2) Direct emasculation : This consists of direct removal of anthers. The usual practice is to remove all underdeveloped, overdeveloped or previously pollinated flowers from the inflorescence. Only flowers at the optimum stage are retained. Depending on the floral structure, the floral parts that cover the sex organs are entirely or partly removed or left intact. The anthers are removed by means of a forceps of suitable size. The forceps may be sharp, pointed or blunt with a curved or straight end. The choice of type of forceps to be used depends on the size of flower and the crop. In some crops a spear pointed needle, pencil, stick or scalpel is used; in other crops no tools are required and the stamens are removed by rubbing them between the thumb and forefinger. For some crops some very specific procedures have been developed.

3) Indirect emasculation methods : i) Thermal inactivation : The hot-water method of emasculation is a standard practice in some countries for crops such as rice and sorghum. An inflorescence that has just begun to flower is selected. All the open flowers are removed and the inflorescence is dipped in hot water. The temperature of the water and the immersion time vary with the crop (rice, 43°C, 5 minutes; sorghum, 47-48°C, 10 minutes). The hot water treatment kills the pollen but does not injure the pistil. Usually a thermos bottle of appropriate size is used to give the hot-water treatment. After treatment, the inflorescence is allowed to dry. Any unopened buds/florets are removed. After drying the inflorescence is bagged. Pollination is done 30-60 minutes after treatment.

ii) Alcohol emasculation : Emasculation in some crops, for example alfalfa, may be done by immersing the whole raceme in 57% ethanol for 10 seconds and then washing it in water for a few seconds. Species differ in susceptibility of the stigma to injury by exposure to alcohol. The seedset is relatively low.

iii) Use of chemical gametocides : Chemical emasculators (for example, monosodium methane arsenate, sodium methyl arsenate), particularly in hybrid seed production of rice, have also been used in China. This method is not commonly used.

iv) Dehiscence control : High humidity is known to delay anther dehiscence and therefore can be used as a tool to facilitate or eliminate emasculation in some grass species in which the flowers open at the time of pollen shed, provided some selfs are not objectionable and the quantity of hybrid seeds needed is more. To facilitate emasculation the florets are held under high humidity during flower opening so that anthers do not dehisce. Anthers are then removed from open flowers much more easily than from immature ones. To eliminate emasculation a panicle is kept in high humidity for several days until the desired number of florets have opened and then hand pollinated.

GENERAL PRECAUTIONS

The following time-tested safeguard measures may be adopted for excluding contamination either through accidental self-pollinations or natural outcrossing.

(1) Inspect emasculated flowers for complete anther removal and accidental anther dehiscence. A magnifier may be used to examine flowers that are very small.
(2) Rinse the forceps and needles, etc. in alcohol or water when changing the parents in a crossing block. A similar rinse is needed when anthers accidentally dehisce during emasculation.
(3) Cover the emasculated flowers and wherever necessary the unemasculated flowers also to exclude contamination through undesired pollinations and dehiscence of flowers.

LABELLING

The emasculated flowers are usually covered by bagging. Stamping the date on which the flower was bagged helps in determining when the flower is ready for pollination. A label is hung on the stem immediately below the flower to record the date of emasculation, parentage, etc. Later on, when pollination is done, the date of pollination and parentage of pollen parent are added.

2. Pollination

POLLEN COLLECTION AND STORAGE

The pollen availability in flowers may be readily seen by striking or stroking the inflorescence and watching for pollen, or by removing an anther from the flower with a forceps and brushing it on the thumbnail to check for pollen shed. Male flowers are usually bagged 12 to 24 hours before they are used for pollination to exclude foreign pollen contamination, which may have been carried by wind or insects from other sources. During this period, foreign pollen grains on flowers, if any, lose their viability. The pollen from male flowers is collected the next day or a few days after as the need may be. The best time to collect pollen and implement pollination is during the peak anthesis hours. Immediate use of pollen assures maximum viability and minimizes the time required for pollination. The methods of pollen collection vary in various crops. Pollen may be collected in bulk through the use of mechanical devices, such as mechanical vibrators, in some species.

In most species the pollen is rather short-lived and needs to be used without much loss of time. In some crop species pollen remains viable for ext ended periods without special treatment. Maintenance of pollen viability du.ing storage may be necessary when male and female flowers are not available at the same time. The usual methods are storage at low temperatures, at low humidity, and/or high relative humidity conditions. The exact procedure varies with the crop species.

POLLEN APPLICATION

1) Direct pollen application : Direct pollen application, that is, transfer of pollen directly onto the stigma is done either by crushing the anther onto the stigma, or using a brush, piece of cotton, artist's brush, rubbing of male flowers over female flowers, or pouring (dusting) pollen collected in a paper bag over the stigmatic surfaces, or a toothpick or a piece of cardboard (in the case of sticky pollen). These are the most commonly used methods of pollen application. The exact procedure varies with the crop and the nature of the pollen. In some instances where indirect pollination is possible, one has to resort to direct pollination if the pollen supply is limited.

2) Indirect pollen application : Indirect pollen application, that is, transfer of pollen indirectly through dissemination of enough pollen close to the female flower is also used in many crops. Examples : injecting pollen with a hypodermic syringe into the paper bag covering the female flower, covering the hole with a tape and agitating the bag to disperse the pollen; inserting the male inflorescence (exposed anthers) into the bag covering the female inflorescence and rotating it vigorously to stimulate anther dehiscence; or positioning female flowers slightly below the male inflorescence and enclosing both in a bag. Detached male inflorescence can be used, which usually would supply pollen for several days, if the cut stem is placed in a container of water. The exact procedure varies with the crop. This method has the advantage of providing pollen over an extended period of time and usually results in a higher percentage of seedset than direct pollen application.

3) Pollen application in self-incompatible species : Inflorescences of similar maturity from each genotype to be crossed are placed together at the same height in a bag or in an isolated enclosure for mutual pollination.

GENERAL PRECAUTIONS FOR POLLINATION

1) Rinsing of pollination equipment including fingers and hands with water

or alcohol when changing male parents is essential to avoid possible contamination.

2) Covering the female flower after pollination is necessary for all species (at least under field conditions). The covering used after emasculation or the bag used for pollen collection is generally used to cover the pollinated female.

3) In some crops covering (bagging) after pollination may reduce seedset. In such instances, after a few days, when the flower has faded, the coverings should be removed.

4) New buds may develop on an inflorescence after hand pollination is completed and set seeds by self- or cross-pollination. Systematic removal of new flowers may therefore be necessary to eliminate errors at harvest.

5) Removal of a part of calyx, corolla, glume or lemma during emasculation helps in differentiating hand-pollinated flowers from naturally pollinated flowers.

Labelling

The hybridized flowers are appropriately labelled. Each label contains such information as parentage, date of emasculation, date of pollination, number and position of pollinated flowers, etc. The label is usually placed on the stem just below the pollinated flowers. As a safeguard the pollinated flowers should also be marked with strings, or plastic ties, or foils or wires that are colour-codes or stamped to identify different male parents.

SELF-POLLINATION

Self-pollinated Crops

In crops in which the extent of natural outcrossing is negligible, for example cleistogamous species, nothing is required to ensure self-pollination. In crops in which the extent of natural outcrossing is not negligible, it becomes necessary to exclude foreign pollen. This is done in one of the following ways.

1) Bagging of inflorescences prior to flower opening so as to exclude chance pollinations either through visits by insects or deposition of foreign pollen by wind.

2) Manual (hand pollination) transfer of pollen from the anther to the stigma of the same flower. Bagging would be necessary to exclude chance outcrossing.

3) Specific procedures have been evolved for crops in which tripping of the flowers is required to ensure pollination.

Monoecious Crops

The procedure is exactly the same as practised for artificial hybridization. The only difference is that pollen for self-fertilization is obtained from the same plant on which the female flower occurs.

Self-Incompatible Crops (Bud Pollination)

Self-pollination in these crops in done at the young bud stage (3-4 days prior to natural flower opening). The usual practice is to open 6-8 young buds at a time and pollinate them with the pollen from an older open flower on the same plant. Pollen can be transferred with a brush or a thumbnail. A fertile flower in full bloom is often used and its anthers brushed across the stigmatic surface to transfer the pollen. This is also known as *bud pollination*. After pollination the female flowers are covered with glassine bags in the usual manner.

Labelling : Labelling is done in a manner similar to that described for artificial hybridization. Information such as date of self-pollination and parentage is written on the label.

TECHNIQUES FOR SYNCHRONIZING FLOWERING DATES

Several methods are employed to achieve simultaneous flowering among genotypes which differ in maturity, in particular the number of days to flowering. This knowledge is particularly essential for obtaining higher seed yields from hybrid seed production fields. The usual methods are described below.

1. Delayed planting/multiple plantings (staggered plantings) of early flowering genotypes : When the difference in number of days to flowering is precisely known, synchronization is rather a simple affair. The early parent is planted late so that both the parents come to the flowering stage at almost the same time. When the precise difference in number of days to flowering is not known, the late parent is planted on a single date and the early parent planted thereafter on several dates. Usually 2-3 plantings at one-week intervals are considered sufficient to ensure pollen availability when the late-flowering parent comes to the flowering stage.

   For hybrid seed production the optimal time (gap between staggered plantings) can be determined by observation and experience at each location (Agrawal, 1995).

2. Manipulation of growing environment : Under controlled environmental conditions, for example growth chambers, manipulation of daylength, temperature, relative humidity conditions etc. may help delay or hasten flowering. Species and varieties differ very widely in the basic vegetative phase (BVP)[2] and daylength requirements. A knowledge of the critical photoperiod[3] is essential when daylength is used to delay the flowering of a genotype. Similarly, a knowledge of optimum photoperiod[4] for a species/genotype is necessary to ascertain flowering and seed

---

[2] BVP refers to the minimum amount of vegetative growth, after which floral induction can take place.

[3] Critical photoperiod is defined as the miximum daylength at which a short-day plant will flower, and minimum daylength at which a long-day plant will flower.

[4] The optimum photoperiod is the daylength at which a plant will flower in the shortest time after BVP is completed.

development. For each crop variety timing, duration, level and quality of supplemental lighting need to be standardized and/or considered for manipulating flowering behaviour.

Cool temperatures generally delay flowering while warm temperatures hasten it. Excessively high temperature conditions may slow development in many species.

3. Cultural practices : Reduced plant population per unit area encourages tillers and branches. Tillers and branches generally flower later than the main stem and thus prolong duration of flowering of a plant. Similarly, removal of the growing point on the main stem promotes tillers and branches that flower later.

In some forage species plants are cut back and the secondary growth produces more flowers and seed than the first growth. The differential number of cuttings before the crop is allowed to flower may help synchronize flowering of late and early genotypes. Additional techniques used to cause small changes in flowering dates are fertilizer application and irrigation. Spraying one per cent urea 2-3 times at an interval of 2-3 days or additional irrigation to the lagging parent may help synchronization of parents.

4. Pruning : In many instances pruning (clipping) of leaves without damaging the main stem can delay flowering. For example, in corn the top leaves of young plants are destroyed by fire to delay flowering of male plants in fields for hybrid seed production. Removal of flowers and seeds from a plant may induce prolonged flowering. Clipping, burning and moisture stress should involve only the male parent, whenever possible.

5. Use of hormones : Many long-day plants can be induced to flower sooner by the addition of gibberellic acid. Some short-day plants can be induced to flowering by ethylene-producing chemicals.

SPECIFIC CROSSING AND SELF-POLLINATION TECHNIQUES

The general principles of artificial hybridization discussed earlier are basic to all crops. The specific procedures applicable to some important crops are described below.

Cereals and Millets

Wheat
Allan (1980) has described the technique in detail.

FLORAL TRAITS
The wheat inflorescence is a determinate, composite spike. Sessile spikelets are alternately arranged on opposite sides of the rachis of the main axis of the spike. Each spikelet has several florets. The number of florets varies from 2-9. The two outer parts of the floret, lemma and palea enclose 3 stamens and a pistil with two short styles and a branched feathery stigma. The main culm flowers bloom first and the tillers later, in the order of their

formation. Flowering begins in the upper part of the spike and proceeds in both directions. Flowering can occur anytime during the daylight with a large peak in the morning and a smaller one in the afternoon. Wheat is normally self-pollinated. The glumes normally open during the flowering process, the anthers protrude from the glumes, and part of the pollen is shed outside the flowers. The anthers may shed their pollen without being extruded, however. The stigma is most receptive up to 3 days.

EMASCULATION

Emasculation is done 1-3 days before anthesis. The right stage for emasculation may be determined by observing the few emerging spikes. At the right stage, the anthers are well developed but still light green in colour, and the feathery stigmas are clearly visible and extended to about one-fourth the length of the floret. By checking spikelets about midway up the spike, the most advanced stage of floral development can be determined.

One to three basal spikelets and upper spikelets may have non-functional flowers. These spikelets and awns are removed with scissors. The primary and secondary florets in the remaining spikelets are retained and all other florets are removed by grasping the spike gently but firmly between the thumb and fingers of one hand, then lightly pressing the thumb below the tertiary floret the unwanted florets are pulled downwards and outwards with the forceps. After this, the anthers are removed from each of the retained florets by carefully inserting the forceps between the lemma and palea and spreading them. The anthers are removed with forceps carefully, so as not to injure the feathery stigmas. The emasculated spikes are immediately covered with glassine bags and fastened with paper clips.

POLLINATION

Emasculated flowers are pollinated 2-4 days after emasculation. Mature pollen is taken from the male parent from those spikes that exhibit a few freshly extruded anthers. A forceps is inserted between the lemma and palea of selected florets and anthers are collected just before anthesis.

For pollination, the glassine bag is removed from the emasculated florets and an anther which has just begun to shed pollen is carefully brushed against the stigma of the emasculated floret. The glassine bag is replaced to cover the pollinated florets. Paper tags on which details of parentage, etc. have been recorded are then affixed on the culm below the bag. The success of pollination can be verified within 3-5 days by observing kernel development.

Barley

Starling (1980) has described the technique in detail.

FLORAL TRAITS

The barley inflorescence consists of a spike with alternating rachis nodes on each side with three spikelets at each node. Each spikelet normally consists of a pair of narrow glumes and one floret. The central floret is largest, with laterals ranging from fully fertile (as in six-row barleys) to vestigial (functionless, sterile, as in two-row barleys). The outer lemma and inner palea of each floret enclose the stamens and pistil. When awns are present, they are an extension of the lemma. The pistil consists of a two-lobed ovule and a two-branched feathery stigma. There are three stamens, each composed of a two-lobed anther and a slender filament which originates at the base of the ovule. Filaments are capable of extensive elongation at flowering and often push anthers outside their enclosures. Two anthers are located towards the palea and the third is toward lemma. At the base are two lodicules which swell and force the lemma and palea apart, exposing the stigma, if pollination does not occur at flowering.

Barley is normally self-pollinated, with anthers often dehiscing before spikes emerge from the leaf sheaths (boots) and usually without flower opening. Pollination often is practically complete before awn tips appear. The first flowers to mature on a spike are those in the central row located midway in the upper part of the spike. Florets mature in both directions from this section, with apical and basal florets maturing up to 2 days later. Lateral florets mature later than central ones, but in the same order. The stigma remains receptive for a few days. The seedset is maximum when pollination is done 2 days after emasculation. The period of pollen viability is short (< 10 minutes). The best seedset is obtained from dehiscing anthers. Early to mid-morning is the most effective period for collecting pollen and implementing pollination.

EMASCULATION

A spike is ready for emasculation when its anthers are plump and light green to yellowish in colour. At this stage the spike is in the boot, but the edges of the flag-leaf sheath are separated to partially expose it. A check of central florets in the middle of the spike suffices to indicate appropriate development. Emasculation may be done anytime of the day; however, it is generally preferred to emasculate during afternoons to avoid pollen shed. For emasculation the flag-leaf sheath is opened by splitting the sheath lengthwise with the point of a forceps. A gentle rotation and forcing the sheath apart exposes the spike. The sheath and its attached flag-leaf florets, and all undeveloped florets at the tip and base of the spike are removed using forceps. From 6-8 florets on each side of the rachis are sufficient for a simple cross. The lemma and palea are clipped at an angle of 30-45° from horizontal at a level just above the anthers with scissors. The three anthers are removed with forceps. Following emasculation, spikes are covered immediately with glassine bags to prevent outcrossing. The coverings should be attached securely at the base of the spike with a paper clip to prevent their blowing off. A tag should be hung on the culm showing the date of emasculation.

POLLINATION

Emasculated flowers are generally ready for pollination 2 days after emasculation. In spikes ready for pollination the lemmas and paleas have separated and the feathery stigmatic branches are exposed. Any closed flowers found between widely opened ones should be removed prior to pollination because selfing has probably occurred. The stigma is receptive at all times of the day.

From early to mid-morning is the most effective period for collecting pollen and concurrent pollination. A spike with suitable pollen is the one near emergence from the flag-leaf sheath and in which anthesis normally would occur that day. It will be slightly more advanced than the stage described for emasculation. Anthers ready to dehisce will be plump and yellow. Such individual anthers (from male parent) are collected after cutting the florets in the proper stage just above the anther tips. In a matter of minutes, anthers ready to dehisce are pushed out of the flower by elongating filaments. These anthers are used directly, after removing the cover of the emasculated spike; single anthers are broken against the lemma and palea edges of each floret, allowing pollen to fall on the stigma. A single good anther may be used to pollinate several florets if pollen is abundant. Following pollination spikes of the female parents are again covered with the same glassine bag that was lifted and secured with a paper clip. Parentage and date of pollination are then added to the tag attached at emasculation time.

Oats

Brown (1980) has described the technique in detail.

FLOWER TRAITS

The inflorescence of oats is an open panicle consisting of a main axis, and lateral axillary branches that arise from alternate sides of the nodes of the main axis. The main axis and each of the lateral branches terminate in a single spikelet. The oat spikelet consists of several florets enclosed in two empty glumes with the tip of one glume extending slightly above the other. Normally, only the two basal florets are fertile, but occasionally the lower three may be. The florets are arranged alternately upon a central axis, the rachilla. Oat flowers are perfect and enclosed in a lemma and palea. Each flower contains three stamens, a pistil and two lodicules. The lodicules swell at anthesis, causing the flower to open. After anthesis, the lodicules collapse.

Anthesis takes place in the afternoon, but the exact time of opening varies with environmental conditions. Anthesis occurs in oat flowers in the order of emergence from the sheath. Primary florets open first, but at times the primary and secondary florets open on the same day. Florets on an individual panicle bloom over a period of 8-9 days. Individual stigmas are usually receptive 1 day before natural anthesis and remain receptive for as long as 3-5 days. The period of stigma receptivity is shorter under high temperature conditions. Excessive mutilation of flower parts during emasculation contributes to floral desiccation and more rapid loss of stigma receptivity.

The pollen of oats is most abundant and of the highest quality just prior to and during natural anthesis. Suitable pollen for effective crossing can be found in individual florets 1-2 hours before the flower opens during natural anthesis. Anthers dehisce most of their pollen just before or during opening of the floret.

EMASCULATION

The optimum time to emasculate oats is when the anthers have attained full size but are not yet ready to dehisce. One can easily observe withered, empty anthers from natural anthesis the previous day hanging from between the glumes of a closed spikelet, and these anthers provide an excellent bench-mark below which florets can be selected for emasculation. Five to eight primary florets near the top are usually emasculated on each panicle. The most important consideration in emasculation is to be sure that anthers are removed before they have dehisced any pollen. It is preferred to do emasculation work in the morning to avoid the period of natural anthesis which occurs in the afternoon. The spikelet is held near the base between the index finger and thumb of one hand with the ventral outer glume facing the person emasculating the spikelet. The outer glumes are separated with forceps. The exposed secondary floret can be removed by snipping it with the forceps. The palea is separated from the lemma by inserting the tip of one prong of the forceps between the palea and lemma and pulling forward on the palea to expose the three anthers of the primary floret. The three anthers are extracted from the floral cavity with the forceps being careful not to break any ripe anthers that are present to damage the stigma branches. The floral parts are replaced in their original positions. A slight rotation of the spikelet helps in tucking the edges of the ventral glume inside the edges of the dorsal glume. The emasculated florets are covered with a glassine bag to exclude outside pollen and secured with a paper clip or stapled. When securing the bag, the paper clip should not be extended across the culm because it will often cause the culm to break at that point. The emasculated panicle is labelled with a tag designating the female parent and date of emasculation. A small tag attached on the main panicle axis just below the last emasculated floret can indicate that all florets above that point have been emasculated. A small length of yarn can also serve the same purpose. Any spikelets above this point that were not emasculated are removed. Only the emasculated florets are covered in the bag.

POLLINATION

Ideal environmental conditions for making successful pollinations are clear days and moderate to low temperatures. High temperatures (30°C and above) reduce the availability of usable pollen and contribute to floral desiccation and loss of stigma receptivity. High temperature on days between emasculation and pollination causes reduced seedset, particularly in cases where 3 or 4 days elapse between emasculation and pollination. The optimum time between emasculation and pollination is 1 to 3 days. Excellent results may also be had from emasculations made in the morning and pollinated in the afternoon of the same day.

Pollination is done between 1:00-4:00 p.m. Anthers with usable pollen will be yellow, plump and should dehisce within one minute when removed from the floret and placed on the hand. Anthers may be collected from primary and secondary florets. The removed anthers should be used for pollination immediately. To pollinate the glassine bag is removed and the lemma and palea of each emasculated floret are separated, and a mature anther held in forceps is gently tapped against the inside of the lemma to be sure that the anther opens and spreads pollen on the stigma hairs. The palea and glumes are returned to their original position. When all emasculated florets on a panicle have been pollinated the glassine bag is replaced and a tag fastened below it for identification of the cross.

Rice

Coffman and Herrera (1980) have described the procedures in detail.

FLORAL TRAITS

Rice is a self-pollinating crop. Inflorescence is a panicle, which bears perfect flowers in single-flowered spikelets. A flower consists of six stamens, each composed of a two-lobed, four-loculed anther borne on a slender filament, and a pistil containing one ovule. The short style bears a feathery stigma with two branches. The flower is fully developed and the stigma is fully receptive at the time of pollen shedding. At this stage, the lodicules become turgid and force the lemma and palea apart. Anther dehiscence and extrusion occur more or less simultaneously after which the lemma and palea close. The blooming of rice normally occurs between 8:00 and 11:00 a.m. The flowers in a single panicle bloom over a period of 7-10 days, but most of the flowers bloom between 2-4 days after emergence of the panicle from the boot-leaf. The stigma remains receptive for 4-5 days, but receptivity drops at rather a high rate after the first day. Pollen normally remains viable for a very short time after anther dehiscence. Fertilization is completed within three hours after pollination.

EMASCULATION

Rice is ready for emasculation when 50-60% of the panicle has emerged from the boot. Individual flowers are ready for emasculation after emergence from the boot and prior to anthesis. Emasculation should be done late in the afternoon. The optimal time can be determined by observation and experience at each location. The selected panicle is separated from surrounding ones to make it easy to emasculate. The flag-leaf sheath is removed carefully to avoid breaking the stem. All florets that have pollinated in the top of the panicle are cut off with scissors. Such spikelets appear translucent and often have anthers clinging to the outside. Young florets from the bottom of the panicle wherein height of the anthers is less than half that of the floret are also cut off. About one-third to one-half of each of the remaining florets is cut away obliquely to expose the anthers. The stigma can be injured if the cut is made too low, and if the cut is too high, emasculation is difficult and the pollen may not reach the stigma. Anthers are removed with forceps; the tip of one prong is used to gently press the anthers against the side of the floret and lift them out. Extreme care must be exercised to avoid damage to the stigma and to ensure that all six anthers are removed. After all the florets on the panicle are emasculated a glassine bag is placed over the panicle, the bottom edge is folded over, and a paper clip is placed on the fold against the stem to hold it securely in place.

Hot-water method of emasculation: The hot-water method of emasculation may be used. When this method is employed, panicles in the second or third day of blooming are chosen as female parents. The florets that have already flowered or are immature are cut. An hour or so before blooming normally begins, the tiller is bent over carefully to avoid breaking, and inserted for 5-10 minutes into a thermos bottle containing hot water at 42–44°C. The water temperature is critical. The thermos bottle may be supported on a trough-like holder at an angle of about 35° to prevent loss of water. This treatment causes the florets to open in a normal manner and the stamens may be removed without injury to the stigma. Any unopened spikelets are cut off. Pollination must be done within 30 minutes, before the glumes close naturally.

POLLINATION

Pollination should only be attempted during the period of peak anther dehiscence. This peak usually occurs in mid-to late morning. The pollen is gathered just before anther dehiscence which is indicated by several florets at the top with anthers showing. Glassine bags of somewhat smaller diameter are placed over the panicles before anthesis. At anthesis the culm is cut at a convenient length below the panicle and the flag leaf is removed and carried out to the female.

The emasculated panicles are checked for any anthers that may have been left and any floret is reclipped where the opening is small. To apply pollen, the glassine bag on the female panicle is removed, and the florets (of the male panicle) nearing the blooming stage are opened and turgid anthers are taken with forceps and broken over the stigma of the emasculated female flowers. The glassine bag is again placed over the pollinated panicle and the bottom edge folded over. A paper clip is placed on the fold against the stem to keep the bag secure. An identification tag is attached to the stem and the tag so positioned under the paper clip at the bottom of the glassine bag that the sides showing the parents of the cross face inwards and the pollination date outwards.

Alternatively, the pollen may also be dusted over clipped florets by shaking a shedding panicle over them. If pollination is done in this manner, the top of the glassine bag of the female parent is clipped and blowing into it forms a cylinder around the panicle. One of the male panicles (enclosed in glassine bags) is held downwind, then clipped off and gently shaped into a circular form. The male panicle bag is placed into the bag containing the female panicle, where it is held by the stem above the female and twirled vigorously for a few seconds to cause pollen shed.

Pearl millet

Burton (1980) has described the technique in detail.

Floral Traits

The inflorescence of pearl millet (5-150 cm long and 1-5 cm diameter) is a false spike. The involucre borne on a stalk up to 15 cm long, consists of a cluster of bristles that are usually inconspicuous in mature heads but may extend several cm beyond the grain. The spikelets, 4-7 mm long in each involucre, usually occur in pairs with a sessile male floret and a short pedicelled bisexual floret. The latter has a single pistil with two feathery stylar branches and three anthers enclosed between the lemma and palea.

Pearl millet is a protogynous species and the stylar branches are exserted from the florets several days before the anthers. Stylar branches are first exserted from upper florets in the upper half of the head and by the third day nearly all styles emerge on heads less than 25 cm long. Longer heads usually require more days for complete exsertion of all stylar branches. Styles are usually exserted after the heads emerge from the boot but in some genotypes they are exserted before head emergence. Seedset is usually very poor if anthesis occurs in the boot. The first anthers generally emerge from the first florets to exsert styles at least 1 day after most styles on a head are exserted. Anthers in the sessile male florets are exserted 2 or 3 days later. Most heads shed pollen for 4-6 days. When temperature exceeds 25°C, anthesis occurs anytime during the day with the greatest flush of anthers appearing soon after sunrise. Stigmas in a floret remain receptive from emergence until 1 day after the anthers have been exserted.

Preparation of Female

The protogynous habit of pearl millet makes emasculation unnecessary. To prepare the female parent, the top one or two leaf blades are removed and the partially exserted head and culm are covered with a glassine bag before styles appear. The bag is fastened by folding the open end at a 45° angle and stapling it close enough to the stem so that it does not blow off. The bag must be loose enough, however, to be moved up the culm by the exserting head. The head is examined by looking through the glassine bag for the presence of exserted styles. When styles are exserted, the head is ready to pollinate. Some genotypes start exserting anthers before the florets at the top and bottom of the head are exserted. On such genotypes the florets without stigmas can be removed at the time of pollination without adversely affecting the pollination and seedset of those remaining on the head.

Pollination

Pollen can be collected anytime that it is dry. Glassine pollen collecting bags allow the worker to see the amount of pollen available. Shaking the pollen-collecting bag vigorously and carefully removing it from the head maximizes the amount of pollen harvested. Crossing can be accomplished anytime but the seedset from crosses made in the morning or later in the afternoon is better. Pollen is usually applied by putting the female head in the bag in which the pollen was collected and shaking it. If the female millet is tall, the culms are bent over so that the head assumes a nearly horizontal position before the pollen containing bag is pulled over it. After pollination, the head is enclosed in a kraft-paper bag on which is written the pedigree of cross, the date, and any other pertinent information.

SELF-POLLINATION

Selfing in pearl millet is easily accomplished by removing the top one or two leaf blades of the culm and enclosing the head and culm in a kraft-paper bag before any styles appear. The bag is stapled shut and left on the head until harvesting.

Sorghum

Schertz and Dalton (1980) have described the detailed technique.

FLORAL TRAITS

The inflorescence of sorghum is a panicle that varies from compact to open. The spikelets are usually in pairs on the branches, one being sessile and fertile and the other pedicelled and male or sterile. The terminal sessile spikelet of each branch has two pedicelled spikelets associated with it. The glumes of the fertile sessile spikelet enclose two florets, the upper perfect and fertile, and the lower sterile. The fertile floret has a membranous lemma, a palea, two lodicules, three stamens, and an ovary with two long styles with plumose stigmas.

Flowering begins 0-3 days after the panicle emerges from the boot. It begins at or near the panicle apex and proceeds downwards for a period of 4-7 days. Stigmas are receptive 0-2 days before blooming and if unpollinated, remain receptive for 5-16 days after anthesis. Anthesis usually occurs in the morning. Viable pollen is shed from the anthers until about noon. Pollen is highly functional for about 30 minutes after dehiscence but remains viable up to about 4 hours.

EMASCULATION

Florets are emasculated the day before anthesis. Florets below and within about 3 cm of opened florets usually will open the next day. Hand emasculations in the field are usually made in the afternoon to avoid contamination by viable pollen from other plants. All open spikelets are removed with scissors and the viable pollen is rinsed from the panicle with water. The anthers may be removed with a forceps, a pointed stick, pencil, etc. The point is entered between the outer glumes of the sessile spikelets and the anthers are teased out. If an anther is broken or punctured that spikelet or its ovary is removed and the pollen is removed from the emasculating instrument. Male fertile, pedicelled spikelets and unemasculated spikelets in the region of those emasculated are removed with forceps or scissors. A small paper/parchment bag is placed over the emasculated florets, folded and secured with a clip.

POLLINATION

Pollen for crossing should be collected in the morning when anthers dehisce, usually between 7:00-9:00 a.m. Any panicle that is flowering can be used as a source of pollen, but pollen is most abundant when taken from the centre of the panicle. Stray pollen on the panicle of the pollen parent usually is not a problem and pollen can be collected from open heads. The male can be covered a day before crossing, if one desires absolute protection from pollen contamination or if the wind may agitate the pollen from the anthers of unbagged panicles before all pollination can be completed. Pollen from bagged heads is shed later and is available for a longer time than is pollen from open heads. To collect pollen, the bag covering the panicle is tilted toward the horizontal and the bag with enclosed panicles is rapped sharply. The bag containing pollen is inverted over the emasculated female and shaken to promote movement of the pollen onto the stigmas. After pollination is completed, the panicles or the individually pollinated panicle branches are covered and the bag secured with a staple or clip. Identification of the cross is written on the bag or a tag is attached to the bag or peduncle.

Corn

Russell and Hallauer (1980) have described the technique in detail.

FLORAL TRAITS

Corn is monoecious, with the male flowers in the tassel and the female flowers on the ear shoots. When the male flower is mature, the anthers are exserted from the spikelet and pollen is dispersed through a pore at the tip of the anther. Anther exsertions usually begin on the central spike a short distance below the tip. Each spikelet has two flowers and the anthers are exserted from the upper flower first, then from the lower flower, either later the same day or the following days. When the anther pore opens, the pollen may be completely dispersed in only a few minutes, or over a longer period as determined by the temperature, humidity, air movement and genotype. Pollen shed for a tassel may vary from only 1 or 2 days to more than a week. Pollen dispersal may begin 3 hours after sunrise and continue for 1-3 hours. Once pollen is dispersed into the atmosphere, it will remain viable only a few minutes because of rapid desiccation. The top ear shoot usually is at the sixth or seventh node below the tassel. The silks that emerge from the tip of the ear husk are the functional stigmas and there is one silk for each potential kernel. The first silks to emerge are usually those near the basal part of the ear. Complete silk emergence may occur in only 2 or 3 days under favourable growth conditions or may require 5-7 days with cool temperatures. The silks usually emerge at the top ear node 1-3 days after anther dehiscence has begun. Silks become receptive upon emergence from the ear husk. The receptivity period for silks up to 10 days after emergence has been reported.

PREPARATION OF FEMALE

The ear shoot is covered by an earshoot bag before the silks emerge from the husk tip. Ear shoots may be covered anytime of the day. The usual practice is to cut the ear shoot to within 2 cm of the cob tip before covering with a glassine/butter-paper bag. The bag should be so placed that it is firmly anchored between the shoot and the auricle of the ear head, or sometimes it is expedient to break off the ear leaf and pull the bag down so that it is anchored between the shoot and the stalk.

POLLINATION

A tassel usually produces its greatest volume of pollen in the second or third days of dehiscence. A kraft-paper bag (tassel bag) is placed over the tassel on the day before pollination to eliminate contamination by other pollen that may adhere to the tassel. The bag is held in place by a paper clip or a staple.

For cross-pollination, pollen from one tassel is used for one or several ear shoots simply by pouring pollen from the pollen bag directly onto the silks of another plant. The pollination operation is begun around 9:00 a.m. and continues up to 5:00 p.m. The tassel bags on plants are shaken to collect the pollen. The silk is exposed for pollination either by lifting the ear-shoot bag or by tearing off the closed end. After the pollen has been dusted onto the silk, the tassel bag is quickly placed over the ear shoot, pulled down towards the ear node, and fastened around the stalk with a staple. Marking the tassel bag is necessary to identify the parents crossed.

SELF-POLLINATION

For self-pollination, the pollen is taken from the tassel and placed on the silks of the same plant in the manner described earlier.

Legumes

All legumes have typical papilionaceous flowers. The flower consists of five sepals, five petals (one standard, two wings and two keels), ten stamens (nine fused to form a staminal column and one free) and a carpel with style borne laterally on the ovary.

Pea

FLORAL TRAITS

Anthesis proceeds sequentially upwards from node to node. Pollen can effect fertilization from the time the anthers burst until several days thereafter. The stigma of pea is receptive to pollen from several days before anthesis until about 1 day after the flower wilts. In the field it is desirable to emasculate and pollinate in the morning.

EMASCULATION

Flowers developed to the stage just before anther dehiscence (petals usually extending beyond the sepals) needs to be selected for emasculation. Younger flower buds are difficult to emasculate and self-fertilization will have already taken place in more mature flowers. When two or more flowers are borne on a peduncle, only one is retained and the others are snipped off. Emasculation is usually carried out immediately before pollination. The sepal in front of the keel is torn away with the tip of forceps. By positioning the forefinger behind the flower and the thumb in front of the flower, a light squeezing pressure is applied to spread the standard and wings and expose the keel. The keel can be slit by carefully inserting one of the sharp tips of the forceps about halfway down and slipping upwards or downwards so as not to damage the pistil. After the keel is slit, it is spread apart by inserting both tips of the forceps. The two halves of the keel can be held down with the thumb and finger to spread the keel. The 10 anthers are removed by grasping the filaments with the forceps and pulling gently.

POLLINATION
Pollen of the intended male parent flowers at a stage just prior to their opening or just after is collected from freshly dehisced anthers and transferred on the tips of forceps or a toothpick to the stigma of emasculated flowers or alternatively, the anthers may be used as a brush to transfer pollen to the stigma of the emasculated flower. Pollen must be carefully applied to the stigma so as to avoid injury to the style. After pollination, it is necessary to fold the petals back into position to protect the stigma from desiccation and to reduce exposure to foreign pollen. The emasculated flowers may or may not be covered with butter-paper bags. A tag is attached to the pollinated flowers for identification of the female and male parent.

Lentil
Muehlbauer et al. (1980) has described the technique in detail.

FLORAL TRAITS
The small size and delicate structure of the flower are major factors preventing rapid and readily successful hybridization. A magnifying glass may be necessary to view parts of the small lentil flowers.

EMASCULATION
For emasculation flowers are chosen when the tips of the petals reach 50-75% of the length of sepals. At that stage of development, almost none of the flowers will have pollinated.

During the emasculation process the selected flower bud is held between the thumb and forefinger with the suture of the keel facing the operator. Care should be taken not to bend or twist the peduncle. Sharply pointed forceps are used to remove the sepals closest to the keel. Remove a notch from the standard; this makes it easier to fold it back to gain access to the wings. After removal of the wings the standard and keel are folded away to expose the staminal column and stigma. The stamens are removed by grasping the filaments with forceps and breaking them free from the staminal column.

POLLINATION
Pollination is usually done immediately after emasculation. Flowers in which the corolla has elongated to three-fourths the length of sepals are selected as sources of pollen. Most flowers at that stage have anthers that have recently dehisced their pollen. Viable pollen suitable for transfer is identified by its bright orange-yellow colour and is contained in a flower with a turgid-looking keel and wing petals. Pollination is accomplished by first removing or

folding back the keel and standard of the male flower to expose the 10 dehisced anthers and the pollen-laden stigma. The anthers and pollen-laden stigma are held with forceps and brushed slightly against the stigma of the emasculated flower. The keel and standard of the pollinated female flower are carefully returned to their original position to protect the stigma. After pollination has been completed, a tag is affixed to the internode directly below the pollinated flower. Pollinated flowers do not need protection from foreign pollen.

Chick-pea

Auckland et al. (1980) have described the technique in detail.

FLORAL TRAITS

Self-pollination takes place 1 or 2 days before the flower opens. The flowers open on 2 successive days. On the first day they open from 9:00 a.m. to 3.00 p.m. and close about 6.00 p.m. The second day they open earlier and close sooner. Hybridization work is not begun until after the first pod has formed.

EMASCULATION

Emasculation is usually carried out in the afternoon of the day previous to pollination between 2:00-6:00 p.m. Buds in which the petals are first showing and the anthers are about half the height of the style are selected for emasculation. The bud for emasculation is held lightly at its base, between the thumb and first finger. The front sepal may be drawn back or snipped off. The keel petal is gently pushed downwards with a fine forceps. This exposes the anthers to be removed from the filaments. The peduncle, stigma and style are fragile and should not be touched during the operation. Coloured cotton thread tied loosely around the stem below the emasculated flower and removed sepal serves for identification. There is usually no need to protect emasculated flowers from cross pollination. New buds emerging after the flower has been emasculated should be removed for a few days thereafter.

POLLINATION

Pollination of emasculated flowers is done the next morning between 8:00 a.m. –1.00 p.m. Half-open flowers are selected from the male parent for pollination. Pollen at this stage is mature, yellow and slightly sticky. Gently holding the keel petal and filaments with forceps, the pollen is liberally but lightly applied to the stigma of the emasculated female parent, avoiding damage to stigma, style or pedicels. The pollinated flower is tagged on the stem below. Usually there is no need to protect the pollinated flower.

Oil-Seeds

Soybeans

Fehr (1980) has described the detailed technique.

FLORAL TRAITS

The structure of the soybean flower is similar to that of other legumes. The stigma is receptive to pollen about 1 day before anthesis and remains receptive for 2 days after anthesis, if the flower petals are not removed. The anthers dehisce on the day of anthesis. Pollen shed is influenced by temperature and usually occurs between 7:00-9:00 a.m.

EMASCULATION

Flowers that are expected to open the following day are selected on the female parent. The flower buds at this stage are swollen and the corolla is just visible through the calyx or has begun to emerge. Usually no more than 2 buds in a raceme are prepared and all others are removed. For emasculation the selected bud is grasped between the thumb and index finger and the location of the stigma determined by examining the sepals. A long, curved sepal covers the keel, and the stigma is on the opposite side of the flower. The sepals are removed with forceps. The exposed corolla is removed by grasping it just above the calyx scar without injuring the stigma. The ring of anthers is visible after the corolla is removed. The anthers are then carefully removed. The emasculated flowers are pollinated immediately after emasculation.

POLLINATION

Pollination is done by removing the stamens and pistils with forceps from a flower (with a recently opened corolla) of the male parent and gently brushing the anthers against the stigma of the female flower. Brushing the anthers on the stigma causes them to rupture. After brushing, the pollen should be visible on the stigma. Pollinated flowers are identified with a tag attached to the internode below the pollinated flower. The pollinated flowers are not protected.

Groundnuts

Norden (1980) has described the detailed technique.

FLORAL TRAITS

The inflorescences consisting of three or more flowers are spike-like. Each flower is subtended by a bract and is borne on a minute branch of the inflorescence which arises in the axil of a second bract. Thus two bracts are found below each flower but the flower bract actually occurs on the axis of the inflorescence. Main stem inflorescences occur in the Spanish and Valencia varieties, while inflorescences occur on lateral branches in the Virginia varieties. The calyx and corolla are located at the top of the hypanthium. The staminal column is surrounded by and runs the entire length of the hypanthium. The calyx is five lobed, one lobe being separate and opposite the keel and the other four fused, except for their tips. The keel closely embraces the stamens and style.

The staminal column is usually composed of 10 filaments, 8 of which normally bear anthers. The filaments are fused through one-half to two-thirds of their length. At the point of separation. the free ends of the filaments are sharply reflexed towards the standard, forming acute angles with their fused hypanthium. The pistil is surmounted by the long style which extends through the hypanthium, bends sharply through the staminal column, the filaments and ends in a club-shaped stigma. The style is clothed with upward-slanting hairs near its summit on the surface facing the standard.

The most prolific period of flowering occurs between 6 and 11 weeks, depending on the variety, with a high proportion of the first flowers producing mature fruits. Few of the flowers that open late produce mature fruits. Groundnuts produce many more flowers than the plants can support in fruit and seed production. Approximately two-fifths fail to begin fruit development and an additional two-fifths produce immature pegs which abort before pod enlargement occurs.

Usually only one flower of an inflorescence reaches anthers on a given day, occasionally two or more in Spanish types. Intervals between successive flowers on an inflorescence vary from one to several days. Twenty-four hours before anthesis, the flower bud is 6-10 mm long. During the day the hypanthium elongates slowly and the buds attain lengths of 10 to 20 mm. At night, elongation accelerates, so that at anthesis flowers are usually 50-70 mm long. The bud usually opens with commencement of daylight. Pollen dehiscence may occur as early as 7 to 8 hours before flower opening, but even upon flower opening the anthers may not have dehisced in some flowers. The stigma is receptive from 24 hours before to 12 hours after flower opening.

Emasculation

Usually only one flower bud per inflorescence is emasculated on the female parent. The flower bud selected should be as near to the main stem as possible. The plants should be in the early stage of flowering. Flower buds appear above the leaf axils during the afternoons of warm, bright days, and emasculation can be accomplished without difficulty as early as 5:00 p.m. On cloudy, rainy days emasculations are easier to accomplish in the late night hours (9:00–10:00 p.m.).

For emasculation the bud is grasped with the thumb and index finger of one hand. The sepal in front of the keel is removed and the sepal on the side of the standard is folded down. The standard is opened with forceps and the wing petals pulled out and down. As the standard is held back with the thumb and index finger, the point of the forceps is used to break the keel and to move up the keel to pull it free of the stigma and anthers. The keel is pulled down and held out of the way with the index finger and thumb while all the anthers and as much of the filaments as possible are removed with forceps. After the anthers are removed, the standard is returned to its original position over the stigma. Emasculated flowers are usually not protected. They can be identified in a number of ways, for example by attaching a small thread to the hypanthium of the emasculated flower.

Pollination

On the morning of the day following emasculation, the standards are usually expanded and the stigmas exposed; thus pollen can be applied. Pollinations are accomplished between 7.00–9.00 a.m. by removing a healthy flower from the male parent, squeezing its pollen onto a forceps and transferring the pollen to the stigma of the emasculated flower. Covering the female plants after pollination with a polyethylene cover (tube) in which a few holes have been punched to allow aeration, reportedly results in better pollen germination and ultimate fertilization, precluding the need for covering the pollinated stigmas individually. After pollinations are completed, all unpollinated flowers on the plant are removed by breaking the hyparthium near the base with forceps by 10:00 a.m. on the day pollination looks to have been completed. If fertilization is successful an aerial peg will become visible in 7 to 10 days or so after pollination. The developing peg, with the withered flower and string still attached, can be identified by means of a colour-coded wire looped around the peg before it penetrates the soil. The other end of the wire can be attached to the stake that supports the label of the female plant.

Rape-seed and Mustard

FLORAL TRAITS
The floral arrangement of *Brassicas* is a typical corymbiform raceme. The flowers are regular, bisexual and hypogynous with four free sepals in two whorls, median and transverse, and four free diagonally placed petals. Flowers have one pair of lateral stamens with shorter filaments and four median stamens with longer filaments. Sutures of the anthers are introse in the bud stage, but the anthers of the four long stamens become extrose after flower opening in all forms of rape except *Brassica campestris* var. *yellow sarson*.

EMASCULATION
Buds which are about to open or will open the following day are selected for emasculation. Six undehisced anthers are removed with forceps and discarded. Emasculation is usually done in the afternoons or evenings. Emasculated buds are protected from foreign pollen for up to 1 week following pollination. All other flowers, open buds and remaining young buds are discarded. Tags for identification are attached to the flower pedicle.

POLLINATION
Pollen transfer is accomplished by picking selected anthers with a pair of forceps and applying them to the stigma of an emasculated flower, or by collecting pollen on a small camel-hair brush and dusting the stigma. Tags for identification are attached to the pedicle of the pollinated flower after pollination.

Flax
Beard & Comstock (1980) have described the detailed technique.

FLORAL TRAITS
Flax flowers are regular, perfect and borne in terminal, multiflowered panicles. The 5 sepals, petals and stamens are attached in whorls below the ovary.

The stamens alternate with the petals and are united at the base of the filaments. There are 5 pistils, each with an ovary, style and stigma. The flower shape may be tubular, funnel form, star-shaped or crimped or large disc-shaped. Flax flowers open about 20 minutes after sunrise and anthers dehisce 20 minutes later. The actual time is dependent on temperature and weather conditions. Cloudy days or cool weather conditions somewhat delay flower opening and pollen shedding.

EMASCULATION

Emasculation and pollination may be done simultaneously or the flowers may be emasculated a day before in the late afternoon. An unopened bud is selected that has the petals protruding about 3-6 cm past the sepals. The petals are removed with forceps with a gentle twisting pull. The anthers are easily removed. The emasculated flowers are identified by jewel tags which are gently tied on fragile pedicles. There is no need to protect emasculated flowers.

POLLINATION

Stigmas are receptive from the day before the flower opens until the morning of the day after pollen shed. The stigma shrivels the day after pollination. Any flower with petals has usable pollen. Flax pollen is short-lived, but adequate pollen is generally available from the time anthers dehisce until the petals start to fall. The petals are removed from the flower to be used for pollen and the anthers rubbed against the stigma of the emasculated flower. No protection is needed for the pollinated flowers. The pollinated flowers are identified by loosely tying a coloured thread and attaching a jewel tag with requisite information.

Sunflower

Dedio and Putt (1980) have described the technique in detail.

FLORAL TRAITS

The head of the sunflower is a compound inflorescence composed of many individual flowers in a large disc subtended by large ray flowers. The ray flowers are normally asexual. The disc flowers are perfect with petals and 5 anthers united in separate tubes. The disc flowers are arranged in concentric circles radiating from the centre of the head. The ray flowers open first and flowering then proceeds from the periphery to the centre of the head at the rate of one to four rows per day.

The individual disc flowers are effectively protandrous and the positioning of the stigma above the anthers makes self-pollination difficult. Early in the morning (about 7:00 a.m.) the staminal filaments rapidly elongate and exert the anther tube from the corolla. Immediately after this stage is reached, the anther locules dehisce, releasing their pollen inside the anther tube. An elongation of the lower portion of the style follows, which forces the two-lobed pubescent stigma up the anther tube. The stigma is not receptive at this stage because the two lobes are held together covering the inner receptive surface. At the same time, the staminal filaments lose turgidity and the anther tube begins to recede into the corolla. The next day, the stigma lobes are fully emerged and have separated to expose the receptive surfaces to pollination. Pollination and fertilization soon occur, after which the stigma withers and recedes. By the morning of the second day, after the flower opens, both the anther and the stigma have receded almost completely into the corolla.

EMASCULATION

The ray flowers and bracts are usually removed before emasculation to make the flowers in the disc more accessible and to eliminate a large surface area on which the pollen could lodge. Usually, only those flowers which open on a single day are emasculated and all other are removed. The flowers opening prior to the day of emasculation can be removed from their ovaries by a simple sideways pulling motion with the thumb and forefinger. In the field emasculation should be carried out very early in the morning, prior to 7:00 a.m. The ideal time to emasculate is in the period when the anther tube is extended sufficiently to be grasped with the forceps but the pollen has not yet dehisced. However, in practice the anther tube is often removed after dehiscence but before the stigma has grown far enough into it to be injured, or for the stigmatic lobes to separate when the tube is removed. Free pollen on the outside of the stigmatic lobes must be blown off. Undeveloped central florets are removed, usually by cutting them off with a knife at a point just above the ovaries. A few flowers closely adjacent to those emasculated are removed with forceps.

The stigma normally remains receptive for 3 to 5 days. Stigmas of flowers which are not fertilized continue growth for several days to form a coil which will allow the stigmatic surface to contact pollen adhering to the outer surface of the stigmatic lobes. Emasculated flowers are protected by covering the head with paper or perforated plastic bags.

Emasculation is not necessary in male sterile or highly self-incompatible plants which can be pollinated directly provided they have been protected to prevent undesirable crossing.

POLLINATION

Pollen for crossing is collected by placing paper bags over the heads of the male parents a day or two before flowering begins. Adequate pollen will be found in the bags 1 or 2 days after flowering commences. It is best to use the pollen shortly after collection. Pollen should be applied when the stigmatic lobes have separated and have exposed their receptive surfaces. Receptive stigmas are brushed with a pollen-laden cotton wad or camel-hair brush, usually a day after emasculation. Alternatively, pollination can be effected simply by rubbing heads from the two parents against each other.

The pollinated head is maintained in isolation by rebagging. The bags also serve as a protection against birds. Identity of the cross may be lettered on the bag or indicated on a tag attached to the plant.

Self-pollination

Self-pollination is achieved by placing a cloth bag over the head before anthesis or anther dehiscence. The amount of seedset depends on the degree of self-compatibility of the lines and can be considerably increased when heads are manipulated by brushing with muslin batting.

Other Crops

Alfalfa

Barnes (1980) has described the technique in detail.

Floral Traits

Alfalfa flowers are borne on recemes, with about 10 flowers per raceme. Each flower contains 10 stamens which form a tube in which 9 filaments are fused. The tenth stamen is nearest the standard and is free. Filaments alternate long and short so that, during development, anthers fit tightly around the stigma in a double ring. The pistil consists of a single carpel that develops a superior ovary, a smooth awl-shaped hollow style and a well-defined stigma. All annual species are cleistogamous and are exclusively self-pollinated. Generally, the perennial species are cross-pollinated. Before pollination can be successful, the sexual column (ovary and staminal column) must be tripped (released). This is usually done by bees; in other instances environmental factors, such as hot dry winds or rains can cause tripping.

Tripping : Two major forces are involved in the tripping mechanism:

i) the pressure exerted by the sexual column from cells under tension at the juncture of the staminal tube and the keel, and

ii) the restraining mechanism of the keel petals that cohere due to interlocking projections of cutinized tissue in the appressed petal surfaces.

Tripping takes place whenever the restraint of the appressed keel petal is reduced or becomes less than the pressure of the staminal column. Plants differ in ease of tripping. Four bud stages are recognized in alfalfa, namely straight bud, pointed bud, hooded bud and erect standard. Anther dehiscence usually occurs in the pointed bud stage. A cuticular membrane forms a continuous film over the stigma, thereby preventing pollination before tripping. When tripping occurs, the stigmatic membrane ruptures as the stigma strikes either the bee or standard petal. Alfalfa pollen is sticky and readily adheres to pollinating insects.

The time flowers remain open varies from 5-16 days. The pollen remains viable and the stigma is receptive from the late bud stage until the flower begins to wilt. Successful self- and cross-pollinations can be made at any time of day or night with any flowers that are not wilted and have fertile ovules and pollen.

Emasculation

The standard petal is clipped and the flower is gently tripped. Anthers and pollen are removed either by vacuum suction or alcohol. With vacuum emasculation, a glass tube, drawn to a 1 mm tip and attached to a vacuum source, is used to remove the anthers. A hand lens may be used to check the thoroughness of pollen removal. When alcohol is used, the whole raceme is immersed in 57% ethyl alcohol for 10 seconds, then rinsed in water for a few seconds. Alcohol is more effective than vacuuming for emasculation but causes more injury to the stigma.

POLLINATION

Pollinations can be made at any time of the day or night. A camel-hair brush or a small piece of cardboard may be used to transfer the pollen to the stigma. Cross-pollination by hand without emasculation is accomplished by alternately tripping flowers of the two parents into a folded cardboard pollen reservoir. By randomly tripping flowers from many plants into the same pollen reservoir, bulk seed can be produced on a population of plants.

SELF-POLLINATION

Self-pollination requires tripping the flowers without introducing foreign pollen. This is done either by inserting a toothpick in the throat of a flower; by inserting a toothpick that has the tip covered with emery-paper to rupture the stigmatic membrane; by depositing the pollen into a folded cardboard which retains the pollen for the stigma to fall into; or by gently rolling or squeezing racemes between the fingers. The last is more efficient than other methods. Alcohol should be used to clean hands between plant treatments.

Sugar-cane

James (1980) has described the techniques in detail.

FLORAL TRAITS

The inflorescence of sugar-cane referred to as an arrow, is an open, branched panicle and may contain as many as 100,000 or more spikelets in pairs alternately at rachis nodes. One spikelet of each pair is sessile and the other pedicellate. At maturity, the lateral axes of the inflorescence break below the spikelets at rachis nodes. The pedicellate spikelets break free, leaving the sessile spikelets attached to the rachis segment and the stalk of the pedicellate spikelet. Each spikelet contains a single flower containing a whorl of three stamens, and a single ovary with two feathery purple stigmas. Flowers that are about to open can be recognized by stigmas protruding from the closed glumes. As anthesis approaches, the filaments lengthen and the anthers are extruded. Often the flowers are with only rudimentary ovary or with anthers devoid of pollen grains. Anthesis may begin as the inflorescence starts to emerge from the flag leaf or as much as several days after full emergence. Anthesis takes place from the top of the panicle downwards and from the ends of panicle branches inwards. Some flowers usually open before sunrise each day, and the process is completed in 3-14 days. After flower opening the stigmas are receptive and anther dehiscence usually takes place after some drying has occurred, if the humidity is high. The flowers open between 5:00 and 6:00 a.m. Degree of anther dehiscence and pollen viability ranges from 0-100%. A few complex species hybrids are completely sterile.

CROSSING TECHNIQUE

Due to the small size and large number of sugar-cane flowers on an arrow, emasculation is not practically feasible. Cross-pollinations are usually obtained by isolating the parent lines and enclosing the arrows of the parent lines within a lantern, or pollen may be collected from the male parent and dusted over the arrow of the female parent between 5:00 and 6:00. a.m. When the arrows of the two parents are enclosed within the same lantern, shaking the arrows occasionally, may help to disseminate the pollen. Since an arrow normally flowers over a period of 5-10 days, the pollination processes must be repeated daily during the period of flowering.

MODIFIED PROCEDURES

Marcotted stalks[5] and detached stalks with arrows (which are kept alive in a weak acid solution: 150 ppm SO_2, 75 ppm H_3PO_4, 37 1/2 ppm H_2SO_4 and 37 1/2 ppm HNO_4 changed biweekly) allow transfer of several stalks to a central crossing area. These are supported in an upright position with the cut end immersed in the aforesaid solution. Pollination of arrows on stalks maintained in solution or on marcotted stalks may be done by enclosing male and female arrows under the same hood, or by dusting the arrows of the female parent with pollen collected as described earlier.

SELF-POLLINATION

Selfing may be ensured by covering the arrow with a cage covered with a closely woven cloth or a polyethylene bag, forming what is commonly called a lantern which is supported by a bamboo pole. The temperature within the lantern may get rather high during midday so the bags are sometimes opened between noon and 4:00 p.m., when pollen dispersal is minimal, in order to reduce the temperature inside.

Cabbage

Dickson and Wallace (1986) have described the technique in detail.

FLORAL TRAITS

Flowers are borne in racemes on the main stem and axillary branches. The flower has four sepals, four petals, six stamens and two carpels, as well as a superior ovary with a false septum and two rows of ovules. The androecium is tetradynamous (2 short and 4 long stamens). The buds open under pressure of the rapidly growing petals. Opening starts in the afternoon and usually the flowers become fully expanded during the following morning. The anthers open a few hours later, the flowers being slightly protogynous.

---

[5] Just prior to flowering a polyethylene strip containing a mixture of moist potting soil is wrapped around a bud of the sugar-cane stalk about 2 nodes above ground level. Roots develop on the stalk where the bud has been marcotted within 10 days. Marcotted stocks are then severed.

Natural pollination is by insects, particularly bees, which collect pollen and nectar (secreted by two nectaries situated between the bases of the short stamens and ovary), because the pollen is sticky and not wind blown. Depending upon the variety a few to most plants are self-incompatible. Few seeds will be set following self-pollination. Pollination can best be performed in the greenhouse or in a large screened cage to eliminate insects. If crosses or self-pollination are desired in the field, cheesecloth bags can be used to enclose the blossom of one or two plants.

CROSSING TECHNIQUES
Select buds that are likely to open in the next 1 or 2 days for emasculation. The buds are opened with a toothpick or forceps and the anthers removed with the forceps. The desired pollen is then transferred to the stigma in the same manner as for bud pollination.

If a male sterile or self-incompatible plant is used as a female parent, then an open flower with protruding pistil can be used to avoid the time and effort required to open the flower bud. However, only certain lines have these protruding pistils.

SELF-POLLINATION
Self-pollination of cabbage can be obtained by brushing or shaking the open flowers if the plant is self-compatible. When the plant is self-incompatible, self-pollination is achieved by bud pollination.

BUD POLLINATION
For bud pollination, buds can be opened 1-4 days before they will open naturally and be bud pollinated. The largest unopen bud is probably too old for successful bud pollination as the incompatibility factor will be biosynthesized by that stage. A younger bud (3-4 days prior to natural flower opening) will have the least self-incompatibility and still be large enough for bud pollination. Thus six to eight buds can be opened at one time with a pointed object such as toothpick or forceps, and pollen from an older open flower transferred to the stigma and seed obtained. Pollen can be transferred with a brush or on a thumbnail. Also, a fertile flower in full bloom is often used by brushing the anthers across the stigmatic surface to transfer the pollen.

Potato
Plaisted (1980) has described the technique in detail.

FLORAL TRAITS
The inflorescences are borne terminally on the stems. The number of flowers per inflorescence and the number of inflorescences per plant depend on the variety. About the time the first flowers are fully expanded, a new shoot develops at the proximal leaf axis and grows beyond the first flower cluster to produce the second inflorescence. The size of the corolla varies with the variety. Flowers of cultivated species usually open in the early morning and remain open for 2-4 days. The stamens are attached to the corolla tube and

bear the erect anthers which form a close column or cone around the style. The stigma generally protrudes beyond the ring of anthers and remains receptive for about 2 days. Most flowers can be used as a pollen source for 2 days after the corolla first opens. Pollen production is most abundant in the morning.

EMASCULATION

Emasculation is done in the afternoons. The opened flowers are removed and mature unopened buds are emasculated. The buds at this stage are plump and the petals appear ready to separate and exhibit their final colour. The petals are opened and the anthers are removed with forceps without damaging the style. The emasculated flowers are then bagged with thick butter-paper bags.

POLLINATION

Pollination is done in the early morning. Flowers with plump, bright yellow anthers with a brownish tip are selected from the intended parents. The pollen is removed from mature anthers by inserting a blunted narrow scalpel at the base of the suture in the anther lobe and scraping it the length of the anther. The pollen is applied by touching the stigma with the tip of the pollen-bearing scalpel or inserting the stigma into the petri dish containing collected pollen. After pollination the flowers are again bagged and labelled for identification.

SELF-POLLINATION

The flowers (unemasculated) are allowed to open and the pollen is removed from an anther with the help of a blunt forceps and applied to the stigma of the same flower. The flower is bagged with a butter-paper bag and labelled for identification.

The crossing techniques for crops, namely castor, cotton, tomato, egg-plant, okra, pepper, cucumber, watermelon, and squash are discussed in Chapter 32, Hybrid Seed Production.

8

Genetic Basis of Plant Breeding

VARIATION

Types of Variation

The variability, that is, the differences observed in individuals of a population or among populations with regard to one or several of its characteristics, is known as variation. This apparent variation, known as *phenotypic variation*, in turn depends upon the genotypic composition of the population and the environment in which it is raised. The individuals in a population may be composed of different genotypes and the different populations may have a different genotype or genotypes. Variation arising due to differences in genotypic composition of individuals in a population is known as *genotypic variation*. The group of individuals belonging to different biotypes[1] may show consistency for some characters and variability for others. A biotype may be homozygous or heterozygous and a population may be composed of individuals of different degree of homozygosity or heterozygosity. The different genotypes, however, react differently to varying environmental conditions. This differential behaviour is known as *genotype* x *environment* interaction. The genotypes which show greater consistency over a range of environments are said to be widely adapted (greater phenotypic stability). The genotypes which give widely varying performance over a range of environments are said to have narrow adaptation. Nevertheless, all genotypes react to the prevailing environment and genotype x environment interaction is always there. The difference between wide adaptation and narrow adaptation is one of degree or extent of reaction of a genotype over a range of environments.

1. Environmental Variations

Variation arising entirely due to differences in environment such as temperature, humidity, rainfall, soil fertility, etc., is known as *environmental variation*, which may also cause *developmental variations or fluctuating variations*[2].

---

[1] Population in which all the individuals have the same genotype is referred to as a biotype.
[2] Observed variation due to differences in factors such as soil fertility, etc.

2. Agroecotypic Variations

Agricultural practices, such as soil fertility, irrigation, soil temperature, atmospheric conditions, etc., greatly affect the expression of various plant characteristics. The resultant variations arising due to these factors are known as agroecotypic variations. A proper understanding of agroecotypic variations is necessary for a plant breeder for evaluation of the characters desired in the breeding material because certain plant characteristics express themselves fully only when the environment is favourable. A few examples are given below.

1) Development of chloroyphyll pigments in plants is both under genetic and environmental control. Absence of manganese, iron, magnesium or light may lead to chlorophyll deficiency.
2) Soil fertility (plant nutrition) and water relation of soils have great influence over developmental variations. Poor fertility leads to poor growth, which may force the plant to flower early, while rank growth delays flowering. Dehiscence of anthers is known to be influenced by atmospheric humidity.
3) Very low soil temperatures retard seed germination. Low atmospheric temperature delays flowering and ripening of the produce.
4) Light affects the earliness and flowering period, expression of chlorphyll and anthocyanin pigments.

By controlling environmental conditions, such as temperature, changes have been produced which closely resemble similar mutants already found in nature. The term *phenocopy* is applied to those individuals of a population in which the newly produced variation mimics an earlier mutant.

Ecotypes

In nature one finds resemblances and differences between individuals of a population and between populations of such magnitude and complexity that it is almost impossible to comprehend the whole. This variation is chiefly governed by genetic recombinations, polyploidy, mutations and environmental modifications. The genetic response to ecological conditions gives rise to *ecotypes or biotypes*. The term '*ecotype*' refers to ecological subspecies arising as a result of the differentiation of the species population in response to a particular environmental condition. Ecotypes represent an intercrossing community whose members have become clustered in groups (ecotypes) primarily because of the differentiating effect environmental factors exert on the genetically heterogeneous population. Accordingly, ecotypes are classified as climatic, edaphic, biotic (synecotypes and agroecotypes) and geographical. Ecotypes are distinguished primarily by their response to environment, and may or may not possess morphological differences that would enable their identification in the field.

When reproductive isolation is not complete, hybridization between borderland species often leads to introduction of new genes into the genome of one species from another. This is known as *introgressive hybridization*. When the reproductive isolation is complete and the populations are separated from one another by barriers which prevent intercrossing, the nature of variation observed depends on the degree to which these barriers have developed and on the diversity in populations so isolated.

Ecospecies

The populations so related that they are able to exchange genes freely without loss of fertility or vigour in the offspring are known as *ecospecies*. Ecospecies so related that they are able to exchange genes among themselves to a limited extent through hybridization are known as *coenospecies*. Coenospecies between which hybridization is possible either directly or through intermediaries are known as *comparium*.

Importance of variation to a plant breeder : An intimate knowledge of the variation existing in nature not only within the species, but also in allied species and genera, is necessary for a plant breeder for the following reasons.

1. It may be possible to find in nature a form which is most suited for introduction into cultivation as an improved type.
2. Evolutionary changes in different geographic regions have progressed in different directions and therefore it is possible to find useful genotypes either within the same species or in allied species and wild forms. These types may be used in the hybridization programme with a view to transferring the desired characters to the adapted variety.
3. Precise knowledge of the variations arising due to genotype x environmental interactions and due to environmental factors alone, for example, agroecotypic variations, is necessary for evaluation of varieties for either specific or wide adaptation.

HEREDITY

Hereditary Material

Gene is the functional unit of heredity (inheritance). Structurally, a gene is a small segment of deoxyribonucleic acid (DNA) which codes for a particular polypeptide, a small chemical factor, which controls a physiological process of the organism and thus its structural development. Genes are located in a linear order on *chromosomes*, the thread-like structures present in the nucleus of a cell. Each crop species has a fixed number of chromosomes. Each gene is located on a specific chromosome and at a specific location called the *locus*. In each cell nucleus there are two identical sets of chromosomes, one set from the male parent, and another from the female parent. The pairs of corresponding chromosomes are called *homologous chromosomes*. A set of chromosomes is called a *genome*.

There may be several alternative forms of a gene, called *alleles*, which occur at the same locus. When the two alleles at the corresponding *loci* on a chromosome are the same, the individual is called *homozygous* for that character, and when they differ, *heterozygous* for that character. A population is called homogeneous when all its individuals are of the same genotype, and *heterogeneous* when the individuals are of different genotypes.

Gene Specializations (Gene Effects)

Gene effects are not simple. A gene may exert influence, that is, it may also affect the expression of other genes as well. A few terms used for gene effects are defined below.

Modifying genes : Genes which have small effects by themselves may nevertheless influence the expression of major genes. Such genes are called modifying genes. Modifying genes may also have a marked effect upon another group of major genes, for example, those which control male sterility. Modifying genes are widespread through many species.

Pleiotropic genes : Some genes influence the expression of more than one character; they are termed pleiotropic genes. They show their presence as dominants through a group of associated characters. The whole character group is either missing or altered in appearance when a pleiotropic gene is present in the recessive state.

Threshold genes: Genes whose expression and/or development depends upon a specific environment are known as threshold characters.

Linked genes: A large number of genes are present on any chromosome, and although chromosome segments (as chromatids) break and rejoin at cell meiosis, genes closely adjacent to one another on a particular chromosome tend to remain together. Such genes are said to be closely linked genes.

Mutant genes: The combined activity of the whole gene complement is responsible for the external appearance of all parts of a plant. Occasionally, non-typical or unexpected forms may be observed among a conventional population or may be revealed in the course of critical genetic experiments. The new forms are frequently the result of a spontaneous change in the gene. Such genes are called mutant genes.

Isoalleles: These are mutant alleles, grossly similar in action, which in homozygous form produce phenotypes similar to that of the standard wild type.

Marker genes: Marker genes are usually major genes which are recognizable by their phenotypic expression and may be dominant or recessive. They have a particular value in genetic studies as well as for varietal identification. Their main purpose lies in aiding the identification of varieties, in showing the degree of contamination within varieties, or the success with which crosses have been made. When dominant, they are best used as male parents in crosses so that true hybrids can be identified in a mixed progeny of selfs and hybrids. Conversely, plants with homozygous recessive marker genes are used as females in crosses in order that self-fertilized plants may be identified in the progeny by the presence of the 'marker character'. Any hybrid plant will have a phenotype other than that of selfed plants. Marker genes are a valuable tool in F_1 hybrid seed production, particularly when the 'marker' character can be identified at a young age (before flowering).

Pseudoalleles (heteroallele) : Alleles that are located at different sites within a complex gene locus are designated heteroallelic, provided this can be verified through recombination or in other ways.

Penetrance and expressivity : *Penetrance* is the ability of a gene to express itself within the individual that is carrying it. When the gene fails to express itself in all the individuals carrying it, penetrance is said to be *incomplete*. *Expressivity* denotes the ability of a gene to express itself in all the individuals carrying it. Expressivity may be *uniform*, that is when the expression is fairly uniform; or *variable*, that is when the degree of expression in different individuals varies.

Physical Basis of Heredity

Diploid organisms result from the union of male with female reproductive cells (male and female gametes, each with *n* number of chromosomes). The chromosome number in a zygote is thus twice that of a gamete. There is remarkable consistency in number of chromosomes generation after generation in a species. This is due to reduction division (meiosis) that takes place in the reproductive cells during gametogenesis. In homozygous types both the alleles are the same at corresponding loci on the chromosomes and hence a single kind of gametes is produced; in heterozygous types, the alleles being different, two types of gametes are produced.

During pollen maturation the pollen mother cells undergo meiotic division to give rise to pollen grains. The pollen grains undergo mitotic division to give rise to one generative nucleus and a tube nucleus. The generative nucleus subsequently undergoes a mitotic division to give rise to two male gametes or sperms, one of which unites with the egg and the other with two polar nuclei.

Similarly, the embryo sac mother cell undergoes a meiotic division to give rise to 4 embryo sac initials (megaspores), of which 3 degenerate and die. The remaining megaspore undergoes three mitotic divisions to produce 8 nuclei. The three nuclei that move to one end are termed antipodal cells, the two nuclei that remain in the centre are called polar nuclei, and the remaining three go to the micropylar end. Of these three, the central nuclei is called the egg cell (female gamete) and the other two one on either side of it, are called synergids. The arrangement of these nuclei is shown in Fig. 8.1.

Mitosis

Mitosis is the process of cell division in which a cell divides into two daughter nuclei which are qualitatively and quantitatively alike. Its primary function is the duplication of chromatic material. The various stages of mitosis are very briefly described below. The detailed description may be found in Swanson (1957).

Interphase : The stage between the two mitotic divisions. The two daughter cells are fully formed. Chromosomes duplicate themselves during interphase.

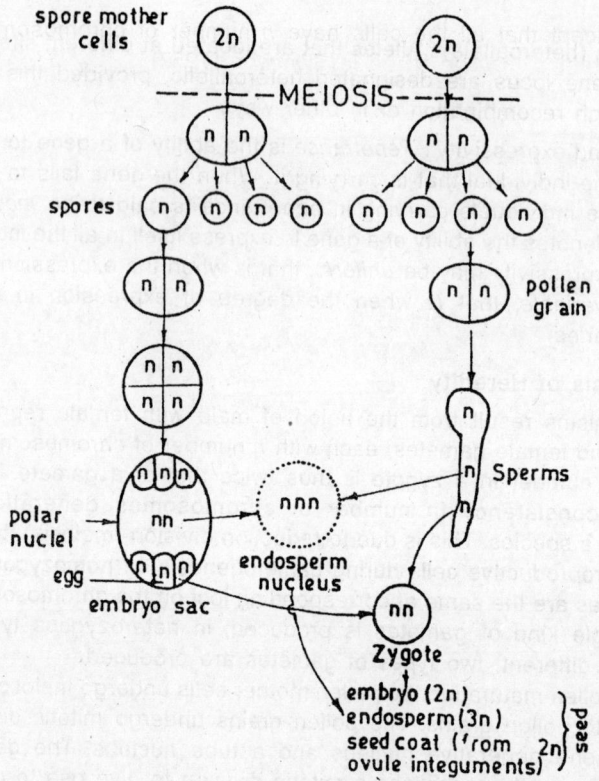

Fig. 8.1. Diagram showing chromosome numbers in cells of the reproductive organs of a seed plant (adopted from Poehlman and Borthakur, 1969).

Prophase : Chromosomes appear as thread-like structures. At the late prophase stage chromosomes shorten and become distinct. Each chromosome can be seen to consist of two chromatids joined at the centromere.

Metaphase : Nuclear membrane disappears. Chromosomes become arranged on an equatorial plate. A spindle is formed and the chromosomes are attached to the spindle in such a manner that centromeres lie along this plate.

Anaphase : Chromatids separate at their centromere and the two daughter chromosomes begin to move to opposite poles.

Telophase : The two daughter chromosomes reach poles. A nuclear membrane is formed. A cell wall is formed between the two daughter nuclei to form two daughter cells.

Meiosis

Meiosis is also known as reduction division since the resultant nuclei have only half the number of chromosomes (*n* number of chromosomes).

Meiosis consists of two phases, in the first phase (reduction division) parent cells divide to form two daughter cells each having only half the number of chromosomes; the second phase (equational division) is similar to mitosis, except that all the cells have *n* number of chromosomes. The detailed description may be found in Swanson (1957).

The various stages of meiosis are described in Fig. 8.2.

THE GENETIC BASIS OF PLANT BREEDING

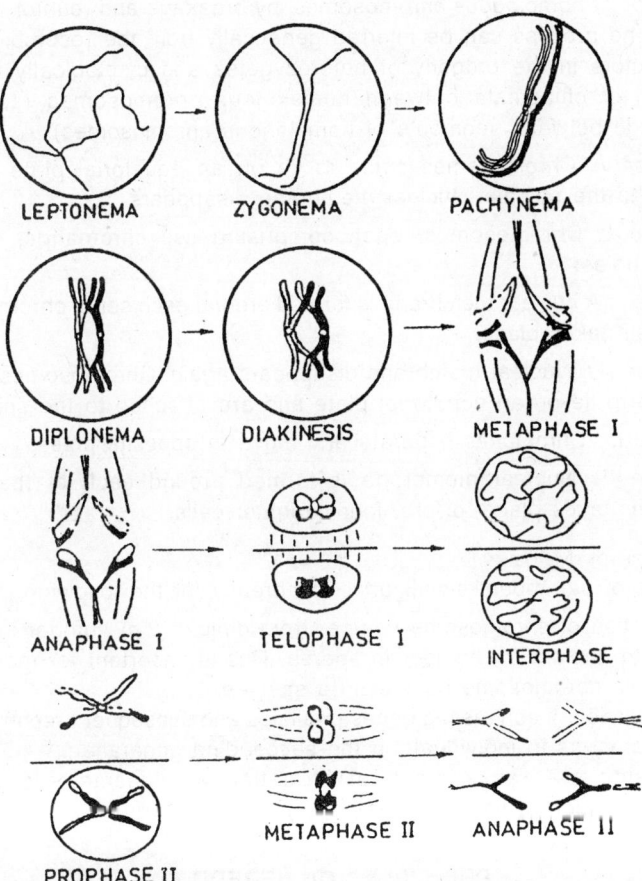

Fig. 8.2. The stages of meiosis (diagrammatic). Only a single pair of chromosomes is followed (Rhoades, 1950)

Prophase I: At the start, the chromosomes become visible, and the homologous chromosomes pair (*synapsis*) lengthwise to form bivalents. Each bivalent has four chromatids in which pairing takes place. Subsequently, each bivalent separate starting with the centromere. Chiasmata (places of exchange) are formed while two chromosomes are still temporarily joined.

Chiasma is the point of exchange of partners in paired chromatids. The hypothesis is that two chromosomes which have paired and remained associated until metaphase do so by virtue of the formation of a chiasma or visible exchange of partners (crossing overs) among their chromatids. Only sister chromatids are associated after chiasma-formation and therefore every chiasma results from crossing over between two individuals. Long chromosomes form more chiasmata than short chromosomes.

Crossing over is the exchange of corresponding segments between chromatids of homologous chromosomes, by breakage and reunion following pairing. The process can be inferred genetically from the recombination of linked factors in the progeny of heterozygotes and cytologically from the formation of chiasmata between homologous chromosomes (segmental interchange between segments of homologous chromosomes).

Metaphase I: Chromosomes come to lie on an equatorial plate and are attached to the spindle. Nuclear membrane disappears.

Anaphase I: Chromosomes, each comprising two chromatids, move to opposite poles.

Telophase I: A nuclear membrane is formed around each set of chromosomes. Cytokinesis takes place.

Metaphase II: Nuclear membrane disappears again. Chromosomes in each cell come to lie on an equatorial plate and are attached to the spindle.

Anaphase II: Chromatids separate and move to opposite poles.

Telophase II: Nuclear membrane is formed around each of the nuclei. Cytokinesis takes place to form four daughter cells.

SIGNIFICANCE OF MEIOSIS

Meiosis is of paramount significance in heredity for the following reasons.

1. Reduction of chromosome number from diploid (2 n) number in mother cells to haploid (n) number in spores. This is important for maintaining a stable chromosome number in a species.
2. Segregation of contrasting genes or alleles and subsequent recombination of characters in individuals in the succeeding generation.
3. Crossing over as a result of formation of chiasmata leads to a recombination of linked genes.

PRINCIPLES OF HEREDITY

The fundamental principles of heredity were enunciated by Mendel. These are known as Mendel's Laws of Inheritance.

1. The Law of Segregation : The law of segregation states that two alleles in a heterozygote (F_1 of a cross between parents having contrasting characters, say, tall or dwarf) segregate, that is, separate from each other during gamete formation, so that only one of the alleles is transmitted to a particular gamete.

Thus, when a plant is homozygous for an allele, all the gametes carry the same allele, and as a result no segregation is apparent. The heterozygous plant, on the other hand gives two types of gametes; 50% gametes thus carry one allele and the other 50% carry the other allele. The resultant generation (F_2 generation) thus segregates into respective allele types.

Dominant versus recessive character : The character (allele) which expresses itself in the F_1 of a cross is called *dominant* and the one which fails to express itself is called *recessive*. The dominance of an allele may be complete or partial.

2. *The Law of Independent Assortment :* The law of independent assortment states that the segregation and assortment of different characters (say tall or dwarf, smooth or wrinkled) are independent of each other. The law of independent assortment is, however, applicable to genes located on a different chromosome. The genes located on the same chromosome tend to go together and are said to be *linked*. The segregation and independent assortment of linked characters, however, also takes place to the extent linkages are broken because of crossing over and chiasmata formation during meiotic division. The extent of crossing over depends on the distance at which two genes are located.

Regardless of the nature of genes and the number of genes involved, inheritance essentially operates per these fundamental laws and there are no exceptions.

Inheritance of Qualitative Characters

The inheritance of qualitative characters is relatively simple since these are governed by a few genes, and are relatively insensitive to environmental variations. In segregating generations, classification of individuals into discrete classes is also simple and easily done. A typical ratio may thus be observed which indicates the nature of gene interaction and inheritance pattern of the gene(s) involved in the expression of a character. Inheritance ratios were first enunciated by Mendel and hence are called Mendelian ratios.

Mendelian Ratios (F_2 ratios)
1. *Monohybrid ratio (3:1) :* When dominance is complete, the F_2 generation of a cross involving one character will segregate into two discrete classes in a 3:1 ratio, dominant expression versus recessive expression respectively. When dominance is incomplete the observed ratio is 1:2:1.

2. *Test cross ratio (1:1) :* A heterozygote when crossed to a recessive parent segregates in the following generation in a 1:1 ratio, dominant (or incomplete dominance) expression versus recessive expression.

3. *Dihybrid ratio (9:3:3:1) :* When dominance is complete, the F_2 generation of a cross involving two characters segregates into four discrete classes in a 9:3:3:1 ratio, both the genes showing dominant expression, single genes individually showing dominant expression and both the recessive genes showing recessive expression respectively.

4. *Dihybrid test cross ratio (1:1:1:1) :* Just as in monohybrids, here 4 discrete classes are observed in equal proportion.

5. *Trihybrid ratio (27:9:9:9:3:3:3:1)* : When three independent factors are taken into consideration, segregation holds good, just as in the dihybrid.

Modification of F_2 ratios (Gene interactions)

The dihybrid ratio described above may not always be 9:3:3:1. This is due to interaction between non-alleles.

1. *Complementary factors (9:7)* : When the presence of two dominant genes is necessary for the expression of a character, the typical F_2 ratio would be 9:7. For example, when both the dominant factors A-B- are present, the flower colour is purple, but when only one dominant gene, A- or B-, is present or both are homozygous recessive the flower colour is white.

2. *Supplementary factors (9:3:4)* : When a gene alters the action of a gene at the other locus, but have no action on their own (i.e., do not express any dominant reaction) the typical F_2 ratio would be 9:3:4. For example when both dominant factors A- and B- are present the flower colour becomes red, but when only A- is present the flower colour is pink, and when only B- is present or only the recessives are present the flower colour is white.

3. *Epistasis (12:3:1)* : When a gene suppresses the action of another gene, the consequent F_2 ratio is 12:3:1. For example, when a dominant gene W- is present the grain colour is red, when it is recessive and another dominant gene Z- is present it is pearly white, and when only the recessives are present the colour is chalky white. (The masking factor is epistatic to the masked factor.)

4. *Inhibitory factors (13:3)* : When one gene inhibits the expression of another gene the consequent F_2 ratio is 13:3. For example, when a dominant gene I- is present the stem colour is green, and when it is recessive and another dominant gene Lp- is present the stem colour is pigmented, and when both the genes are recessive the stem colour is green.

5. *Duplicate factors (15:1)* : When a dominant gene governing the same character is present at either locus or at both loci the phenotype is the same. The F_2 ratio is 15:1.

6. *Polymerism (9:6:1) (additive genes)* : When the presence of a dominant gene at either locus, or at both loci gives the same effect, but the intensity is increased when both the dominants are present, the consequent F_2 ratio is 9:6:1.

Inheritance of Quantitative Characters

The plant characters which are the concerns of modern plant breeders are generally not under the control of major genes, but of genes which, by themselves, have only a small effect. When aggregated together in one plant these minor, or *polygenes*, are able, nevertheless, to produce a very marked response upon the character in question. The effects of these genes are *quantitative* as opposed to *qualitative* major genes. Population, defined as a community of individuals which shares a common gene pool constitutes the unit of inheritance for studying quantitative characters. The genetic behaviour of quantitative characters in a population is studied through well-designed experiments and working out certain statistics, namely, means, variances, covariances and partitioning of the variances. The concepts of populaion mean and variances in turn are based on gene frequencies and gene effects. Estimated parameters refer to a specific population only. It is for this reason that the inference cannot generally be translated from one population to another.

Hardy-Weinberg Equilibrium

Hardy (1908) and Weinberg (1909) independently discovered that gene frequency and genotype frequencies in a Mendelian population (random mating population) remain constant generation after generation, provided there is no selection differential, mutation, migration or random drift. Subject to these conditions any Mendelian population attains equilibrium after one generation of random mating. This is known as *Hardy-Weinberg Equilibrium*.

BASIC TERMS USED IN INHERITANCE OF QUANTITATIVE CHARACTERS

a) Means

Phenotypic value, is the measured value of a character in an individual.

Genotypic value, is the mean phenotypic value of a genotype in a population.

Average effect of a gene, is the average effect of a gene at a locus resulting from substitution of one allele for another.

Breeding value, is the value of an individual judged by the mean value of its progeny. The property of an individual and the population mean are equal to the average effect of the genes it carries. Also referred to as *additive effects* of genes.

Dominance deviation (for a single locus), is the difference between the genotypic value and breeding value. (i.e. additive effects of genes).

Interaction deviation, is the epistatic deviation due to non-allelic interaction.

b) Variances

Phenotypic variance (σ^2_p), is the total variance observed in an experiment in respect to a character; i.e. the total variance of all phenotypic values in an experiment.

Genotypic variance (σ^2_G), is the variance arising due to genotypes; the total variance of all genotypic values.

    (i) Additive genetic variance (σ^2_A), is the variance of breeding values, the average effect of genes. It is the primary measure of resemblances between parents and their offspring. It is the fixable component through selection.

    (ii) Dominance variance (σ^2_D), is the variance due to intra-allelic interaction of genes at segregating loci. It measures the behaviour of alleles in a heterozygote. This component is not fixable by selection and is exploited through hybrid vigour.

(iii) Epistatic variance (σ^2_I), is the variance due to interallelic interaction of genes at two or more segregating loci. It may be further partitioned into additive x additive (σ^2_{AA}, σ^2_{AAA} ...), additive x dominance (O^2_{AD} X O^2_{ADAD} ...) and dominance X dominance (σ^2_{DD}, σ^2_{DDD} ....). The interactions involving only additive effects are the fixable component through selection; the remaining are not fixable by selection and are exploited through hybrid vigour.

Heritability, is the proportion of genetic variance and total variance. Broad-sense heritability is equal to (σ^2_G/σ^2_P) x 100; narrow-sense heritability is equal to (σ^2_A/σ^2_P) x 100.

A character with high heritability is more likely to respond to selection compared to a character with low heritability.

Relationship between Genes and Genotypic Value

The relationship between genotypes and genotypic value for a single locus may be represented as follows:

Genotype A_2A_2 A_1A_2 A_1A_1

Genotypic $-a$ (μ)   d $+a$
value

where,

μ = mean of the value of two homozygotes`(+a and −a; o)
d = value of the heterozygote, which depends upon the degree of dominance (d/a)

Degree of Dominance (d/a)

The degree of dominance is worked out as follows:

when d = 0	No dominance (d/a = 0)
o < d, < a	Partial dominance (0 < d/a)
d = +a or −a	Complete dominance (d/a = 1)
d > +a or < −a	Overdominance (d/a > 1)

Methods of Estimating Genetic Parameters[3]

There are several ways in which genetic parameters of a population are worked out.

1. By partitioning of means : The genetic estimates (Kempthorne, 1957; Hayman and Mather, 1955) obtained through partitioning of means provides evidence of the presence of different types of gene action. This method is applicable in cases wherein gene frequency is previously known. The positive and negative effects may, however, cancel each other or may even result in negative estimates.

---

[3] Detailed procedure for analysis and interpretation may be seen in the references cited in the parenthesis.

2. By partitioning of variances and covariances : The genetic estimates obtained by partitioning of variances and covariances are more useful since they are always expected to be positive. Voluminous literature has developed. The interested reader is referred to Kempthorne (1957), Falconer (1960) and Hanson and Robinson (1963).

(a) *Estimation based on segregating generations from crosses of two pure lines* (Mather 1949).

(b) *Variance and covariance of half-sibs, full-sibs* (Comstock and Robinson, 1948, 1952).

A *half-sib* family is obtained from seeds produced by one plant (female parent) that was pollinated by a random sample of pollen from the population. A full-sib family is obtained by crossing a random pair of plants from the population.

(c) *Line X tester analysis* (Comstock and Robinson, 1952; Kempthorne, 1957).

(d) *Powers partitioning method* (Powers, 1951)

(e) *Nested designs* (Horner et al., 1955; Brim and Cockerham, 1961).

(f) *Diallel analysis*
    i) *Numerical approach* (Jinks, 1954; Hayman, 1954, 1958).
    ii) *Graphical approach* (Hayman, 1954; Jinks, 1954).

(g) *Partial diallel approach* (Kempthorne and Curnow, 1961; Fyfe and Gilbert, 1963)

(h) *Combining ability studies :*
    i) *for homozygous parents* (Griffing, 1956).
    ii) *For heterozygous parents* (Matzinger et al., 1959; Gardner and Eberhart, 1966).

(i) *Triallel and Quadriallel analysis* (Rawlings and Cockerham, 1962; Ponnuswamy, 1974).

Genotype-Environment Interactions

The relative ranking of genotypes grown over a wide range of environments differs very widely. This interplay of genotype-environment interaction reduces the correlation between genotype and phenotype which in turn reduces confidence in inferences from experimental data to both plant improvement and inheritance mechanisms.

Measurement of Genotype-Environment Interaction

1. Replicated Yield Trials Over Locations and Years
Variance analysis over locations/and or years provides the mean sum of squares due to genotype x environment interaction, the significance of which can be tested by a simple F test (MS_2/MS_3) (Table 8.1).

Table 8.1. Analysis of Variance Over Environments

Source of variation	d.f.	Mean sum squares	Expectations
Environments	(e−1)		
Genotypes	(g−1)	MS_1	$\sigma^2 + r\sigma^2ge + re\sigma^2g^{\cdot}$
Genotype x environment	(e−1) (g−1)	MS_2	$\sigma^2 + r\sigma^2ge$
Error	ge (r−1)	MS_3	σ^2

* If the experiment is so designed, the σ^2g can also be partitioned into σ^2_A, σ^2_D and σ^2_I.

2. CORRELATION ANALYSIS

Another measure of genotype environment interactions involves the correlation of performances of an array of genotypes in one environment with their performances in other environments (Stuber et al., 1973). Large positive value for this type of correlation coefficient indicate little effect of genotype-environment interactions, whereas the converse is true when evaluating the magnitude of variance components attributed to such interactions.

3. REGRESSION ANALYSIS

Although, in general, genetic effects are not independent of environmental effects, a number of authors have observed that the relation between the performance of some genotypes in various environments is often linear, or nearly so. From these observations, Freeman and Perkins (1971) concluded that there is strong evidence indicating a genuine underlying linear relationship between performances of specific genotypes and environmental conditions, even though this relation does not always account for all interactions observed.

Because of the linear relationship as discussed above, several authors have used regression techniques to characterize responses of genotypes in varying environmental conditions. In particular, regression analyses have been used to provide measures of phenotypic stability[4] (Finlay and Wilkinson, 1963; Eberhart and Russel, 1966).

HETEROSIS

Heterosis or hybrid vigour is regarded as the superiority of a hybrid over its parents, that is, the superiority of an F_1 cross between two individuals, which are neither closely nor distantly related.

Quantitative Measure of Heterosis

(a) *Relative heterosis* : The heterosis expressed in relation to the average of two parents involved in the cross is termed relative heterosis:

$$h = \overline{F}_1 - \left(\overline{P}_1 + \overline{P}_2\right)/2$$

---

[4] Refers to the performance of genotype over a wide range of environments.

(b) *Heterobeltiosis :* The heterosis expressed in relation to the better parent of the cross is termed heterobeltiosis:

$$h = \left(\overline{F_1} - \overline{P_1} \ \text{or} \ \overline{P_2}\,(\text{whichever is higher})\right)$$

(c) *Standard heterosis :* The heterosis expressed in relation to a standard variety (check) is termed standard heterosis.

$$h = \left(\overline{F_1} - \text{average performance of a check variety}\right)$$

where h is the heterosis observed, $\overline{F_1}$ is the averge performance of a cross, and $\overline{P_1}, \overline{P_2}$ is the average performance of the parents.

Genetic Basis of Heterosis

The genetic basis of heterosis is still conjectural, although the following hypotheses have been discussed for a long time.

Non-Mendelian Hypotheses

(a) *Physiological stimulus or heterozygosity hypothesis :* Shull (1914) gave a non-Mendelian physiological-stimulation explanation of heterosis. He stated that hybridity, that is, the union of unlike elements, the state of being heterozygous in itself has a stimulating effect upon the physiological activities of the organism which results in hybrid vigrour.

(b) *Greater initial capital hypothesis :* Ashby (1930) suggested that heterosis is due to the maintenance of initial advantage in the embryo size and not to an acceleration of metabolic processes. East (1936), however concluded that seed size, or the size of any part of the seed cannot be the cause of heterosis.

Mendelian Hypotheses

a) Dominant Favourable Alleles Hypothesis
Keeble and Pellew (1910) and Bruce (1910) suggested that hybrid vigour was due to the action of favourable dominants (dominant alleles). This hypothesis assumes that in general dominant factors contributed by each parent of the hybrid are desirable and recessive factors are harmful. Thus a hybrid is more vigorous than its parents because it has more dominant factors than recessive factors. The salient features of Bruce's hypothesis are: (i) that heterosis would occur if the parents differed in gene frequency and (ii) dominance was present.

b) Overdominance Hypothesis
East (1936) attributed the vigour of F_1 to its heterozygous condition. Thus the greater the number of genes in which a plant is heterozygous, the greater the heterosis. A type of allelic interaction at a single locus with dominance lacking was proposed. Hull (1945) defined the superiority of a heterozygote over either homozygote as '*overdominance*'.

Complementary genes produce an effect which differs from that of any of them separately. Complementary interaction of alleles at a single locus has been proposed as a basis for heterosis. These phenomena have been interpreted as interallelic interaction. Sakai and Utiyamada (1957) suggested that vigorous heterotic growth may be due to independent competitive ability arising from overdominance in the heterozygous state.

c) Validity of Dominant Favourable Alleles Hypothesis

Proponents of the overdominance hypothesis have raised objections against the dominant favourable alleles hypothesis on the following grounds.

i) If the heterosis is due to accumulation of dominant favourable alleles, then it should be possible to obtain inbreds as good in yielding ability as the hybrids. This has never been accomplished.

ii) F_2 population does not show skewed (asymmetrical) distribution.

iii) There is no evidence of the occurrence of dominance in the expression of quantitatively inherited traits.

Proponents of the dominant favourable allele hypothesis have held their ground by rebutting these objections as follows:

i) If the number of gene pairs is more than ten for a single trait the chances of isolating a line homozygous for all the ten pairs are indeed remote. Since the number of quantitatively inherited genes involved is usually very large in all instances it is practically not feasible to obtain homozygous inbreds for all the pairs of genes involved in the expression of a character and hence the criticism of the dominant favourable allele hypothesis is not valid. Also, the relative vigour, yielding ability of inbred lines continues to improve in successive cycles (Collins, 1921).

ii) Failure to obtain skewed distribution could be due to linkage between favourable and unfavourable genes (Jones, 1917). Also, when the number of factor pairs is large, even in the absence of linkage skewing would not appear (Collins, 1921).

iii) The most evidence gained so far indicates that partial to complete dominance is the primary mode of inheritance.

The only objection raised to data against the overdominance hypothesis is that the heterosis in respect to quantitative traits cannot be a single locus heterosis.

Definitive proof for either of the hypotheses proposed for the genetic basis of heterosis is difficult to establish due to the complexity of the inheritance of quantitative traits. It appears that the evidence gained so far supports that heterosis results from an accumulation of dominant favourable growth factors. Epistasis, that is, non-allelic gene interaction, also may contribute to the heterosis expressed in crosses, although studies indicate epistasis does not seem to be a major component of the genetic variability. Most likely, all types of gene action, both inter- and intra-allelic are involved. Heterosis, however, can be utilized without knowledge of the exact genetic basis of its occurrence.

Biometrical Concepts of Heterosis

For practical purposes what is important is a knowledge of the predominant types of gene action which are operative (in a population being used) in designing effective and efficient breeding schemes for continued progress. Falconer (1960) showed that heterosis is expressed when,
   i) dominance is present (at some level); and
   ii) there is relative difference in gene frequency in the two parents. If either or both conditions do not exist, heterosis will not be manifest.

Example : Taking two populations, A and B, let us consider genotypic values and frequencies in two populations in Hardy-Weinberg equilibrium for one locus with two allels (Table 8.2). Genotypic values and frequencies in a cross between these two populations are given in Table 8.3.

Table 8.2. Genotypic values and frequencies in a population in Hardy-Weinberg equilibrium for one locus with two allels

Genotypes	Gene frequency		Coded genotypic values	No. of A_1 alleles involved
	Population A	Population B		
A_1A_1	p^2	r^2	$+a$	2
A_1A_2	$2pq$	$2rs$	d	1
A_2A_2	q^2	s^2	$-a$	0

Table 8.3. Genotypic values and frequencies in a cross between two populations in Hardy-Weinberg equilibrium.

Genotypes	Frequencies	Coded genotypic values
A_1A_1	pr	$+a$
A_1A_2	$ps + qr$	d
A_2A_2	qs	$-a$

Where p and r the frequencies of allele A_1 and $q(q = (1 - p))$, and r and $s(s = 1(1 - r))$ are the frequencies for allele A_2, and a and d are the genotypic effects.

The mean values of population $A(P_A)$ and $B(P_B)$ and their F_1 shall be as follows

$$\overline{P}_A = (p - q)a + 2pq \cdot d$$

$$= \{p - (1-p)\}a + \{2p(1-p)\}d$$

$$= (2p - 1)a + 2(p - p^2) \cdot d$$

$$\overline{P}_B = (r - s)a + 2rs \cdot d$$

$$= \{r - (1 - r)\}a + \{2r(1 - r)\}d$$

$$= (2r - 1)a + 2(r - r^2) \cdot d$$

$$\overline{F}_1 = (pr - qs)a + (ps + qr)d$$

$$= \{pr - (1 - p)(1 - r)\}a + \{p(1 - r) + r(1 - p)\}d$$

$$= \{pr - (1 - p - r + pr)\}a + (p - pr + r - pr)d$$

$$= \{pr - 1 + p + r - pr\}a + (p + r - 2pr)d$$

$$= (p + r - 1)a + (p + r - 2pr)d$$

The heterosis (h) as deviation from the mid-parent values shall be

$$h = \overline{F}_1 - (\overline{P}_1 + \overline{P}_2)/2$$

$$= (p + r - 1)a + (p + r - 2pr)d - \left(\frac{1}{2}\right)\{(2p - 1)a +$$

$$2(p - p^2)d + (2r - 1)a + 2(r - r^2)d\}$$

$$= (p + r - 1)a - \left(\frac{1}{2}\right)\{(2p - 1)a + (2r - 1)a\} +$$

$$(p + r - 2pr)d - \left(\frac{1}{2}\right)\{2(p - p^2)d + 2(r - r^2)d\}$$

$$= (p + r - 1)a - \left(\frac{1}{2}\right)\{a(2p - 1 + 2r - 1)\} +$$

$$(p + r - 2pr)d - \frac{1}{2}\{2d(p - p^2 + r - r^2)\}$$

$$= (p + r - 1)a - \left(\frac{1}{2}\right)\{a(2p + 2r - 2)\} +$$

$$(p + r - 2pr)d - \left(\frac{1}{2}\right)\{2d(p + r - p^2 - r^2)\}$$

$$= (p + r - 1)a - \frac{2(p + r - 1)a}{2} + (p + r - 2pr)d -$$

$$(p + r - p^2 - r^2)d$$

$$= d(p + r - 2pr - p - r + p^2 + r^2)$$

$$= (p - r)^2 d$$

It can be readily seen from the above that if there is no dominance (i.e. $d = 0$) the mean F_1 equals to the mean of mid-parent and there is no heterosis.

Now when we consider that in population A, $p = 0$ or 1 and in population B, $r = 0$ or 1 for same locus, depending on whether the allele is in homozygous recessive or dominant condition the heterotic response will be observed. The heterotic response in the first generation cross is due to the loci where $p = 1$ and $r = 0$ or vice versa. Thus the heterotic effect depends on the number of such contrasting loci and also on the level of dominance at each locus. The heterosis will be greatest when one allele is fixed in one population and the other allele in the other population. If the populations crossed do not differ in gene frequency there will be no heterosis.

In any case, heterotic response is expected to occur whenever there is difference in gene frequencies and some degree of directional dominance at one or more loci is involved in the control of the character.

Gene Action and Heterosis

Explaining in terms of gene action, if the gene action is purely additive there would be no heterotic response as the average response would be equal to that of mid-parent values. If the gene action is dominant (intra-allelic interaction is such that one allele manifests itself more or less, when heterozygous, than its alternative allele), and/or epistatic (dominance of one gene over a non-allelic gene; non allelic interaction whereby manifestation at any locus is affected by the genetic phase at any or all other loci) only then can a heterotic response be expected, either because of the coming together of a greater number of dominant alleles and/or complementary interaction of favourable dominant alleles, and interallelic interaction.

From the plant breeder's point of view, the basic issue is whether the best genotypes are homozygotes or heterozygotes. If non-allelic interaction is important, the best genotype would be a heterozygote. If there is partial or complete dominance, the best genotype would be a homozygote, and wherever feasible it might be desirable to isolate transgressive segregates (plants ouside the parental range in respect of some characters in segregating generations of a cross).

Geometric Model for Yield

Grafius (1963) proposed the novel view that genes for a complex trait, such as yield, do not exist; there are only genes for components of yield. He proposed a geometric model for yield, wherein the volume of a parallelepiped with three edges corresponds to the component traits, heads per plant, seeds per head, and average seed weight. Grafius provided data showing that changes in yield, indeed heterosis, can be explained by epistatic interactions among the components. To the extent that these are additive, they are fixable in inheritance. This corresponds to what we know about progress in breeding for yield. Grafius' views are important from the viewpoint of relationships to selection pressure on components in his model.

Utilization of Heterosis

Heterosis has been commercially exploited in the form of hybrids or synthetic varieties in a large number of sexually propagated crops (Chapter 32). In asexually propagated crops it has been utilized in almost all crops of economic interest.

9

Selection

What is Selection?

Selection in plant breeding means, selection of individual plants having desirable characteristics from a group of plants (populations, breeding materials etc.). The critical factor, however, is that only the selected plants be taken to the next generations, that is, allowed to reproduce. This is necessary because if not done, the selection will not be effective. Selection thus brings changes in gene frequencies of different combinations of genes and hence the genetic value of the population in the desired direction. It may be noted, however, that selection can act effectively only on heritable differences and that it cannot create variability but acts on that which already exists.

1. NATURAL SELECTION

a) *Stabilizing selection*: The process of selection is continuously ongoing in nature. Under natural conditions, directional selection favours plants adapting to new environmental conditions, namely, new georgaphic areas and changing environmental conditions in the original habitat. For characters such as viability, fertility, etc., which directly affect *fitness* once nature has produced the optimal phenotype for a given habitat, further selection will operate to perpetuate this phenotype as long as the habitat itself remains stable. Under these circumstances extreme expression, whether positive or negative, is at a disadvantage as nature selects for a population mean. For example, at flowering time, natural selection in such circumstances will favour neither the very early nor the late types in the population. This type of natural selection is known as *stabilizing selection*. Characters which have been subjected to stabilizing selection show quite a different architecture compared to those subjected to directional selection. Here, dominance will be low or absent, and if present may be ambidirectional in nature, whereas epistasis will not generally occur. Stabilizing selection promotes largely additive variation.

b) *Disruptive selection*: Natural habitats encompass a number of distinct 'ecological niches'. Among these, the differences may be spatial (microniches) or temporal (seasonal or long-term cycles). Each of these ecological conditions favours selection of different optima in form of function which often involves several related processes, such as density-dependent and/or frequency-dependent interactions between genes or genotypes. Inevitably, the process favours diversity (polymorphism) rather than uniformity within the populations. This is known as disruptive selection. The effect of disruptive selection on population structure depends on the following two factors (Mather, 1973).

1) Whether the different optimal phenotypes are independent of each other or dependent on each other for maintenance or functioning. For example, coexisting differences such as male and female forms in dioecious species are interdependent in reproduction and so gene exchange is 100 per cent ensured. Thus they share a common genetic background which has to ensure that both channels of development are equally adapted.

2) The rate at which genes are exchanged between the differentially selected genotypes.

Genetic Control of Polymorphism

A precise genetically controlled polymorphism is built up in response to selection for consistent interdependent functions governing the reproductive processes. A few examples:

1. *Incompatibility alleles :* In species in which outbreeding is controlled by incompatibility genes, the rarer the allele at a locus, the greater its chances of securing a compatible mating and vice versa for the common alleles. Such frequency-dependent selection is capable of building up a large number of incompatibility alleles in a population.

2. *Disease- and pest-resistance genes :* In wild populations a relatively low proportion of resistant plants curtails the spread of disease and so offers protection to the population at large. Challenge by different pathogen races thus builds up a large number of alleles which buffer the population against epidemic attack.

3. *Stable mixtures :* Density- and frequency-dependent interactions can result in stable mixtures of coadapted (mutually interdependent) genotypes having different and complementary requirements for resources, such as nutrients and light, which enable them to escape competitive elimination.

Coadaptation

Coadapted genotypes have been shown to maintain high levels of variability in wild populations of self-fertilizing plants and in land races of cereals. The faculty to develop coadapted associations is also possible for outbreeding species provided they are perennials and long-lived or are propagated asexually, as with perennial pasture species. The coadapted genotypes, however, cannot be preserved as such over sexual generations but add to the genetic heterogeneity of the seed population, and thus increase the population's capacity to respond to further selective change in a heterogeneous environment.

A further dimension of coadaptation occurs at the interspecific level in mixed plant communities, for example grass and legume associations. If spatial and temporal differences are on a sufficient scale to fragment the habitat into independent niches, then integrated disruptive selection becomes two-way directional selection for independent phenotypic optima. Differences can occur over very short distances and despite considerable gene flow from the adjacent sites if the selection pressures are high enough. If continued long enough, genetic barriers to crossing may arise and the new populations will become genetically isolated and may even become distinct species.

2. DIRECTIONAL SELECTION

In plant breeding the aim of selection is to change existing populations, varieties, etc. to achieve specific objectives. Selection is therefore 'directional' towards maximal expression of the targeted character(s), or on optimal expression. This requires reassociation of genes affecting a number of connected characters in producing a fully balanced phenotype. Directional selection for a character leads to the establishment of dominance and epistasis. Several selection methods have been devised to help the plant breeder make accurate decisions. These methods are based on genetic concepts, namely progeny test or family evaluation and types of gene action, and experimental procedures that attempt to minimize the effect of environments on the expression of genotypes. The genetic basis of each of the selection methods has been discussed together with selection methods in subsequent chapters. Here we discuss some important general features of selection and the ways through which gains from selection can be increased.

Systems of Mating and Their Effect on Selection

Each mating system has advantages and disadvantages for particular goals. The number of genes involved and heritability greatly influence the effect of different mating systems. (Allard, 1960).

1) Random Mating (random mating with selection)

Random mating means, individuals are mated at random or by chance. Strictly speaking, it means that each member of the population has an equal chance to mate, that is, any female gamete is equally likely to be fertilized by any male gamete of the same plant or with any other member (plant) of the population, and there is equal chance for it to produce seeds. Random mating brings no change in either gene frequencies, existing variability in the population or genetic correlation between close relatives. It really does not fix genes, regardless of selection or no selection. This theoretical concept is not met, however, and cannot be met in plant breeding, since selection is involved; therefore it is more appropriately known as random mating with selection. Random mating with selection, on the other hand, effects change in gene frequencies and the mean of the population while at the same time there is no appreciable or little effect on the homozygosity, population variance (variability) or genetic correlation between close relatives, provided the population is large enough.

When the population size is limited (small) variation in gene frequencies due to *genetic drift*[1] and inbreeding reduce its heterozygosity. After a sufficient number of generations the population ultimately becomes homozygous for all loci. The effect of drift will clearly be more visible when intense selection is practised.

Random mating is of utmost use in the preservation of desirable alleles, such as in germplasm composites, which otherwise might be lost by chance under systems, namely inbreeding. It also has a place in progeny testing.

2) Inbreeding (genetic assortative selective mating)

Inbreeding means, mating of individuals which are more closely related by ancestry or self-fertilization. This is also known as genetic assortative mating. Inbreeding uncovers genetic variability concealed in heterozygotes and makes it accessible to action by selection.

Close inbreeding (continued self-fertilization for a number of generations) breaks populations up into numerous non-interbreeding groups and automatically brings about fixation of types (homozygosity) and an increase in prepotency. But inbreeding with selection favouring only one type of homozygote reduces genetic variance to zero as the type becomes fixed. If the goal is production of homozygous lines, inbreeding is the obvious choice.

3) Phenotypic Assortative Mating

Phenotypic assortative mating means, mating of plants which resemble each other more closely (phenotypically) than the rest of the population. It tends to concentrate the population towards the two extremes. Total variability is increased if the two extremes are maintained.

This type of mating is inefficient in fixation of type or in changing prepotency. Most of its effect is produced in only a few generations after it is begun and is likely to disappear once random mating is resumed; the population will then return to its original composition, provided selection has not altered gene frequencies.

If the goal is development of an extreme phenotype, assortative mating with selection is appropriate.

4) Disassortative Mating

Genetic disassortative mating means, mating between individuals which are less closely related than they would be under random mating. Its real application in breeding lies in crossing different strains.

Phenotypic disassortative mating, on the other hand, means mating of individuals on the basis of contrasting phenotypic characters. It is practised to compensate for defects by choosing contrasting parents, each of which compensates for the weaknesses of the other. It is most useful in plant breeding to maintain diversity in populations which serve as sources of genes. Mating of unlike types in the population counteracts the erosion of diversity and helps to maintain the usefulness of the population as a source of genes.

---

[1] Random fluctuation in gene frequency; unintentional variation in sampling of alleles say *A* or *a* owing to small population size.

Genetic Effects of Phenotypic Disassortative Mating

a) This system of mating tends to maintain heterozygosity in a population. When highly heritable characters are governed by few genes, it may even increase heterozygosity slightly above the level expected with random mating.

b) It tends to decrease population variance, since the offspring of opposing extreme types tend to be nearer the population mean than the offspring of random mating.

c) Mating of unlikes tends to reduce the genetic correlation between relatives.

Disassortative mating is therefore the most conservative of the mating systems and the one which best holds a population together. It has a role to play, for example, in maintaining a source population in which maximum stability is the goal.

Genetic Advance under Selection

Voluminous literature has developed on the subject. Interested readers are referred to Falconer (1960). Genetic advance through selection may be enhanced in the following ways.

a) Increasing selection pressure : Selection differential (k) is a direct function of the proportion of selected units, that is, selection intensity. As the value of k increases, the proportion of the selected group decreases in size, leading to greater expected progress. Very high selection pressure, however, may drastically reduce genetic variability, which might not be desirable. The selection intensity therefore should be carefully chosen while attempting to increase genetic advance through increased selection pressure.

b) Adjusting the coefficient of σ^2_A (C) : Control over female and/or male sex is an important way to increase progress through selection; for example, $C = 1/2$ for mass selection (no control over male sex), and $C = 1/8$ for half-sib family selection if only female gametes are selected; but if both sexes are selected at the same selection intensity then $C = 1$ and 1/4 for mass selection and half-sib family selection respectively.

c) Increasing genetic variability: The greater the genetic variability, the better the chances for selection of superior types. Development of germplasm complexes from genetically divergent sources is also one of the very important ways for increasing gains through selection.

d) Controlling environmental effects : Gains from selection increase directly by decreasing phenotypic variance among selection units. This could be achieved by improving experimental techniques, namely, use of appropriate experimental design, increase in number of replications, etc. Selection of levelled fields, uniform fertility and irrigation system besides other good agronomic practices, as well as careful recording of data and analysis help to minimize environmental variations and increase genetic gains.

Correlation between Characters and Correlated Response

Correlation, measured by a correlation coefficient, is a measure of the degree of association, genetic or non-genetic between two or more characters. If genetic association exists, selection for one trait will cause changes in other traits. This is called *correlated response*. Estimation of phenotypic, genotypic and environmental correlations are based on the components of variances and covariances from analyses of variance and covariance respectively.

a) Genetic Causes of Correlated Response

i) Pleiotropism : Pleiotropism, that is, when a gene affects simultaneously several physiological pathways, its influence is observed on several characters. This is known as the pleiotropic effect.

ii) Linkage disequilibrium : Linkage refers to genes located on the same chromosome, that have a tendency of being transmitted together. For example if two non-allelic genes A and B with gene frequency P_A and P_B population are included in the gametes, the probability of their being transmitted together is $P_A \cdot P_B$, but if these are linked the probability is more than $P_A P_B$ because of the tendency of two genes to go together. The amount of linkage disequilibrium tends to be dissipated over generations in a random mating population. The rate of dissipation depends on the rate of recombination between the two genes. It is slower for closely linked genes and relatively faster for not so closely linked genes. When genes are not closely linked linkage disequilibrium is not an important cause of correlation between characters in a random mating population. In such cases the existence of genetic correlations is mostly attributed to pleiotropic effects.

b) Non-genetic Causes (Environmental causes)

Environmental correlations also exist because measurements in respect of several traits are taken for the same individual or from the same family. For example, a positive environmental correlation is expected to occur between plant height and ear height in the same plant because a microenvironment that favours plant height also increases ear height and vice versa. When two traits are evaluated by the average (or total) of a family, the environmental deviation in a given plot affects all individuals of a family and causes an environmental correlation among plots.

IMPORTANCE OF CORRELATED RESPONSE

Genetic correlations (correlations due to genetic causes) have a wider use in homozygous self-fertilized species and apomictic species. In cross-fertilized species, genetic correlations involving only additive effects are more appropriate because the information from the correlations is used in connection with recurrent selection. Additive genetic correlation is important in selection programmes because it gives information about the degree of association between two traits by way of additive or breeding values of individuals, which are the effects that can be changed by selection. Thus, selection for one trait will cause a change in the mean value of selected individuals, and if another trait is correlated additively to the first, selection will cause an indirect change in the mean of the second trait. This indirect change, called 'correlated response' can also be predicted in the same way as direct response for one trait. The merit of indirect selection relative to direct selection for the second trait is measured by the ratio of expected correlated response over direct response.

$$r_A \, K_1 \, h_1 / K_2 \, h_2$$

where, r_A = additive genetic component, K_1 and K_2 = selection differentials, and h_1 and h_2 = square roots of heritabilities of trait 1 and trait 2.

Indirect selection has an advantage over direct selection when $r_A \, h_1 > h_2$, that is when the secondary trait has a higher heritability than the desired character and the additive correlation between them is high.

Selection for More Than One Trait
Selection for more than one trait may be done in one of the following ways.

1) Tandem Selection
Tandem selection involves selection of a trait for a number of generations, which is followed by selection for another trait for a few generations and the process thus goes on. The major problem here is in determining the number of generations for which selection for a given trait is to be practised. The twin related issues are that if a high selection intensity is practised, there is a rapid loss of genetic variability and consequently the expected genetic gains from selection are reduced after a few generations; on the other hand, if the selection intensity is low, it will take many more generations to achieve the desired progress. In view of this, it is important to set the desired limit for each trait and the importance of the trait besides determining the heritability before the tandem selection is initiated. Information about correlated response is also important because a long-term change in one trait may cause an indirect but undesirable change in another important trait. An alternative procedure would be to alternate the selection cycle. For example M_1 cycle for trait X followed by N_1 cycle for trait Y and so on; and again M_2 cycle for trait X and so on.

Tandem selection is useful when the relative importance of each trait changes throughout the years. If genetic correlations do not exist, such as yield and disease resistance, tandem selection can be used effectively to increase the level of disease resistance before selection for yield is initiated.

2) Independent Culling
This method involves selection at a given intensity for several traits in the same generation. For example, selection for yield, ear height and lodging resistance may be practised in the same generation first by selecting say 40% plants/families for yield; of this 40% then select 50% plants/families for ear height and then out of these selected plants/families select 50% plants/families for lodging resistance. The total selection intensity thus would be .4 × .5 × .5 or 0.1 (10%).

3) Selection Index

This is the most widely used selection method for selection for more than one trait. In most applied selection procedures breeders use an *intuitive selection index* for simultaneous selection for several traits. This is based upon relative weightage placed on each trait, visual acuity and experience. The inherent subjectivity of the selection process enables breeders to recognize the desirable genotypes. The genotypic evaluation, however, is based on individual plant observations by phenotypic selection. The selection efficiency thus depends upon the manner empirical weights are given to several traits. Even the selection based on family evaluation in replicated trials requires accuracy in the evaluation of several characters to allow selection to be efficient. For example, characters such as plant height or ear height could be measured directly but a metrical evaluation for disease or lodging resistance is not feasible and these are measured on a relative score (0-10). The accuracy of evaluation on score basis depends upon experience and the amount of compromise made for different traits. When the progeny is especially high yielding the breeder may select despite the fact that the level of some other trait may not be as high as desired. The consistency of family performance over replications is an important criteria for selection. Also, the variation in plant stand needs to be looked into. Thus, the breeder's decision regarding selection is based on his ability to give to the several traits the appropriate weights he has visualized.

Usefulness of the selection index relative to selection based on only one trait and over tandem selection or independent culling is well established. The superiority of index increases with increasing number of traits under selection but decreases with increasing differences in relative importance. The superiority of the selection index is maximal when the traits considered are equally important. It is therefore imperative that the choice for traits be done carefully and with objectivity. Interested readers are referred to Henderson (1963) and Hanson and Johnson (1957).

METHOD OF MAKING SELECTION INDEX

The basis of the selection index is to assign relative weights to a combination of traits (the traits for which selection is to be practised). Thus each individual has an index value (score) and the selection is based on this value. It is possible that highest yielding individuals will not get the highest scores in the use of the selection index.

The genotypic value of an individual considering several traits is:

$$H_j = a_1 G_{1j} + a_2 G_{2j}, + a_i G_{ij}$$

where, a_i (i-1 to n) is the relative weight, G_{ij} is the genotypic value of the *j* th individual for the *i* th trait, and H_j is the combination of characters.

The objective is to make a selection so that H_j is the most desirable. Because G_{ij}, are not known but are evaluated through P_{ij} (phenotypic values), H_j is evaluated by an index called I_j, which is based on phenotypic values such that the correlation between H_j and I_j is as great as possible.

$$GA = PB$$

where, G = matrix of genetic variances and covariances,
A = a vector of values (relative weights),
P = matrix of phenotypic variances and covariance,
B = a vector of unknown values.

Restricted selection index : Tallio (1962) generalized the theory of restricted selection index. It is used when it is desired to improve n_1 traits without limit and n_2 traits only to a predetermined limit.

Base Index (Williams, 1962). For the base index the traits are weighted directly by their economic values. It is 95% as efficient as conventional indexes.

10

Interspecific and Intergeneric Hybridization

Interspecific hybridization refers to hybridization between two species belonging to a genus, while intergeneric hybridization, refers to hybridization between species of two different genera. These are also referred to as wide crosses.

APPLICATION AND USES OF WIDE CROSSES

Though the individual opinions among plant breeders may vary certain plant breeding problems can be solved only by using a wide cross, one involving representatives of different species or genera. The specific uses of wide crosses are (Briggs and Knowles, 1967).

1) Transfer of a desirable gene or genes from a wild progenitor or related species and genera to cultivated varieties. These include disease- and insect-resistance gene(s), cytoplasmic-genetic male sterility system, restorer gene(s), hardiness, etc.

2) To exploit hybrid vigour in certain species, for example, interspecific hybrids in cotton; forage species, for example sorghum x Sudan grass hybrids; asexually propagated crops, for example, sugar-cane, potato.

3) To produce new alloploid species, for example, triticale.

4) To determine the evolutionary relationship among species. Plant evolutionists have used interspecific hybridization as a means to determine the species relationship.

Major Issues

The major issues in the utilization of wide crosses are reproductive barriers (Hadley and Openshaw, 1980). The external barriers, namely spatial, ecological or seasonal or mechanical isolation of species, however, pose no problem for plant breeders. Individuals from separated populations can be brought together and freely crossed to produce fertile F_1 hybrids which allow free flow of genes (genetic interchange) from one population to another. Strictly speaking, such populations are not different species, but more appropriately the subspecies or races. Internal barriers, on the other hand, are due to disharmonies between physiological and or cytological systems of plants which prevent genetic interchange between two species. In such instances wide crosses are difficult to make for the following reasons.

1. Cross Incompatibility (failure to produce F_1 zygote): Cross incompatibility may be due to disharmonies in physiological systems of plants from the different species or genera, which prevents F_1 zygote formation. For example, pollen does not germinate on the stigma, or the pollen tube does not completely traverse the style, or the male gamete does not combine with the egg even though the pollen tube reaches the ovary.

2. Non-viability of F_1s: Some of the F_1s of interspecific crosses are either non-viable or too weak to be of any use to the plant breeder. This may be due to the following reasons.

    i) *Chromosomal sterility :* The structural differences between chromosomes of two species interfere with pairing and disjunction at meiosis due to the disharmonies between two different genomes involved in the interspecific hybrid.

    ii) *Disharmonies between the genes of one species and the cytoplasm of another :* The resultant interaction may be deleterious or even lethal.

    iii) *Disharmonies between the genotype of F_1 zygote and the genotypes of endosperm or the maternal tissue with which the developing F_1 embryo is associated :* Probably the combination of two genomes, the female parent genome and the very different genome from the male parent, may produce unfavourable dosage effect.

3. F_1 sterility: Chromosomal sterility or disharmonies between parental genomes, or the genome of one parent and the cytoplasm of the other, as described above may cause the F_1 to be sterile. When sterility is caused by specific gene complexes, it is known as genic sterility (genic-hybrid sterility). When chromosome pairing is absent in interspecific hybrids, sterility results from abnormal distribution of the gametes of different numbers and combinations of chromosomes from the parental genomes. In some interspecific hybrids, even though the pairing and disjunction of chromosomes is normal, the F_1 is still sterile. This may be due to structural differences in chromosomes. Also, intergenomic interaction can result in the elimination of chromosomes from one of the parental genomes in the interspecific F_1 zygote during early mitoses in the developing embryo. The resultant mature plant is highly sterile. Chromosome elimination may also be due to incompatibility between a genome and cytoplasm. Sterility is also induced by some combinations of cytoplasm of one species and genomes of another. The cytoplasmic-genetic type of male sterility is a phenomenon that has been recorded for numerous interspecific crosses by plant breeders.

4. Hybrid breakdown: Some interspecific hybrids are both vigorous and fertile, and yet give rise to F_2 plants that are weak or sterile. Such a

situation is known as hybrid breakdown or genetic disability. This may be due to recombination of chromosome segments during meiosis in the F_1 hybrid involving small structural differences that can lead to gametes with small but significant deficiencies and duplications which may render the gametes non-viable (F_1 hybrid sterility). Recombination can, however, produce abnormal F_2 or later generation sporophytes, if the abnormal gametes escape elimination in the gametophytic stage of the life cycle. This is what is known as F_2 hybrid breakdown.

From a plant breeder's point of view, unless the plant breeder is able to overcome the reproductive barrier(s) he cannot induce any genetic interchange between the populations involved.

Methods to Overcome Reproductive Barriers

Genetic exchange between two species is brought about by either overcoming the reproductive barrier(s) or circumventing them. The plant breeder may employ the following approaches for successful interspecific or intergeneric hybridization.

A. Prefertilization Barriers

1) Selection of species for hybridization: A plant breeder has three different streams of gene pools (germplasm resources) at his disposal, namely, cultivated varieties, wild progenitors and companion weeds, and alien germplasm (other species in the genus). The plant breeder should carefully screen potential germplasm for expression of the desired character. If the choice is possible, the germplasm most closely related to the cultivated variety should be selected. The alien germplasm should be exploited only when potentialities of the cultivated and wild pools have been exhausted (Zohary, 1973). The best way to determine closeness of relationship between a cultivated variety and its relatives is on the basis of data from crossing experiments. Biosystematic studies made by cytotaxonomists with an appreciation of plant breeding problems is the best way to determine relationships. The plant breeder has to attempt to obtain different hybrid combinations.

2) Number of crosses: After making a choice of germplasm (for example, wild progenitors or alien germplasm), the plant breeder should attempt a sufficiently large number of crosses. This is important from the viewpoint that considerable genetic variation exists in some cultivated varieties, as well as in their wild relatives. Restriction of crossing attempts to a combination of only a few individuals from the wild species and a few from the cultivated variety may not serve the purpose of arriving at a sound conclusion regarding crossability of the two populations involved. In fact, the plant breeder should try to cover a broad range of variation in the material he is working with (Allard, 1960; Briggs and Knowles, 1967)

3) Reciprocal crosses : The plant breeder should invariably make reciprocal crosses, more particularly, when he has no knowledge for the parental

combinations. When he is successful in making reciporcal crosses, he should grow and observe them for possible post-fertilization barriers caused by cytoplasmic-genomic disharmonies, and also their F_2 generation as there might be some useful differences.

4) Increase in ploidy levels : Chromosome doubling has been successfully used in crosses between species that differ in number of genomes, as well as total chromosome numbers. The most commonly used technique is the application of colchicine in an aqueous solution. Numerous other techniques to induce chromosome doubling include temperature shock, wounding, exploitation of certain mutant genes, etc.

5) Indirect crosses : When a plant breeder fails to cross two species he may adopt an indirect approach. Such an approach consists of a series of crosses involving the cultivated variety and two or three related species. Although the cultivated variety and a particular cross-incompatible species are never directly crossed, chromosomes of their genome are brought together in the same organism through indirect crosses. This is not all that easy, however and may require several crossing attempts and the screening of large segregating populations to maintain the desired genes under transfer.

6) Modification in crossing technique : Crossing techniques for wide crosses are the same as for intraspecific crosses (chapter 7). Suitable modifications in the techniques of emasculation and pollination or both, however, may help overcome internal prefertilization barriers in specific cases. A few such techniques are described below.

  i) *Grafting of embryos :* This involves grafting the embryos of germinating seeds of one species onto the endosperms of germinating seeds of another species. For example, wheat embryos grafted to rye endosperms gave wheat plants. When such plants were crossed with rye pollen they produced 5 times more number of hybrids compared to control (ungrafted plants) (Hall, 1954).

  ii) *Mixing of pollen :* mixing viable incompatible pollen with non-viable compatible pollen has been successful in overcoming certain incompatible reactions in some species, for example, *Poplar* species (Knox et al., 1972).

  iii) *Use of organic solvents :* In some species (*Poplar* species) incompatibility between pollen grain and foreign stigmatic surfaces could be broken by application of organic solvents to the stigma or treatment of the incompatible pollen with organic solvents.

  iv) *Physical removal of stigma :* Failures of certain interspecific crosses (e.g. *Solanum* species) could be overcome by removing the stigma and replacing it with a drop of an agar-sucrose gelatin medium known to support pollen tube growth in many species (Swaminathan, 1955).

  v) *Application of growth regulators :* In some interspecific crosses the pollen does germinate on a foreign stigma, but the resultant pollen tubes may grow so slowly through the style that the egg degenerates, or the flower drops before fertilization can take place. This problem may be solved by application of growth regulators that promote pollen-tube growth or fruit development. Such regulators can also be used to increase embryo survival after fertilization has occurred.

vi) *Surgical procedures* : surgical procedures have also been used to shorten styles of one species, so that the pollen tube from the other could reach the egg. Some workers have successfully overcome cross incompatibility by injecting pollen suspensions into surgically opened ovaries in different species.

vii) *Protoplast fusion* : Protoplast fusion has interesting possibilities and may become potentially useful to plant breeders. The various aspects of protoplast fusion are discussed in Chapter 29.

viii) *Attempting wide crosses in large numbers* : Internal prefertilization barriers might also be overcome in some cases by attempting a very large number of crosses. Any method of bulk emasculation or pollination that eliminates working with individual florets or flowers permit attempts to make a very large number of crosses which otherwise would not be practically feasible. If marker genes are available (the wild relatives may differ so much from the crop that no marker gene is required for distinguishing hybrids by their general morphological characteristics), the plant breeder can cross-pollinate without emasculation, harvest seed from the female seed parent in bulk, and screen the progeny to separate hybrids from offspring resulting from sib crosses or selfs. Examples of such procedures for obtaining interspecific crosses are numerous.

Some form of male sterility, if available, may also suffice as long as the plants are female fertile and may be used for making open-pollinated crosses in isolated blocks where insect and wind can function as pollen vectors.

ix) *In vitro pollination* : *In vitro* ovular and placental pollinations, wherein the stigmatic, stylar and ovary wall tissues are almost completely removed from the path of the pollen tube, are potentially very useful in inbreeding and hybridization programmes when the zone of incompatibility lies in the stigma, style or ovary. Placental pollination has proven useful in at least three different areas: overcoming self-incompatibility, overcoming cross incompatibility and haploid production. Zenkteler and Slusarkiewicz-Jarzina (1986) have shown that fertilization and development of young hybrid proembryos can be achieved in extremely wide crosses through placental pollination. The limitation, however is that it is difficult to screen for the presence of genetic material of different origins in the derived plants.

B. Post-fertilization Barriers

Hadley and Openshaw (1980) have considered following techniques for overcoming post-fertilization barriers.

1) Use of growth hormones : In many instances the few F_1 hybrid zygotes or young developing embryos are lost either due to very low seedset, or

failure of embryos to develop. Several workers have successfully retained flowers and fruits by applying growth regulators immediately after pollination or stimulating fruit set.

2) Mixed pollinations : Mixing of pollen from different species and cross-compatible species may also be used for obtaining fruit set. For example, if the flowers of a species have three stigmas per flower and two of these are pollinated by pollen from different species and one with pollen from a compatible plant in the same species, fruit set may result.

3) Embryo culture : In instances wherein a zygote fails to develop normally due to the inability of the endosperm to nourish 'it, the embryo can sometimes be cultured and germinated on an artificial medium. The various aspects of embryo culture are discussed in Chapter 28.

4) Grafting : Grafting the F_1 seedlings onto normal plants of either parent may be resorted to in some cases when the F_1 is not viable due to abnormal chlorophyll development or inadequate root system.

5) Producing alloploids : Numerous interspecific hybrids are highly sterile. The chromosomal sterility may be overcome by doubling the chromosome numbers of F_1, that is, by producing an alloploid.

PART II

METHODS OF CROP BREEDING

11

Pure-line Selection

Pure-line selection, also referred to as individual plant selection, head-to-row selection, progeny selection, pedigree selection, single-line selection and inbred selection, consists of selection of desirable plants on a phenotypic basis from a genetically variable population, for example, local collections, introduced germplasm pools or in segregating populations of crosses. It involves raising the progeny of selected plants to determine the genetic behaviour of the selected plants, evaluation of the selected true breeding progenies and their release as pure-line varieties. Pure-line varieties are highly uniform in appearance and performance.

Goals of Pure-line Selection
Pure-line selection seeks to develop true breeding strains, that is, varieties that exhibit similar traits generation after generations.

Genetic Basis of Pure-line Selection
Pure-line selection method of breeding is based upon the concept of a pure line and application of the pure-line theory.

Concept of a Pure Line
Johannsen (1903) defined pure line as the progeny of a single self-fertilized homozygous individual. Jones (quoted from Hayes et al., 1955, p. 96). defined pure line as the descendants of one or more individuals of like germinal constitution that have undergone no germinal change. This definition is in common use today. Thus, pure line is a strain made up of the progeny of a single self-fertilized homozygous individual or individuals of the same genetic constitution. Unless the genetic constitution is altered, pure lines breed true in respect of all characters.

Pure-line Theory
The pure-line theory states that,
1) selection within a population of mixed genetic types may be effective in isolating lines that are inherently (genetically) different, and
2) within a pure line the variations are due to environmental factors only; hence once a pure line has been isolated, further selection within the line is ineffective.

PURE-LINE SELECTION METHOD OF DEVELOPING A VARIETY

First year: Plant the source population (local collections, introductions, germplasm pools, segregating generations of crosses, etc.). Select a few to several plants on the basis of desired characteristics. A preliminary selection may be made prior to flowering and the final selection made at the time of maturity. Harvest and shell each plant separately for examination of seed characteristics of each selection. Discard selections that have undesirable seed characteristics and retain the desirable ones. Maintain the identity of each selection.

Second year: Grow progeny of each selection made in the first year. Critically observe the variations in each progeny. Record in detail the desired characteristics of each progeny. Variants, if any, in the selected progeny or progenies must be rogued out. Select the best progenies on the basis of their performance. Harvest and shell each of the selected progeny separately and maintain its identity. Seed produce from each progeny should also be examined for seed characteristics and the undesirable types, if any, should be discarded. Each selected progeny now becomes an 'experimental strain'.

Third year: Grow the experimental strains in preliminary yield trial (station trial) comparing with a standard variety as a check. Record the data in respect to all the desired characteristics. Reject inferior progenies, and harvest superior strains after roguing out any aberrants noticed in the experimental strain and maintain the identity of each strain. Record the yield and examine the seed characteristics as done in previous years. Based upon data of the yield trial make a final selection of experimental strains and discard the inferior ones.

Fourth year: Grow the experimental strains in multilocation yield trials, comparing with standard varieties as checks. Record the data of yield trials as in previous years.

Fifth year: Continue yield trials as in the fourth year.

Sixth year: Conduct prerelease advance yield trials at several locations.

Seventh year: Start breeder's seed increases of promising strains for subsequent handing over to seed-production agencies. Move proposals for release.

Application and Uses of Pure-line Selection

Success of the pure-line selection method depends on the amount of genetic variability present in the population in which selection is practised.

Self-pollinated crops: Pure-line selection is usually practised in self-pollinated crops to develop pure-line varieties. Improvement is strictly limited to isolation of the best genotypes already present in the population. Other uses of pure-line selection include purification of old varieties.

Cross-pollinated crops: In cross-fertilized crops pure-line selection is used to develop inbred lines through selfing for several generations.

Limitations

There is no possibility of introducing new characteristics in the population, that is, no new genotypes are created, only the best are isolated. Despite the change in viewpoint during recent years that extreme uniformity may not be all that desirable from the viewpoint of adaptation, pure-line varieties have unbeatable merits, namely, the mixtures are generally lower yielding than the best pure-line variety in the mixture and the mixtures are less attractive than a uniform variety.

Tat e 11.1. Distinguishing Features of Mass Selection* and Pure-line Selection

Mass Selection	Pure-Line Selection
There is no control over pollination	Pollination is controlled (self-fertilization)
No progeny testing is done	Performance of each progeny is tested
Several cycles of selection may be necessary to effect desired improvement	One-time selection of a pure line is enough
Improvement is slow to moderate in each cycle of selection but can go on for a rather long time	Improvement is restricted to isolation of the best genotypes already available in the population
Varieties developed are neither homozygous nor homogeneous.	Varieties developed are homozygous and and homogeneous

*Mass selection is described in Chapter 15.

12

Pedigree Method

The pedigree method of selection consists of selecting individual plants in the segregating generations from a cross on the basis of their desirability, judged individually, and on the basis of a pedigree record.[1] The resultant varieties are pure-line varieties.

Goals of Pedigree method

Pedigree selection seeks to improve an adapted variety in respect of a certain specific feature, for example, disease resistance.

Genetic Basis of Pedigree Selection

Pedigree selection is based on application of the concept of *pure-line theory* to segregating populations. When two self-fertilized lines, say an adapted variety and disease-resistant line (donor parent) are crossed, the F_1 is a heterozygote. Self-fertilization leads to segregation in F_2 and subsequent generations. As a result the segregating population comprises both the homozygous and heterozygous individuals. The proportion of homozygous and heterozygous individuals in each segregating generation can be worked out by expanding the binomial $[1 + (2^m - 1)]^n$, where n is the number of gene pairs involved and m is the number of selfed generations. In the expanded binomial the first exponent in each term gives the number of homozygous loci and the second the number of heterozygous loci. The net effect of continued self-fertilization is a drastic reduction in the frequency of heterozygous plants and a corresponding increase in the number of more homozygous types in which the breeder is interested. Differential survival ability of the homozygote and heterozygotes and the linkages, if any, however affects the rate of return to homozygosity. The final result is a heterogeneous population of homozygous individuals/families. The breeder selects desirable homozygous families on the basis of the individual's pedigree.

Factors Affecting Genetic Advance under Pedigree Method of Selection

Choice of parents: The purpose of hybridization is to breed better varieties by combining desirable characters of two or more lines, varieties or species,

---

[1] Pedigree record means the record of an individual (selected plant and its family).

into a single variety or varieties. Occasionally, recombination of genetic factors leads to production of new and desirable characters not found in either of the parents (transgressive segregates). The parents for hybridization should therefore be carefully selected on the basis of their desirability. A simple way to do this is to list the various weaknesses in the adapted varieties, for example, susceptibility to certain diseases, late maturity, lodging, grain quality, etc., for which improvement is needed. On the basis of identified goals for improvement look for various source materials (donor parents) available in the germplasm, having the desired characters for which improvement is being sought. Record the agronomic features of the source materials. Accordingly, select the appropriate parents for hybridization.

Screening of breeding material : Use of artificial inoculation in screening the material for disease resistance, artificial disease epidemics under greenhouse or field conditions, quality tests, etc. is necessary for selection of desirable plants, which otherwise would be difficult and not very reliable. The plant breeder should seek the aid of specialists, such as plant pathologist, entomologist, biochemist, physiologist, etc. in developing particular tests that could be used in the evaluation of the segregating material in early generations.

PEDIGREE METHOD OF DEVELOPING A VARIETY

First year : Make sufficient number of crosses (50-100) between selected parents. The parents may be varieties, single crosses or multiple crosses and/or backcrosses. For each cross, usually one parent is an adapted variety with a good agronomic base and the other (donor parent) is such as to complement the specific weaknesses of the adapted variety.

Second year : Grow F_1 plants of each cross at low plant density along with its parents for comparison. This comparison is necessary to ensure that F_1 plants being raised are in fact hybrids. Harvest and thresh the F_1 plants of each cross separately and maintain proper identity. Record the usual data and F_1 yield of different crosses for comparison.

Third year : Grow F_2 population of each selected cross at low plant density/ m^2 (7.5-15 cm apart for cereals and 30-60 cm apart for other species). The number of plants may vary from a few thousand to several thousand (2000-10,000) depending on the facilities available. Plant infector rows all along the field. To facilitate selection for resistance to disease and pests artificial epidemics may be created for major disease(s). Use of special techniques for determining specific features may be necessary (Fig. 12.1).

The F_2 population affords the first opportunity for selection in the pedigree method. At this stage all undesirable plants must be eliminated as soon as they are identified. Select 200-500 most vigorous plants among the remaining plants that show the sought characteristics in the new variety. The breeder must have a clear perception concerning the morphological, physiological and other factors of the plants he saves for accurate judgment of the worth of selected plants. Harvest and thresh produce of each plant separately.

Fig. 12.1. Screening material for disease resistance in a disease nursery.

Examine physical characteristics of seeds of each seed pile. The piles unacceptable or found unsuitable should be discarded and the remaining bagged separately. Maintain the identity of each selection retained.

Fourth year : Grow progeny of each of the selected F_2 plants (F_3 generation) at low plant density per square metre. There must be approximately 30 or more plants per progeny. Selection in this generation continues to be on an individual plant basis, but emphasis is on selection of superior progenies. Use of artificial epiphytotics, etc. may also be made, since there may be progenies segregating for disease resistance. The number of selections made in the F_3 generation usually varies from 100 to 150. Harvesting, threshing and handling are done in the same manner as in the previous year. 25-50 seed piles may be discarded after examining the physical characteristics of seed and the rest retained for planting in the next season.

Fifth year: Grow progeny of each of the selected F_3 plants (F_4 generation) in the same way as the F_3 generation was raised. Selection in this generation also continues to be on the same basis, that is, individual plant basis due to the fact that in the F_4 generation homozygosity is not reached to the point where most breeders are willing to propagate progenies *per se*. The number of plants selected, however, are less than the F_3 generation. The selection is based on a combination of pedigree and visual evaluation. Some of the similar appearing material may be discarded. Approximately 100 plants are selected which are reduced to about 90 after examining the physical characteristics of seeds for planting in the next season.

Sixth year : Grow progeny of each of the selected F_4 plants (F_5 generation) on larger plots and at normal field spacings. If seed supplies permit a preliminary yield trial may also be conducted. Selections in this generation are based on visual examination of the desired characteristics, progeny performance and plot yields (if plots are large enough). If a preliminary yield trial has been conducted, yield data is used as an additional criterion of selection. As in F_4, about 100 plants may be selected and subsequently reduced to 80 after examining the physical characteristics of the seeds. In making selections it is assumed that the plants are homozygous and that they will breed true. Therefore, selection should be made from progenies that appear uniform.

Seventh year : Grow progeny of each of the selected F_5 plants (F_6 generation) in a manner similar to the F_5 generation. The basis of selection in this generation shifts however, from single plants to single progenies. Also, the plots are harvested in bulk. Approximately 15 most promising progenies are retained after visual evaluation, progeny performance (yield data) and seed examination. These are now termed 'selections'.

Eighth year : Grow the selections in station trials including check varieties and selections from other crosses. Record data on various aspects, including lodging, disease resistance and/or any other specific quality attributes and yield. Retain a few most promising selections.

Ninth year : Grow the promising selections in yield trials at several locations. Give one set each to a pathologist and an entomologist associated with the crop for making precise evaluation of disease and insect resistance, agronomic response, etc. Discard the inferior ones, if any.

Tenth to twelfth year : Grow the promising selections in multilocation tests to determine the superiority of selected strains over check varieties and also to determine the areas of adaptation. Record the data in the standard manner.

Thirteenth year : Conduct prerelease advance yield tests at several locations.

Fourteenth year : Start breeder's seed increases of promising selections for handling over to seed-production agencies. Move proposals for release of the varieties.

The scheme outlined above is not rigid and depending on the situation, the time required can be slightly reduced by taking off-season crops at suitable locations.

Application and Uses of Pedigree Method

The pedigree method has been the most widely used basic method of handling segregating materials in self-fertilized crops. The method can be used most advantageously if the characters to be combined in the cross are such that they can be seen easily and used as the basis for selection during the early generations. Since this method permits the plant breeder to see the progeny behaviour and exercise his skill in selection to a greater degree than any of the other methods of breeding self-pollinated crops, it is the best choice when the plant breeder is trying to get the most out of one or a few crosses in the shortest possible time.

Limitations of Pedigree Method

1) Relatively much longer time is required to breed a variety through the pedigree method of selection.
2) Much painstaking work and record taking is required during the early generations of segregating material, in addition to considerable resources in terms of land, material and manpower.

13

Bulk-population Method

The bulk-population method of selection in self-fertilized crops consists of raising the early segregating generations (F_2 to F_6) from a cross in a bulk plot, and selecting individual plants in F_6. The progenies of selected plants are handled in a manner similar to pure-line selection. The term *mass* is sometimes used as a synonym for bulk population. The resultant varieties are pure-line varieties.

Goals of Bulk-population Method

The bulk-population method of selection seeks to develop improved varieties of self-fertilized species through hybridization followed by the bulk-population method of selection.

Genetic Basis of Bulk-population Method

Stricto sensu, like the pedigree method, bulk selection is also based on application of the pure-line theory to segregating populations, as the objective is to develop pure-line varieties of self-fertilized crops. However, since no selection is practised in the early segregating generations, natural selection exerts a dynamic influence on the composition of the population at each generation, resulting in changes in gene frequencies as the hybrid moves towards homozygosity. It is therefore necessary to understand population dynamics. (Allard, 1960).

Population dynamics: There are three major expectations of population dynamics when bulk populations of self fertilized crops are perpetuated over a period of generations:

    i) that homozygosity at all loci of the component plants in the bulk population is approached;

    ii) that natural selection will improve the mean population performance; and

    iii) that genotypes with good agricultural fitness will be retained in the population; further, these genotypes can be isolated anytime by mass selection for developing as pure-line varieties.

Genetic shifts during bulking: Genetic shifts in bulk populations take place in favour of highly competitive types. This is due to higher seedset (fecundity) and reduced seed weight as compared to poor competitors. Because of unequal fertility, highly competitive types are represented in higher frequencies in successive generations resulting in decreased variability in the population and increased frequencies of closely related genotypes. Intrapopulation competition could thus be detrimental in isolating homozygous individuals with good performance in pure stands, if they are poor competitors.

Factors Affecting Genetic Advance under Bulk-population Method

1. Natural Selection

Natural selection presumably eliminates some of the weaker types, leading to genetic shifts in favour of good competitive types.

a) Short-time bulks (6-10 generations)

Natural selection operates principally on individuals that are heterozygous at many loci. Here the competition is primarily among heterozygotes. Since complete or near complete homozygosity is the ultimate goal, natural selection is halted when it might become effective in selecting among homozygous lines.

b) Long-time bulks (20-25 generations)

Upon attainment of near homozygosity by F_6–F_8 generation, bulk populations are composed of a vast number of mostly homozygote types. Also the proportion of homozygotes with obvious agricultural defects is reduced. The long periods of bulk handling are more desirable since natural selection may be more discerning than the plant breeder in selecting subtle differences related to adaptation and certain other desirable characteristics. Hence natural selection may be essential, if the very best types are to be increased to the point where they make up a significant proportion of the population. Selection differentials are likely to be small and not constant from generation to generation. Thus long periods of bulk handling are more desirable.

2. Artificial Selection in Bulk Populations

Though natural selection exerts strong selective pressures in bulk hybrid populations, particularly in respect of intangible characteristics that provide for good adaptation, other important characteristics, for example disease resistance, appear to be neutral in competition. Therefore opportunities exist for practising artificial selection to shift the population towards an agriculturally desirable type. It is therefore important that the population be purged of unwanted genotypes, especially when the unwanted types are highly competitive, for example determinate versus indeterminate types. Besides, there is also good reason to apply artificial selection against undesirable seed colours, pubescence, awns, barbs or other such characteristics that are undesirable in commercial varieties but which may be neutral in survival.

3. Use of Artificial Aids

Use of artificial disease and insect epidemics provides the opportunity to screen resistant types. Similarly, mechanical sieving for desired seed shape and size provides a rapid way of eliminating undesirable types. Bulk harvesting when only a part of the plant is mature is an effective and inexpensive way to select for earliness.

4. Duration of Bulking Period

When the primary objective is to achieve homozygosity, the duration of bulking depends on genetic considerations and can thus be specified quite precisely. Average percentage of homozygosity is very high by F_6. Almost 100 per cent homozygosity is achieved by the F_{10} generation even though a large number of genotypes are segregating. If heterozygotes are more productive than homozygotes the rate of return to homozygosity will be delayed.

BULK-POPULATION METHOD OF DEVELOPING A VARIETY

First year: Make sufficient number of crosses between selected parents. The parents may be varieties, single crosses. or multiple crosses or bulk crosses.

Second year: Grow F_1 plants of each cross at low plant density per square metre of area along with its parents for comparison. This comparison is necessary to ensure that F_1 plants being grown are in fact hybrids. Harvest the seeds of each cross in bulk.

Third year: Grow F_2 population at normal spacing or thin spacing (if selection for disease resistance is to be done). Harvest and bulk seeds from all the retained plants. Allow natural selection to exert selective pressure in respect of intangible characteristics that provide for good adaptation. Some other characters, e.g. disease resistance, specific quality attribute, etc., appear to be neutral in competition, however. Artificial selection may therefore be applied here for shifting the population towards agriculturally desirable types in respect of such characters. Infector rows and artificial epiphytotics may be used to remove the susceptible types from the population.

Also, the population may be purged of unwanted genotypes, for example determinate versus indeterminate types, seed colours or other such characteristics that are undesirable in commercial varieties. Subpopulations may also be made, if correlated responses are expected. As mentioned earlier, bulk harvesting when only a part of the plant is mature is an effective and inexpensive way to select for earliness, and mechanical sieving for desired seed shape and size provides a rapid way of eliminating undesirable types.

Fourth to sixth year: The procedure for handling F_3, F_4 and F_5 generations is the same as for F_2. The population is grown in such a field, if possible, where differentiation for insect or disease attack will show up. In both F_3 and F_4 about 10,000 plants are grown. In F_5 only about 5,000 plants are grown. About 1,000 plants may be selected in F_5 and each is threshed individually. Seed piles are examined on seed tables and about 500 best seed piles are kept individually for carrying on F_6 progeny tests in the nurseries.

Seventh year: Grow progeny of selected F_5 plants at low plant density per square metre. Select individual plants. Due emphasis should be given to the performance of progenies, however. Individual selections should be from superior progenies. All the plants are harvested individually and seed examined before the F_6 seed piles are pronounced uniform and bulked as new experimental strains.

Eighth year: Grow the experimental strains in station trials including check varieties and other selections. Record data on various aspects including lodging, disease resistance and yield at maturity.

Ninth year: Grow the promising selections in yield trials at the state level and at several locations. Give one set each to the pathologist and entomologist associated with the crop for making precise evaluation of disease and insect resistance, agronomic response, etc. Discard the inferior ones.

Tenth to twelfth year: Grow the promising selections in multilocation yield trials to determine the superiority of selected strains over check varieties and the areas of adaptation. Record the data in a standard manner.

Thirteenth year: Conduct prerelease advance yield trials at several locations. Start breeder's seed increases of promising selections for handing over to seed-production agencies. Move proposals for release of the varieties.

The scheme outlined above is not rigid and depending on the facilities, time and other considerations the period of bulking can be increased or decreased and, too, the use of artificial selection aids.

Application and Uses of Bulk-population Method

The bulk method is an inexpensive, simple and convenient method of carrying populations in large numbers than is generally possible in the pedigree method. It precludes the painstaking bookkeeping required in the pedigree system. The method is operationally best suited to crops in which harvesting of individual plants is cumbersome. It is particularly well suited for use whenever a heavy elimination of easily recognizable undesirable plants can be made in the early segregating generations by growing them in special nurseries.

Limitations

1. This method should not be resorted to when most of the characters, for example, straw strength, shattering resistance etc. are not easily judged from an individual plant.
2. This method cannot be expected to produce outstanding new varieties as frequently as the pedigree method.
3. Since no progeny testing is done in early segregating generations the effectiveness of the plant breeder's selection work depends entirely on the extent to which undesirable characters, such as disease susceptibility, show up in individual plants.
4. Genetic shifts and natural selections due to inadequate sampling and intergenotypic competition can reduce genetic variability and alter gene frequency in an undesirable direction.
5. This method takes much longer time to derive pure lines.
6. Unlike pedigree selection, the bulk method depends on the nature and outcome of mass trials by natural selection acting on a heterogeneous mixture of competing genotypes. The key point is therefore the type of correlation that exists between the agronomic productivity of a genotype and its competitive ability.

COMPOSITE CROSS-BULK-POPULATION METHOD (EVOLUTIONARY METHOD OF BREEDING, AFTER SUNESON, 1956)

The composite cross-bulk-population method consists of systematically crossing a large number of varieties. These multiple crosses are produced by crossing a pair of parents, then crossing pairs of F_1s until all parents enter into a common progeny to produce ultimately a single hybrid stock. Subsequent handling of the material is done as in the bulk method with no conscious selection. The method seeks to develop varieties with very wide adaptability.

Genetic Basis of Composite Cross-Bulk-population Method

There exists a relationship between the yielding ability of an individual and its *fitness*, that is, the ability to survive under natural selection. This relationship is essential for this method to be useful to the breeder. Here natural selection indirectly favours the better yielding genotypes along with better adaptation through the survival of the fittest. Fitness selection, however, is made up of three components, namely seed germinability, survival to reproductive age and number of seeds produced. Each of these components, in turn, might be governed by a large number of genetic and developmental factors. Natural selection achieves its goal by taking various alternative pathways depending on the components of fitness involved in the process. Accordingly, the array of genotypes representing the adaptive norm of a population is the characteristic of its initial structure and evolutionary history. Since such composite cross bulks have a great genetic diversity present at all stages, a number of significant adaptive changes may be expected under natural selection. During its propagation the bulk hybrid population encounters a succession of different environments. This may lead to many more adaptive peaks. The intermediate values are more likely to coincide with optimum fitness in respect of metric traits. Another important factor related to slowing down selection response is the interenvironmental slippage due to fluctuations in the optima selected for.

In general, genetic changes in response to selection should give a consistent and directional change. As the inbreeding generations are proceeding, the relative amount of additive genetic variance may be expected to increase due to dissipation of some of the non-additive gene effects.

COMPOSITE CROSS-BULK-POPULATION METHOD OF DEVELOPING A VARIETY

First year: Cross the selected varieties in pairs, for example 16 varieties; A × B, C × D, E × F, ... O × P, etc. Any other suitable crossing scheme may also be adopted. H.V. Harlan and his coworkers in their composite cross-bulk-population method intercrossed a fairly large number of barley varieties of diverse geographic origin in a diallel cross or some other scheme and all hybrids were bulked together for propagation by the usual bulk method.

Second year: Cross F_1s in pairs (AB × CD, EF × GH ..... MN × OP, etc.)

Third year: Cross these F_1s in pairs (ABCD × EFGH; IJKL × MNOP, etc.)

Fourth year: make composite cross (ABCDEFGH × IJKLMNOP)

Fifth year: Grow fist selfed generation (F_2 generation). Harvest and bulk seed from all the plants.

Sixth to twentieth year or more : Grow F_3 and the following generations on large plots (1/20 to 1/15 ha) from random seed samples. Harvest and bulk seed from all the plants. Allow natural selection to exert pressure. By this time the bulk population is composed of homozygote types.

Twenty-first year and onward : The bulk populations thus developed may be tested *per se* as experimental varieties, or some selections can be made and tested as experimental varieties in the same manner as described earlier for bulk method.

Application and Uses of Composite Cross

The use of bulk populations for commercial production might have great potential as shown from the findings of H.V. Harlan and his coworkers. They obtained a number of improved strains of barley from composite cross populations. The great genetic diversity in these populations is likely to have a more stable performance level over different environments (population homeostasis) than any single pure line or mixture of pure lines. Another important feature is the economy of the entire programme.

Limitations
1) This method takes a much longer time to derive pure lines.
2) The yielding ability observed under competitive circumstances of such heterogeneous bulk populations may or may not correspond to yielding ability in pure stands. It is thus necessary to investigate the nature of the relationship between their performance in bulk and in pure stands to derive superior strains by this method.
3) The extent of coincidence between the adaptive peaks of the population and the interest of the plant breeder determines the usefulness of this method.

SINGLE-SEED-DESCENT METHOD (SSD)

SSD is designed to maintain the total range of variation in a population by precluding loss of noncompetitive plants by taking a single seed (or equal no. of seeds) from each individual of the population, starting from F_2, and compositing the seeds to propagate the next generation. Selection is not practised until F_5 or F_6. At this stage the individuals in the population are reasonably homozygous. SSD is a modification of the bulk method of breeding (Brim, 1966).

Computer simulation studies revealed that at high heritabilities the pedigree method is more effective while at low heritabilities SSD is more effective. It has also been reported to be more effective in situations in which competition effects are important.

SINGLE-SEED-DESCENT METHOD OF DEVELOPING A VARIETY

First year: Make a sufficient number of crosses between selected parents. The parents may be varieties, single or multiple crosses, etc.

Second year: Grow F_1 plants of each cross at low plant density along with parents for comparison to ensure that F_1 plants being grown are in fact hybrids. Harvest the F_1 plants in bulk for each cross.

Third year: Grow F_2 generation of each cross. Harvest the plants of each cross in bulk. Take single or equal number of seeds from each plant and composite them for raising the next generation.

Fourth to sixth year: F_3, F_4 and F_5 generations are raised in a manner similar to the F_2 generation. About 1000 plants may be selected in F_5 and threshed individually. The seed piles are examined on seed tables and about 500 best plants are kept individually for carrying out F_6 progeny tests.

Seventh to thirteenth year: Same as described for bulk method.

Application and Uses

The SSD method, as a modification of the bulk-population method, has features that overcome the problem of natural selection and inadequate sampling in the conventional bulk-population method. This method minimizes natural selection without eliminating it. Thus, if population size is limiting, it is expected that the SSD method will maintain more genetic variability. The SSD method has obvious advantages in that gene frequencies are stabilized.

Another advantage is that the segregating generations can be advanced with the maximum possible seed, wherever facilities such as greenhouse and off-season nurseries are available. This can be extremely rapid in low-nutrient, continuous-light environments (Grafius, 1965). Depending on the crop plant, the breeding cycle can be reduced from about 8 years with mass selection to about 4 years with SSD. The SSD introduces economics of time and labour and offers good possibilities of isolating superior genotypes. In crops such as lentils, where poor growth habit and lack of synchronous maturity, etc. make it difficult to practice the pedigree method, SSD should be preferred.

Limitations

1. The extent of plant loss due to lack of germination, plant death, and failure of plants to produce a single seed from generation to generation affects the genetic make-up of the SSD population. Some superior genotypes may thus be lost.

2. No genetic advance is realized in a population when raised in greenhouse or off-season nurseries. A seed increase generation is required after which testing can be done in one or more environments.

3. Each plant in the population must be handled at harvest time.
4. No form of selection is possible during the segregating generations.

BULK AND PEDIGREE METHOD

The bulk and pedigree method seeks to combine the advantages of the pedigree and bulk methods. It consists of following the pedigree system up to the F_3 generation, bulk handling of F_4 and F_5 generations, selection of individual plants in F_5 and of progenies in F_6 (Sneep et al., 1979).

BULK AND PEDIGREE METHOD OF DEVELOPING A VARIETY

First year: Make crosses among selected parents. The parents may be varieties, single crosses or multiple crosses. In the case of single crosses only those crosses are included as parents which are well adapted and complement each other with respect to desirable characters. In instances wherein the single crosses involve primitive or unadapted parents, the F_1s are first backcrossed once or twice with adapted varieties before selection as parents.

Second year: Grow a limited number of F_1 generation plants of each cross along with their parents (8 F_1 plants for single crosses and 24 F_1 plants from multiple crosses). In order to determine that F_1 plants are in fact F_1 hybrids, compare them with the female parent. All the F_1 plants from a cross are harvested in bulk. In the case of multiple crosses a weak negative selection is practised in F_1s mostly for straw strength and disease reaction.

Third year: Grow each F_2 population resulting from a cross at very low seeding rate in one standard plot of about 10 m^2 (about 500 plants). The population resulting from crossing multiple cross is expected to show greater genetic and morphological variability and is therefore sown in two plots at the same seeding rate and infector rows are planted all around. Plots with control varieties, sown at the same seeding rate, are also included at regular intervals.

Negative selection is effected several times during the growing season by cutting back single plants which show undesirable characters. After ripening, positive selection is carried out by harvesting a number of single plants from each population. The number of plants harvested from a population depends on the information available about the parents and on the observations made on the F_2 material during the growing season. Selected plants are threshed individually and examined for seed characteristics. Undesirable piles are rejected. A record of each selected plant is maintained.

Fourth year: Grow F_3 progeny of each selected plant in a single 3-m row. Infector rows and control varieties are sown at regular intervals. During the growing season information on important characters such as disease resistance, straw characters, earliness, etc. is collected. No roguing is carried out. At the end of the growing season the best plant progenies of each cross are selected. The variability within the population of each cross is preserved to the maximum possible extent. In the selected plant progenies, single ears are harvested by hand from the best plants, again preserving variability within the progeny to the extent possible. The total number of ears harvested from one cross population is fixed (being determined by the quantity of seed needed for sowing the F_4 generation), but the number of plant progenies from which these ears are harvested may vary from one cross population to another. In this way, after fairly rigorous selection for agronomic characters in the F_3, narrowed F_4 populations are produced.

Fifth year: Grow the F_4 generation. The F_4 populations are sown in two plots of different soil types, together with control varieties; one plot (120 m^2) consists of clay soil (artificially inoculated with disease inoculum) sown at 50% of the normal seed rate while the other plot consisting of sandy soil (5 m^2) meant for observation is sown at the normal seed rate. During the growing season, no selection is practised in either of these plots. As much information as possible is recorded for the performance on the two soil types. The clay soil plot is harvested by combine and seed yield recorded. The sandy soil plot is not harvested but rather left standing as long as possible so as to obtain a good impression of straw characters and susceptibility to ear diseases. The observation plot is then harvested. The seed yield is determined. The selection for grain size is effected by means of rigorous grading. A sample of the seed is reserved for quality tests, for example protein content, flour yield, kernel hardness and dough properties. F_5 populations result from the foregoing procedure.

Sixth year: Grow the F_5 generation. F_5 populations are planted in two plots, in the same way as the F_2 populations from multiple crosses, that is, space planted with infector rows. As in the F_2 generation, roguing is carried out during the growing season and positive selection is practised after ripening. The number of plants to be harvested in a population is now determined on the basis of yield and field observation in the F_4 generation, field observation in the F_5 generation, quality tests carried out in the previous years and population variability (number of F_3 lines which have contributed to the population and observations in the F_4 and F_5). The selected plants are harvested as in the F_2.

Seventh year: Grow the F_6 generation. Plant progenies are sown in the same way as in the F_3. Besides selection for agronomic characters, now plant progenies are also selected for homogeneity. Promising non-segregating plant progenies are harvested in separate progeny bulk. In promising plant progenies which are still segregating, two or more single plants are harvested, which are then re-entered with the F_6 plant progenies in the next year. To avoid later problems during the purification of lines, it is essential that no segregating lines be included in yield trials. About 0.5 kg seed is harvested from each plant progeny in the F_6, this being sufficient for two locations in non-replicated yield trials and a third location on sandy soil for observation. Yield trials are carried out on different soil types and in different climatic conditions. All the lines in yield trials are also sown in disease nurseries for evaluating resistance to diseases.

Eighth to thirteenth year: Conduct yield trials in a manner similar to that described earlier for pedigree or bulk methods.

Seed increase: In deciding upon the generation in which line purification should be started, the breeder has to make a compromise. If the purification programme begins at a late selection stage the material will have reached a very high level of homozygosity, so that little genetic segregation can be expected. This means a considerable amount of pure seed will have to be produced in a very short time. Starting at an earlier stage has the advantage that the purification work can be spread over several generations, but it also has two disadvantages, viz.,

(1) a greater likelihood that segregation will occur in the material, and

(2) purification programmes will have to be undertaken for a large number of lines which are still in trial. Most of this work will prove useless because most of the lines will be discarded in the course of selection anyway.

Application and Uses

While the pedigree method is very labour intensive and losses of good combinations readily occur with the bulk method, the bulk and pedigree method attempts to combine the advantages of both methods.

14

Backcross Method

Backcross

In breeding, a cross of a hybrid to either of its parent is termed backcross. The parent with which successive backcrosses are made in backcross breeding is termed the *recurrent parent*, and the parent from which one or a few genes are transferred to the recurrent parent in backcross breeding is termed the *donor parent*.

Backcross Breeding

Backcross breeding is a system of breeding whereby recurrent backcrosses are made to one of the parents of a hybrid, accompanied by selection for a specific character or characters in the following generation(s). Usually the cycle is repeated for 5-6 generations. Self-fertilization (selfing) at this stage produces homozygosity in respect of the character under transfer. By this time all other characters are homozygous as well. Selection in this population for the desired character gives a variety which is very much like the recurrent parent, but superior in regard to the transferred character, for which the backcross programme was undertaken.

Goals of Backcross Breeding

The backcross method seeks to improve an adapted variety (varieties) with respect to certain specific attributes in the shortest possible time.

Genetic Basis of Backcross Method of Breeding

1. Transfer of Qualitative Characters (major genes)

Recovery of recurrent parent: The F_1 of a cross between the recurrent parent and the donor parent has 50% germplasm from each of the two parents. With every backcross with the recurrent parent, the proportion of the germplasm of the recurrent parent is increased by one-half, while that of the donor parent is decreased by one-half. Thus with 5 or 6 backcrosses the resultant population has 98.4 and 99.2% of the germplasm of the recurrent variety respectively.

Recovery of character under transfer: No difficulty is experienced in identifying plants carrying the desired trait, when it is governed by a dominant gene

(Table 14.1). When it is governed by a recessive trait it becomes necessary to raise an F_2 generation of alternate backcrosses (Table 14.2).

Table 14.1. Transfer of a dominant gene

| Year | Genotypes | | Resultant | | Germplasm |
	Female parent	Male parent	Gener-ation	Genotypes	of recurrent parent (%)
1.	rr	× RR	F_1	Rr	50
2. F_1	Rr	× rr	BC_1	1Rr:1rr	75
3. BC_1	Rr	× rr	BC_2	1Rr:1rr	87.50
4. BC_2	Rr	× rr	BC_3	1Rr:1rr	93.75
5. BC_3	Rr	× rr	BC_4	1Rr:1rr	96.87
6. BC_4	Rr	× rr	BC_5	1Rr:1rr	98.44
7. BC_5	Rr	× rr	BC_6	1Rr:1rr	99.22
8. BC_6	Self-fertilized (Rr)		BC_6F_2	1RR:2Rr:1rr	Select RR/Rr
9. BC_6F_2 progenies	Self-fertilized (RR/Rr)		BC_6-F_3	(i) All RR	Select & mix the seeds
				(ii) 1RR:2Rr:1rr	Discard segregating progenies.

Note : 1) Recurrent parent (rr), Donor parent (RR).
2). In female parent select plants for crossing which more closely resembles the recurrent parent.

Linkages, if any, between the desired gene(s) and an undesirable gene or a block of genes would need to be broken. Where the effect of an undesirable gene is conspicuous, a selfing series following a cross is more effective in breaking the linkages because the opportunity exists for breaks in linkage in both the male and female gametes; in backcrossing contrarily this cannot occur in the recurrent parent. Nevertheless, where selection is not possible for breaks in linkages due to effects of undesirable genes which do not allow expression under selfing and backcrossing, or heritability is low, backcrossing is more effective in breaking the linkages.

2. Transfer of Quantitative Characters

Where a quantitative character is under transfer, each backcross generation is grown through to F_3 lines before the next backcross is made. Such a procedure provides greater opportunities for selecting plants with a larger number of favourable alleles governing the characters under transfer. Also, in all such instances wherein heritability is low and several genes are involved, the number of crosses (backcrosses) of F_2 and F_3 populations would have to be large.

Table 14.2. Transfer of a recessive gene

Year	Genotypes		Resultant		Germplasm of recurrent parent (%)
	Female parent	Male parent	Generation	Genotypes	
1. Initial cross	RR (recurrent parent) ×	rr (Donor parent	F_1	Rr	50
2. F_1	Rr ×	RR	BC_1	1RR:1Rr	75
3. BC_1	Self-fertilized RR/Rr		BC_1-F_2	(i) All RR (ii) 1RR:2Rr:1rr	–
4. BC_1-F_2	rr ×	RR	BC_2	Rr	87.50
5. BC_2	Rr ×	RR	BC_3	1RR:1Rr	93.75
6. BC_3	Self-fertilized RR/Rr		BC_3-F_2	(i) All RR (ii) 1RR:2Rr:1rr	–
7. BC_3-F_2	rr ×	RR	BC_4	Rr	96.87
8. BC_4	Rr ×	RR	BC_5	1RR:1Rr	98.44
9. BC_5	Self-fertilized (RR/Rr)		BC_5-F_2	(i) All RR (ii) 1RR:2Rr:1rr	–
10. BC_5-F_2	rr ×	RR	BC_6	Rr	99.22
11. BC_6	Self-fertilized (Rr)		BC_6-F_2	1RR:2Rr:1rr	–
12. BC_6-F_2	Self-fertilized progenies (rr)		BC_6-F_3	rr	–

Note : In female parent select plants for crossing which more closely resemble the female parent. In self-fertilized generations select plants of rr genotype which more closely resemble the recurrent parent.

Factors Affecting Advance in Backcross Method

1. *Heritability of the characters* : Characters of high heritability governed by single gene are easiest to handle as this method enables their identification with a degree of certainty.

2. *Maintenance of character under transfer*: It is important to maintain the worthwhile intensity of character (expression) throughout the series of backcrosses. Due to the effects of modifier genes in the new genetic background, very often the intensity is somewhat lost despite most stringent selection throughout the backcross programme. This is so, even though the genetic control is monogenic.

3. *Recovery of recurrent parent* : Whereas a properly executed backcross breeding programme allows recovery of all the desirable characteristics of the recurrent parent, the possibility of some characters being modified inadvertently due to the effect of some of the genes tightly linked with the transferred gene cannot be ruled out.

4. *Number of backcrosses made and number of backcross generations*: The number of backcross plants should be sufficiently large so that the breeder is able to select a larger number of plants with desired genotypes. The population size, number of backcross generations, and self-fertilization of some of the backcross generations depend on the number of genes involved and the heritability of the character under transfer.

5. *Use of artificial aids*: Use of aids, such as disease epiphytotic, etc. in the screening of plants for desired character(s) greatly facilitates the selection of desired types.

METHOD OF BACKCROSS BREEDING

A. Transfer of a Dominant Gene

Year-wise handling of the material is done in the following manner.

First year: Make 10-20 crosses between selected parents, that is, recurrent parent and the donor parent.

Second year: Grow F_1 plants. Cross 10-20 F_1 plants with recurrent parent.

Third to seventh year: (BC_1 to BC_5 generation). Grow the backcross generation. Select 30-50 heterozygote plants which most resemble the recurrent parent and cross them with the recurrent parent. Use appropriate screening techniques, if available. A total of six backcrosses are made in rapid succession to produce plants which closely resemble the recurrent parent except for the gene/genes transferred from the donor parent.

Eighth year: Grow BC_6 generation. Use screening aids to select 400-500 desirable plants and harvest them separately to grow individual plant progenies next year.

Ninth year: Grow individual progenies of the plants selected in the previous year at low plant density (spaced planting). Use screening aids to select approximately 100 desirable progenies, which are not segregating. Bulk the seeds of desirable progenies.

Tenth year: Make yield tests to compare the bulk of selected progenies with recurrent variety to determine its equivalence in agronomic values with that of recurrent parent and superiority with respect to character under transfer.

Eleventh year: If the purpose is to release the newly developed line as a variety it would be desirable to conduct multilocation yield trials before release. In all other cases extensive testing of backcross-derived lines is unnecessary, provided the recurrent parent type has been recovered to a satisfactory extent.

B. Transfer of a Recessive Gene

Year-wise handling of the material is done in the following manner.

First year: Same as described earlier for transfer of a dominant gene.

Second year: Same as described earlier for transfer of a dominant gene.

Third year: Grow BC$_1$ generation. Harvest and bulk.

Fourth year: Grow BC$_1$-F$_2$ generation at low plant density (spaced planting). Use appropriate screening techniques to select desirable plants. Backcross 10-20 selected plants with recurrent parent.

Fifth year: Grow BC$_2$ generation. Select 10-20 plants which resemble more the recurrent parent and cross them with the recurrent parent.

Sixth year: Grow BC$_3$ generation. Harvest and bulk.

Seventh year: Grow BC$_3$-F$_2$ generation at low plant density. Use appropriate screening techniques to select desirable plants. Backcross 10-20 selected plants with recurrent parent.

Eighth year: Grow BC$_4$ generation. Select 10-20 plants which resemble more the recurrent parent and cross them with the recurrent parent.

Ninth year: Grow BC$_5$ generation. Harvest and bulk.

Tenth year: Grow BC$_5$-F$_2$ generation at low plant density. Use appropriate screening techniques to select desirable plants. Backcross 10-20 selected plants with recurrent parent.

Eleventh year: Grow BC$_6$ generation. Harvest and bulk.

Twelfth year: Grow BC$_6$-F$_2$ generation. Use appropriate screening techniques to select desirable plants which resemble more closely the recurrent parent. Select 400-500 plants and harvest them separately to grow individual plant progenies next year.

Thirteenth year: Grow individual progenies of the plants selected in the previous year at low plant density. Use appropriate screening aids to select approximately 100-200 desirable uniform progenies. Harvest and bulk the seeds of selected progenies.

Fourteenth and fifteenth years: Same as described for the tenth and eleventh years for transfer of dominant gene.

MODIFICATIONS IN THE BACKCROSS METHOD OF BREEDING

The method of developing improved varieties through backcross is not rigid and plant breeders have often modified the general procedure outlined above. One of the simplest modifications consists of restricting the number of backcrosses to two to three only. Further handling of the material is done in the manner of the pedigree method. This can be adopted when the donor parent is also acceptable as a cultivar. This modification leaves enough heterozygosity for transgressive segregates to appear subsequently. Another modification consists of raising F$_2$ and F$_3$ generations of backcross 1 and backcross 3 for carrying out a rigid selection for the character under transfer and for characters of the recurrent parent. The fourth, fifth and sixth backcrosses are then made in succession. This method may be used for transfer of both dominant and recessive genes. It is understood that effective selection in F$_2$ and F$_3$ generations is equivalent to one or two additional backcrosses. Yet another backcross scheme may involve the use of two or more recurrent parents. Each recurrent parent is used for one or two backcrosses. The underlying objective in this scheme is to combine good genes from each of the recurrent parents besides those from the non-recurrent parent.

ALTERNATIVE BACKCROSS PROGRAMME (Brown and Ellis, 1976)

An alternative backcross scheme was suggested by Brown and Ellis (1976) for selecting genetically fixed characters in selfed generations, namely, F_2, F_3 and F_4. After making the intitial cross F_1 is selfed. F_2 generation is screened for homozygous plants (in respect of desired character) and selected plants are selfed. F_3 generation is screened for confirming homozygous individuals selected earlier. After confirming homozygosity of the selected plants, selection among them is made for the plant type of recurrent parent. Selected plants are selfed again and the F_4 generation is raised. In the F_4 generation selection for characters such as kernel type is carried out. The cycle is repeated with increased selection pressure after crossing the selected plants with the recurrent parent. The second cycle again introduces all the genes of the recurrent parent into the line already selected for desirable attributes of the recurrent parent.

The obvious advantage is that alternation of three selfed generations between backcross cycles eliminates the work of hand-crossing and yet provides ample seed in which to select for homozygous types also. Screening for the desired character (say disease resistance) is confined to F_2 lines, and to small samples (10-20 plants) of F_3 lines to check for homozygosity of the plants selected in F_2. The loss in regaining the genetic background of the recurrent parent is compensated for by the breeder's selection in F_3 field plots and F_4 seed while the desirable character selected in the F_2 is proven to be fixed and homozygous in the F_3.

APPLICATION AND USES OF BACKCROSS METHOD

Self-fertilized Crops

The backcross method has been extensively used in breeding for disease resistance. Nevertheless, it is also a suitable method for the adjustment of morphological characters, colour characteristics and simply inherited quantitative characters, such as earliness, plant height, seed size and seed shape. In fact the method can be used to adjust any character that is moderately to highly heritable.

Cross-fertilized Crops

The backcross method practised for cross-fertilized crops does not differ from that for self-fertilized crops. However, the following two features of cross-fertilized crops should invariably be kept in mind:

(1) A large number of recurrent parent plants must be used in each backcross due to inherent heterozygosity and heterogeneity in cross-fertilized crops.

(2) It may be difficult or rather impossible to grow backcross generations as selfed F_3 lines to identify more accurately the plants carrying the character(s) under transfer.

The backcross method has also been extensively used to transfer cytoplasm in production of isogenic lines and improvement of inbred lines used for the production of hybrid varieties.

Limitations

1. This method is suitable for achieving very narrow breeding objectives, that is, transferring one or few simply inherited characteristics into an adapted variety. When the number of independently inherited desired characters is large or they are polygenic, other breeding methods are more appropriate.

2. The expression of the character which is to be introduced should be sufficiently pronounced and independent of environment to allow the breeder to select for it at all the stages of backcrossing and subsequent selection, and it should be possible to detect the character before flowering.

3. Recessive characters are more difficult to introduce.

4. A satisfactory recurrent parent and donor parent must exist.

5. In cross-pollinated crops particular care is necessary to ensure that the sample of gametes taken from the recurrent parent represents the same gene frequencies which characterise the recurrent parent.

6. The number of backcrosses should be sufficient to reconstitute the recurrent parent.

7. Backcrossing does not permit the achievement of unusual combinations of genes from two or more varieties. If such a combination of genes is obtained from a programme using the pedigree or bulk population methods, the performance of such a variety may be on a plane different from that of the recurrent parent in a backcross programme. The backcross method in such situations is adaptable in part only.

8. Some undesirable genes closely linked to the gene being transferred may also be transmitted to the new variety.

9. Hybridization has necessarily to be done for each backcross.

10. By the time the recurrent variety (after transfer of a gene) is released its agronomic superiority may be lost.

15

Population Improvement

Crop varieties of cross-fertilized crop species are populations of random mating individuals which share a common gene pool. Unlike self-fertilized crops individual plant selections to establish uniform pure-line varieties are not feasible in these crops due to segregation, which causes the progeny to deviate from the parental type. Also, upon self-fertilization vigour and productivity are adversely affected. Hence, in place of developing individual plant selection(s) based varieties, emphasis is given here to improvement of populations. Genetic advance is made through selection of a large number of plants, followed by interbreeding. Several cycles of selection may be necessary to achieve the desired objectives. Other approaches include development of hybrid varieties and synthetics, which involves development of superior inbred lines and their use in hybrid varieties and/or development of 'synthetics'. In this chapter we shall deal with some population improvement methods. The various aspects of recurrent selection are dealt with in Chapter 16.

METHODS OF POPULATION IMPROVEMENT

The various methods of population improvement can be discussed under the following headings:
1. Intrapopulation methods
   —Mass selection and its modifications
   —Ear-to-row selection and its modifications
   —Recurrent selection and its modifications
2. Interpopulation methods
   —Reciprocal recurrent selection and its modifications.

MASS SELECTION

Mass Selection
Mass selection consists of selecting desirable plants on a phenotype basis in a population, harvesting them and mixing their seed for seeding the next generation. The process is repeated until desired improvement has been attained. No progeny test or crossing work is involved in this method. Also, there is no control over pollination either.

Goals of Mass Selection

In general, mass selection aims at improving the general level of the population by selecting and compositing the superior genotypes already present in the population, such as local varieties.

Genetic Basis of Mass Selection

Mass selection in a heterogeneous population shifts the mean of the trait/ traits in the direction selection is practised. This is due to increased frequencies of desired alleles in the population resulting from selection. Shifts in the means of other traits may also occur simultaneously, depending on the linkage between the traits, if any, for which selection is done. Consequently, population variance of the selected and/or unselected traits is reduced.

Factors Affecting Efficiency of Mass Selection

a) *Heritability and gene action of the trait*: According to Romero and Frey (1966) the success of mass selection is directly related to the heritability of the trait. In terms of gene action, Gardner (1961) reported that it is most effective for additive type gene action and hence the greater the amount of additive genetic variance in the population for traits under selection, the greater shall be the success. When indirect mass selection is made (selection for one trait through the expression of another), both the heritability of the selected trait and the genetic correlations between the two traits determine the effectiveness of mass selection.

b) *Selection coefficient,* that is, the extent to which a genotype is removed from the population (selection coefficient), considerably influences the genetic advance.

c) *Number of genes involved*: The more the number of additive genes, the greater the efficiency of mass selection.

d) *Extent of variability in the population*: The greater the variability in a population for the desired character, the better the efficiency of mass selection.

e) *Breeder's selection ability*: The success of mass selection depends on the breeder's ability to judge suitable plants. There should be closer relationship between phenotype and the genotype.

f) *Crop management*: It is necessary to provide full scope for the full expression of each individual plant. Good agronomy and care of the mass selection block is essential in order to provide a chance to the various genotypes to express themselves fully. If needed, the crop should be timely irrigated.

g) *Size of population*: The size of population should be large enough to permit selection of desired types in adequate number. This should be based upon the kind of crop, heritability of character under selection, genetic make-up of the variety, etc. and the ease with which the desired character can be determined.

METHOD OF DEVELOPING A VARIETY THROUGH MASS SELECTION

The general procedure of mass selection is given below :

First year: a) Planting : Plant the source population (local variety, germplasm complex, synthetic variety, bulk population/or an inbred, (if purification is to be done) in which improvement is being sought on a sufficiently large plot so as to provide a sufficiently large population.

The selection plot needs to be isolated to avoid genetic contamination from neighbouring fields of the same kind. This is important for deriving the maximum benefit.

b) *Minimum size of population*: The size of population however, should depend on the kind of crop, genetic make-up of the variety, heritability of the trait, linkage relationship, etc. Obviously, if mass selection is being done only for purification purposes the size of population shall be relatively smaller (5000 plants) than for improvement purposes (about 10,000 to 20,000 plants).

c) *Roguing*: In cross-pollinated crops roguing of undesirable plants prior to flowering/and at flowering should be done to avoid genetic contamination in the selection plot.

d) *Selection of plants* : Select a few to several hundred plants having desired characteristics on the basis of phenotype. Selection must be carefully done. Efficiency will be better if the relationship between the phenotype and genotype is closer. It is desirable to make a preliminary selection prior to flowering or at the time of emergence of the inflorescence, and the final selection at the time of maturity. In cross-pollinated crops, e.g. corn, the rejected plants may then be detasselled to avoid their gene contribution to the next generation. In other cross-pollinated crops the rejected plants may be removed from the population so that only selected plants mate and contribute genes to the next generation.

e) *Selection intensity (percentage of plants selected)* : There should be a proper balance between the percentage of plants selected and the size of population. Selection of a small number of plants would amount to close sib-mating or inbreeding (in cross-pollinated crops), while selection of a large number of plants will result in inclusion of inferior genotypes, which should otherwise be excluded.

Several modification of mass selection (described later) can be adapted to improve its efficiency.

f) *Harvesting* : The selected plants should be harvested and threshed individually for table examination of seeds. If selection is based on individual plant yield, ears should be dried to similar moisture content before weighing. Any seed pile showing undesirable seed characteristics should be discarded. The remaining piles should be mixed and composited into one bulk.

Second year: a) Second cycle of mass selection : Repeat the process adopted in the first year with selected bulk to complete the second cycle of mass selection.

b) *Yield trial*: 1. Grow the selected bulk in a preliminary yield trial (station trial) and compare with a standard variety as a check. Or, if the objective is to improve the population, use original (unselected) population as a check. Record the data in respect of all the desired characteristics.

c) *Genetic advance*: Based upon data of the yield trial determine the advance made in the first cycle of mass selection.

Third year: a) Third cycle of mass selection : Repeat the process adopted in the first year with selected bulk of the second year to complete the third cycle of mass selection.

b) *Yield trial*: 1. Grow the selected bulk of the second year in a preliminary yield trial (station trial) and compare with a standard variety as a check. Record data of yield trial as mentioned in the previous year.

c) *Genetic advance*: Determine the advance made in the second cycle of selection on the basis of yield data.

Fourth year: a) *Fourth cycle of mass selection*: If the desired advance has been achieved, there is no need to initiate the next cycle of mass selection. Otherwise one more cycle of mass selection may be initiated.

b) *Yield trial*: Grow the selected bulk of the third year in a preliminary yield trial (station trial)/and or multilocation trials and compare with standard varieties as checks. Record the data of yield trial as in the previous year.

Fifth year: If no further cycle of mass selection was initiated in the fourth year, grow the selected bulk (from third cycle of mass selection) in yield tests at several locations and compare with standard varieties as checks. Record the data of yield trials as in the previous years to determine the performance and adaptation of the newly selected strain.

Sixth year: Continue yield tests as in the fifth year.

Seventh year: Conduct prerelease advance yield trials at several locations to determine the performance and adaptation of the variety.

Eighth year: Start breeder's seed increases for handing over to seed-production agencies. Move proposals for release of the variety.

MODIFICATIONS OF MASS SELECTION

Stratified or Grid System (Gardner, 1961)

In this system the entire field is divided into squares of equal size. An equal number of best plants from each square of the field is retained. At harvest, the whole field is subdivided into several grids or subplots (of 2 rows each) with 40-60 plants, depending on the selection intensity to be used. A selection intensity of 5 to 10 per cent is usually practised.

Honeycomb Selection (Fasoulas, 1973)

Fasoulas (1973) proposed the honeycomb method of selection in a population to ensure closer relationship between the phenotype of a selected plant and its genotypic value with a view to selecting plants with a superior genotypic value. This can by achieved through a regular triangular pattern of plant positions instead of regular square pattern. With the triangular honeycomb pattern each plant is to be compared with six equidistant neighbours. According to Fasoulas, there are two main factors responsible for masking the inherent genotypic value in an observation of the individual phenotype, that is, competition and soil heterogeneity. He suggested that competition could be avoided by growing plants in the field at low density per sq. m and the effect of soil heterogeneity reduced to the minimum or precluded if the plants are grown under optimal growing conditions and are compared with the plants grown near to each other. The method is said to be efficient in a population which is grown under a pattern of environmental conditions which is too fine textured for efficient grid selection. Bos (1981), however, maintained that only modest results are to be expected from the application of honeycomb selection.

Application and Uses of Mass Selection

Mass selection is both genetically and operationally efficient in exploiting valuable germplasm in a gene pool. Responses to mass selection within bulk gene pools have ranged from 2 to 15% per cycle. Even for characters of low heritability responses between 2 to 7% per cycle were obtained when selection was practised within stratified areas of a field.

Self-pollinated Crops

1. Improvement and purification of local varieties.
2. Handling of crosses and advance generations of bulk hybrid populations. Mass selection being the most effective for additive type gene action permits inexpensive propagation of both a large number of crosses and a large number of plants within crosses. It is especially useful for application to advance generations of bulk hybrid populations.

LIMITATIONS

1. There is little or no opportunity for genetic recombinations to occur in successive generations. Thus the opportunity for rare recombinations to occur is remote.
2. Mass selection creates no new variability but acts only on the variability already present in the original population.
3. The rate of advance is rather slow to moderate compared to other breeding methods.
4. It is not possible to know whether the plants being grouped are homozygous for specific dominant characters. Since heterozygous plants will segregate in the following generation, phenotypic selection may need to be repeated.
5. It is not possible to know whether the selected phenotype is superior in appearance due to hereditary characters or due to environment.

Cross-pollinated Crops

1. *As a breeding method*: Mass selection has been extensively used as a method of breeding in cross-pollinated crops.

2. *As a method of population improvement and development of base material* : Mass selection may be successfully used for improving populations which may well serve as the base material for the development of hybrid varieties. The importqnt reasons for this are :

i) Quantitative genetic studies carried out during the last 30 years or so have provided evidence that sizeable additive genetic variance is present in the populations of many crops.

ii) A large amount of genetic variability is available as a result of extensive germplasm collections. Germplasm pools have been built in many crops, largely as a result of international co-operation and efforts by international crop research centres.

LIMITATIONS

i) It is often difficult to select superior plants on the basis of phenotype alone.

ii) Uncontrolled pollination results in pollination of selected plants by both superior and inferior pollen parents.

iii) Strict selection may reduce population size, which in turn, may lead to inbreeding depression.

iv) The rate of advance is rather slow to moderate compared to other breeding methods.

In both self- and cross-fertilizing species the purpose of mass selection is population improvement which results from modifying the frequencies of desirable genes. In a self-fertilizing species, the selected genes are locked in a certain genotypic combination whereas in cross-fertilizing species gene frequencies are changed more directly.

EAR-TO-ROW METHOD OF SELECTION

The ear-to-row method as originally practised in corn (Hopkins, 1899) consisted of selection of plants in a population on the basis of phenotypic score for the characters for which improvement was desired, harvesting the ears of selected plants and cholling them individually, examining the seed piles, discarding obviously undesirable types and planting ear-to-row progenies for progeny-row evaluation. The selection unit in the ear-to-row method is the progeny (a half-sib family of an ear) rather than individual plants. The ear-to-row method has not been found effective for yield improvement in adapted varieties for various reasons.

MODIFIED EAR-TO-ROW METHOD

a) *Modified ear-to-row selection in corn (Lonnquist, 1964)* : In this method a sample of ear is taken from the population. Progeny row tests are replicated over 3 environments which permits estimation of genotype × environment interaction, and combining data across environments to determine superior families for recombination. A single replication of families (ear-to-row progenies) is planted in isolation, detasselled and cross-pollinated with males consisting of a bulk of all the families. This provides seed from each of the progenies.

b) *Modified ear-to-row selection in corn (Compton and Comstock, 1976)*: In this method all environments in season I are used to identify and select superior families. There is no isolation planting for recombination. In season II, only selected families are planted ear-to-row, detasselled and cross-pollinated with males consisting of a bulk of selected families. In this system parental control is more because selected families are mated with only males formed from a bulk of selected families. Modified ear-to-row selection may be effective in populations having adequate genetic variability. It may be used initially to improve a population or after mass selection has been practised for highly heritable traits.

16

Recurrent Selection

Recurrent selection means repeated cycles of selection with interbreeding of selects to provide for genetic recombination. Selection in a population is not recurrent until selects are interbred and a new cycle of selection is initiated. This selection system as used in corn involves self-pollination of selected individual plants (S_0 plants) in a heterozygous population and/or crossing them with a tester parent, evaluation of selfed progenies or test cross progenies, propagating remnant S_0 seed of selected superior progenies and making all possible intercrosses among them or, alternatively, growing them in isolated plots for recombinations. The resultant intercross population serves as source material for additional cycles of selection and intercrossing. Cycles of selection are repeated as long as improvement in desired traits is shown.

TYPES OF RECURRENT SELECTION

There are four types of recurrent selection distinguished by the manner in which plants with desirable attributes are identified. (Allard, 1960).

1. *Simple recurrent selection*: In this selection scheme plants are identified on the basis of phenotypic scores. Test crosses are not made.
2. *Recurrent selection for general combining ability*: In this selection scheme a tester parent with a broad genetic base is employed to produce test cross progenies. Selection is based on performance of test cross progeny in replicated trials. Variations in test cross performance are ascribed to differences in general combining ability.
3. *Recurrent selection for specific combining ability*: In this selection scheme a tester parent with a narrow genetic base is employed to produce test cross progenies. Selection is based on performance of test cross progeny in replicated trials. Variations in test cross performance are ascribed to differences in specific combining ability.
4. *Reciprocal recurrent selection*: In this selection scheme two heterozygous populations are involved, each of which is a tester for the other. This scheme provides for selection for both general and specific combining ability with certain limitations.

APPLICATION AND USES OF RECURRENT SELECTION

Recurrent selection will not and is not intended to replace classic breeding methods. Recurrent selection methods are in fact population improvement methods and supplement other breeding methods for developing still better varieties and hybrids. Populations resulting from recurrent selection may be used in any of the following ways:

1) as source populations for the development of inbred lines for use in the development of synthetics and hybrid varieties; and
2) as one of the parents in hybrid varieties.

The populations selected for recurrent selection should have adequate genetic variability, high mean value in respect of characters for which selection is to be practised and manifest heterosis in crosses, more particularly so for interpopulation improvement. The choice of appropriate recurrent selection scheme should be based upon defined objectives.

GENETIC BASIS OF RECURRENT SELECTION

Recurrent selection involves interbreeding of selfed seed of a large number of selected plants as a result of which the frequency of desirable alleles and gene combinations is increased while the high genetic variability in the populations involved is retained. This permits effective selection over a longer period.

In terms of gene action the various recurrent selection schemes are quite effective in exploiting additive, partial dominance to dominance and epistatic (overdominance) types of gene action. In simple recurrent selection test crosses are not made and hence effective use is restricted to characters of high heritability. In this scheme therefore additive gene action is operative in the selection of a trait or traits. In all other schemes test crosses are made which permit selection for general combining ability, selection for specific combining ability and/or for both general and specific combining abilities.

When additive gene effects are more important, recurrent selection for general combining ability (RS_{gca}) is more effective than any other scheme, when overdominance is operative than recurrent selection for specific combining ability (RS_{sca}) is more effective than any other scheme and when both additive and overdominance effects are important reciprocal recurrent selection (RRS) is more effective. However, the advantage of RS_{gca} or RS_{sca} over RRS is not very great and hence RRS is as effective as RS_{gca} or RS_{sca}, but when additive gene effects with partial to complete dominance are operative all the three schemes—RS_{gca}, RS_{sca} and RRS are equally effective. The expected genetic advance can be calculated as per formulas given in Table 16.1.

Table 16.1. Expected genetic advance under recurrent selection schemes (from Sprague and Eberhart, 1977).

Recurrent selection scheme and control over sexes (female and/or male) (1)	Years per cycle (2)

1. Mass selection
   (i) one sex (female)

$$k(1/2)\sigma^2 A \Big/ \sqrt{\sigma^2 W + \sigma^2 DE + \sigma^2 AE + \sigma^2 D + \sigma^2 A}$$ 1

   (ii) both sexes

$$k\sigma^2 A \Big/ \sqrt{\sigma^2 W + \sigma^2 DE + \sigma^2 AE + \sigma^2 D + \sigma^2 A}$$ 1 or 2

2. Modified ear-to-row
   (i) one sex (female)

$$k(1/8)\sigma^2 A \Big/ \sqrt{\sigma^2 / re + (1/4)\sigma^2 AE / e + (1/4)\sigma^2 A}$$ 1

   (ii) both sexes

$$k(1/4)\sigma^2 A \Big/ \sqrt{\sigma^2 / re + (1/4)\sigma^2 AE / e + (1/4)\sigma^2 A}$$ 2

3. Half-sib
   Remanent half-sib seed (when parental population is used as tester)

$$k(1/4)\sigma^2 A \Big/ \sqrt{\sigma^2 / re + (1/4)\sigma^2 AE / e + (1/4)\sigma^2 A}$$ 2

   selfed seed (when unrelated population is used as tester)

$$k(1/2)\sigma^2 A \Big/ \sqrt{\sigma^2 / re + (1/4)\sigma^2 AE / e + (1/4)\sigma^2 A}$$ 3

4. Full-sib

$$k(1/2)\sigma^2 A \Big/ \sqrt{\sigma^2 / re + \{(1/2)\sigma^2 AE + (1/4)\sigma^2 DE\} / e + (1/2)\sigma^2 A + (1/4)\sigma^2 D}$$ 2

5. S_1 2

$$k\sigma^2 A \Big/ \sqrt{\frac{\sigma^2}{re} + \frac{[\sigma^2 AE + (1/4)\sigma^2 DE]}{e} + [\sigma^2 A + (1/4)\sigma^2 D]}$$

6. S_2 3

$$k(3/2)\sigma^2 A \Big/ \sqrt{\frac{\sigma^2}{re} + \frac{[(3/2)\sigma^2 AE + (3/16)\sigma^2 DE]}{e} + [(3/2)\sigma^2 A + (3/16)\sigma^2 D]}$$

7. Reciprocal recurrent selection   3

$$\frac{k_1(1/4)\sigma^2 A_{12}}{\sqrt{\dfrac{\sigma^2_{12}}{re} + \dfrac{(1/4)\sigma^2 AE_{12}}{e} + (1/4)\sigma^2 A_{12}}} + \frac{k_2(1/4)\sigma^2 A_{21}}{\sqrt{\dfrac{\sigma^2_{21}}{re} + \dfrac{(1/4)\sigma^2 AE_{21}}{e} + (1/4)\sigma^2 A_{21}}}$$

Contd.

Table 16.1. Contd.

(1)	(2)

8. Reciprocal full-sib

$$\frac{k(1/2)\sigma^2 A_{12}}{\sqrt{\dfrac{\sigma^2}{re} + \dfrac{(1/2)\sigma^2 AE_{12} + (1/4)\sigma^2 DE_{12}}{e} + \left[(1/2)\sigma^2 AE_{12} + (1/4)\sigma^2 D_{12}\right]}}$$

2

9. RRS based on test cross of half-sib families

$$\frac{k_1(1/16)\sigma^2 A_{12}}{\sqrt{\dfrac{\sigma_{12}^2}{re} + \dfrac{(1/16)\sigma^2 AE_{12}}{e} + (1/16)\sigma^2 A_{12}}} + \frac{k_2(1/16)\sigma^2 A_{21}}{\sqrt{\dfrac{\sigma_{21}^2}{re} + \dfrac{(1/16)\sigma^2 AE_{21}}{e} + (1/16)\sigma^2 A_{21}}}$$

3

10. RRS based on test cross of half-sib families of prolific plants

1 or 2

$$\frac{k_1(1/8)\sigma^2 A_{12}}{\sqrt{\dfrac{\sigma_{12}^2}{re} + \dfrac{(1/8)\sigma^2 AE_{12}}{e} + (1/8)\sigma^2 A_{12}}} + \frac{k_2(1/8)\sigma^2 A_{21}}{\sqrt{\dfrac{\sigma_{21}^2}{re} + \dfrac{(1/8)\sigma^2 AE_{21}}{e} + (1/8)\sigma^2 A_{21}}}$$

k = Standardized selection differential; $\sigma^2 A$ = addive genetic variance; $\sigma^2 D$ = dominance variance; $\sigma^2 AE$ = additive $\times$ environment interaction variance; $O^2 DE$ = dominance $\times$ environment interaction variance; σ^2 = error variance; r = number of replications and e = number of environments. When mass selection is practised within ear rows for the primary trait then add additional component, $K(3/8)\sigma^2 A/\sqrt{\sigma^2 W} + (3/4)\sigma^2 AE + \sigma^2 DE + (3/4)\sigma^2 A + \sigma^2 D$.

Definition of additive genetic variance changes slightly with inbreeding and dominance variance is difficult to define unless $p = q = 0.5$.

$\sqrt{\sigma^2 A_{12}}$ and $\sigma^2 A_{21}$ are homologous of $\sigma^2 A$ (additive genetic variance; $\sigma^2 A_{12} = (1/2)$ $(\sigma^2 P_{12} + \sigma^2 A_{21})$ and is the additive genetic variance in the crossed population.

Procedure of Simple Recurrent Selection in Corn

Original Selection Cycle (C_0)

First season: Grow source population (heterozygous population). Select number of plants (S_0 plants) on the basis of phenotypic score for characters under consideration and self-pollinate them. Practise selection among the S_0 plants at harvest as well. Harvest S_1 ears and shell them separately. Discard seeds of undesirable ears or those having undesirable seed characteristics.

Second season: Grow (S_1) ear-to-row progenies from selfed seed of selected S_0 plants[1] in an intercross block. Discard undesirable S_1 progenies[2], if any, before pollen shedding. Make all possible intercrosses among S_1 progenies. Alternatively, grow S_1 progenies in an isolated block for recombination. Bulk equal amount of seeds from all intercrosses or bulk all the harvested seed to form C_1 cycle population.

---

[1] S_0 plants, means plants selected in original population for self-pollination.

[2] S_1 progeny, means the selfed seed of S_0 plants.

First Recurrent Cycle (C_1)
Third season: Grow C_1 cycle population (formed by bulking intercross seed) and repeat the process as in the first season.

Fourth season: Repeat the process as in the second season to form C_2 cycle population.

Second Recurrent Cycle (C_2)
Repeat the process as in the first cycle (C_1) using C_2 cycle population (formed by bulking seeds of intercrosses or bulk harvest) at completion of the first recurrent selection cycle (C_1).

Procedure of Recurrent Selection for General or Specific Combing Ability[3] in Corn

Original Selection Cycle (C_0)

First season: Grow source population (heterozygous population). Select number of plants (S_0 plants) on the basis of phenotypic score for characters under consideration and self-pollinate them, and also cross them to a tester parent with a broad genetic base (for RS_{gca}) or narrow genetic base (for RS_{sca}) to identify such S_0 plants as have good gca or sca respectively. Harvest and shell the seed of each S_0 plant (S_1 progeny) and test cross separately.

Second season: Evaluate the test cross progenies (S_0 × tester parent) made in the first year in a replicated yield trial. Identify superior progenies.

Third season: Grow selfed seed (from S_0 plants) of superior progenies (S_1 progenies) in an intercross block. Make all possible intercrosses among S_1 progenies. Alternatively grow S_1 progenies in an isolated block for recombination. Bulk equal amount of seeds from all intercrosses or bulk all the harvested seed to form C_1 cycle population.

First Recurrent Selection Cycle (C_1)
Fourth season: Grow C_1 population and self-pollinate and cross selected plants as in the first season.

Fifth season: Repeat the process as in the second season.

Sixth season: Repeat the process as in the third season to form C_2 cycle population.

Second Recurrent Selection Cycle (C_2)
Repeat the process as in the first cycle (C_1) using C_2 cycle population formed at completion of the first recurrent selection cycle.

Procedure of Reciprocal Recurrent Selection (Fig. 16.1) in Corn

Original Selection Cycle (C_0)
First season: Grow two source populations, say population A and population B (heterozygous populations) separately. Select number of plants

---

[3] The only difference is the tester parent used.

(S_0 plants) on the basis of phenotypic score for characters under consideration in each population (population A and population B) and also cross them with randomly selected plants from the other population, that is, S_0 plants of population A shall be crossed with randomly selected plants from population B and vice versa. Harvest and shell selfed seed (S_1 progenies) and test cross progenies of each population individually and keep the seeds separately.

Second season: Evaluate the test cross progenies of each population in a separate replicated yield trial. Identify superior progenies of each population.

Third season: Grow selfed seed (from S_0 plant) of superior (S_1) progenies of each population separately in two intercross blocks. Make all possible intercrosses among S_1 progenies in each intercross block or grow in two different isolated blocks for recombination. Bulk the seed of each block separately (at harvest) to form two C_1 cycle populations.

First Recurrent Selection Cycle (C_1)
Fourth season: Grow two populations (C_1 population of A and B) as in the first season and repeat the process.

Fifth season: Repeat the process as in the second season.

Sixth season: Repeat the process as in the third season.

Second Recurrent Selection Cycle (C_2)
Repeat the process as in the first cycle (C_1) using C_2 cycle population formed at completion of the first recurrent selection cycle.

MODIFICATION OF RECURRENT SELECTION PROCEDURES IN CORN

1. S_1 Progeny Recurrent Selection Scheme

In this method the selection units are S_1 progeny means compared with the grand mean of all S_1 progenies. Selection studies comparing S_1 progenies evaluation and test cross progenies evaluation are, however, not in agreement. Lonnquist and Lindsey (1964), Horner et al. (1973) concluded that S_1 and S_2 progeny recurrent selection as compared to test cross progeny evaluation were less effective than expected, while Genter and Alexander (1962), Burton et al. (1971) and Genter (1973) concluded that improvement through S_1 progeny recurrent selection was greater than the test cross method.

In nearly all instances inbreeding depression was less for S_1 lines extracted from populations improved by S_1 progeny selection than the population improved by test cross selection. Theoretically, S_1 progeny recurrent selection is more effective in changing frequencies of genes having additive effects. This method has been used for improvement of traits other than yield.

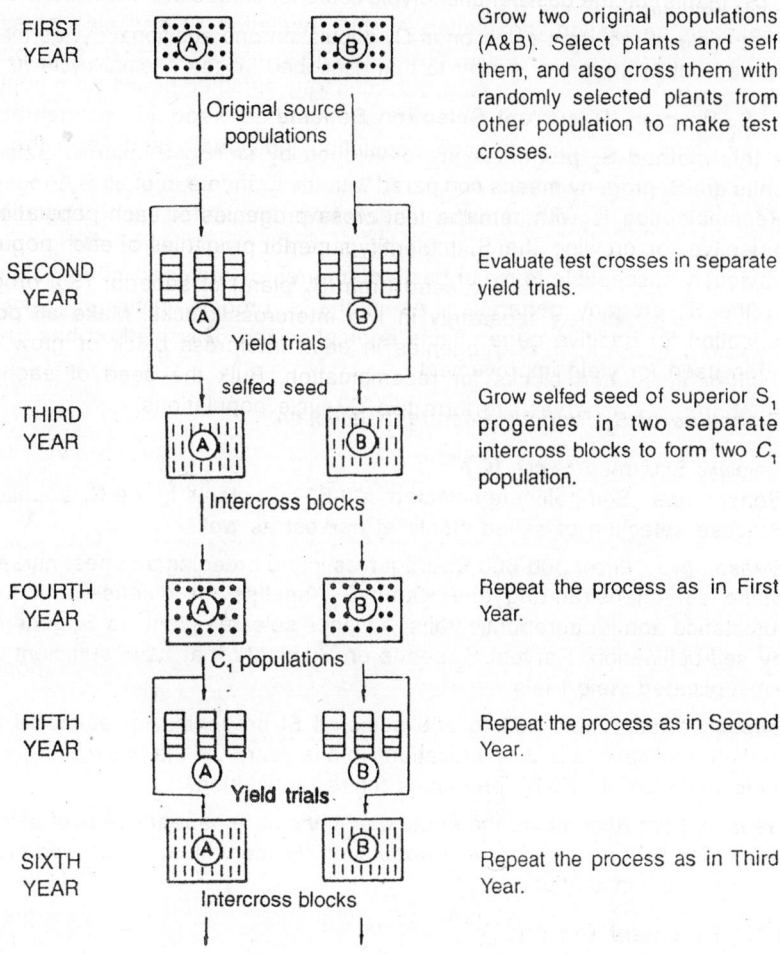

| FIRST YEAR | | Grow two original populations (A&B). Select plants and self them, and also cross them with randomly selected plants from other population to make test crosses. |

Original source populations

SECOND YEAR — Evaluate test crosses in separate yield trials.

Yield trials

selfed seed

THIRD YEAR — Grow selfed seed of superior S_1 progenies in two separate intercross blocks to form two C_1 population.

Intercross blocks

FOURTH YEAR — Repeat the process as in First Year.

C_1 populations

FIFTH YEAR — Repeat the process as in Second Year.

Yield trials

SIXTH YEAR — Repeat the process as in Third Year.

Intercross blocks

May start new cycle of recurrent selection

Fig. 16.1. Reciprocal recurrent selection.

Procedure of S_1 Progeny Recurrent Selection in Corn

ORIGINAL SELECTION CYCLE (C_0)

Season one: In C_0 population, self-pollinate 300-600 S_0 plants. Practise selection among the S_0 plants at pollination and harvest. Harvest S_1 ears that have sufficient seed for replicated trials.

Season two: Evaluate S_1 progenies in replicated trials (3 or 4 locations with 2 replications) in the same year. On the basis of replicated trials select 20 to 30 S_1 progenies for recombination.

Season three: Recombine the selected S_1 progenies to form C_1 cycle population.

FIRST RECURRENT CYCLE (C_1)

First cycle of recurrent selection in C_1 population and additional cycles may be carried out in a manner similar to that described for the original cycle (C_0).

2. S_2 Progeny Recurrent Selection Scheme

It this method S_2 progenies are developed by selfing S_1 plants. Selection units are S_2 progeny means compared with the grandmean of all S_2 progenies. Recombination is with remanent seed from the selfed ears. The primary objective for growing the S_1 progeny generation is to eliminate progenies obviously susceptible to major pests of an area before conducting yield trials in the S_2 progeny generation. As in the S_1 progeny recurrent selection, selection for additive gene effects is emphasized. This method has usually been used for yield improvement.

Procedure of S_2 Progeny Recurrent Selection

ORIGINAL SELECTION CYCLE (C_0)

Season one: Self-pollinate selected 300-600 S_0 plants in the C_0 population. Practise selection of selfed plants at harvest as well.

Season two: Grow 300-600 S_1 progenies in the breeding and pest nurseries. Make selections among and within S_1 families for disease and insect resistance and for agronomic traits. Advance selected plants to S_2 generation by self-pollination. Harvest S_2 seeds on S_1 plants that have sufficient seed for replicated yield trials.

Season three: Grow 100 to 200 selected S_2 progenies in replicated yield trials (2 replications at 3 or 4 locations in one year). On the basis of replicated yield trials, select 20-30 progenies for recombination.

Season four: Recombine the equal quantities of remanent seeds of selected progenies to form the C_1 cycle population. Remanent S_1 or S_2 seed can be used for recombination.

FIRST RECURRENT CYCLE (C_1)

First cycle of recurrent selection in C_1 population and additional cycles may be carried out in a manner similar to that described for the original cycle (C_0).

3. Half-sib Family Recurrent Selection Using S_1 Test Crosses (S_1 × Tester Parent)

Half-sib recurrent selection scheme involves making S_1 plant test crosses and their evaluation for identification of superior progenies. In applied breeding programmes S_1 plant test crosses are preferred, because this affords screening of S_1 progenies and use of selected S_1 plants to produce the test crosses.

Half-sib Recurrent Selection Using S_1 Plant Test Crosses

ORIGINAL SELECTION CYCLE (C_0)

Season one: In the C_0 population self-pollinate 300-600 S_0 plants. Practise selection among S_0 plants at pollination and harvest.

Season two: Grow 300-600 S_1 progenies in the breeding and pest nurseries. Make selections among S_1 progenies whenever possible before pollination. Produce test crosses (half-sibs) of selected S_1 plants in selected S_1 progenies. Self-pollinate the S_1 plants used to produce test crosses.

Season three: Evaluate 100-200 test crosses in replicated yield trials (2 replications, 3-4 locations). On the basis of combined replicated yield trials, select 20-30 progenies for recombination.

Season four: Recombine selfed progeny of remanent half-sib seed to form C_1 cycle population. Remanent S_1 or S_2 seed can be used for recombination.

FIRST RECURRENT CYCLE (C_1)

First cycle of recurrent selection in C_1 population and additional cycles may be carried out in a manner similar to that described for the original cycle (C_0).

4. Full-sib Family Recurrent Selection (Biparental Crosses)

Full-sib families in a population are established by crossing one plant with another. Reciprocal crosses can be made between the two plants included in a cross. Full-sib families are evaluated in a replicated trial for identification of superior full-sib families. Remanent full-sib seed of superior families is recombined to initiate the next cycle of selection.

Method of Full-sib Family Recurrent Selection

ORIGINAL SELECTION CYCLE (C_0)

Season one: Produce full-sib progenies in the population (C_0) under selection. Make 150 to 200 full-sib crosses by reciprocal pollination of two selected plants. Practise selection among S_0 plants at pollination and harvest. Harvest full-sib crosses that have sufficient seed for replicated yield trials.

Season two: Evaluate full-sib progenies : Conduct replicated trials of 100 to 200 full-sib progenies with 2 or 3 replications at 3 or 4 locations.

Season three: Recombine the selected 20-30 full-sib progenies to form the C_1 population.

FIRST RECURRENT CYCLE (C_1)

First cycle of recurrent selection in C_1 population and additional cycles may be carried out in a manner similar to that described for the original cycle (C_0).

5. Reciprocal Recurrent Selection (RRS) Based on Test Crosses of Half-sib Family

In the original RRS individuals are related as half-sibs among progenies and as full-sibs within female progenies. In this method the main difference with original RRS is in the type of parentage among individuals and in test crosses. A cycle of the modified method starts with 200 or more open-pollinated ears (half-sib families) from population A that are planted ear-to-row as females in an isolation block. Male rows are planted with seeds of population B with a ratio of 3:1, 2:1 or 4:2. In a separate isolation block a similar number (200 or more) of open pollinated ears (half-sib families) of population B are planted as females and population A is used as male. The second phase is evaluation of test crosses, half-sib family × population B and half-sib family × population A in replicated field trials. Selection is on the basis of yield means and agronomic traits. Recombination is performed in population A and B for the selected half-sib families, using remanent seeds. Simplicity of the scheme and good seed supplies are major advantages. In genetic terms this method uses a smaller portion of the genetic variability but permits higher selection pressure because the effective population size is about four times the number of families tested.

Method of RRS Based on Test Cross of Half-sib Families

ORIGINAL SELECTION CYCLE (C_0)

Season one: Ears from isolated open pollinated fields (A and B) are planted ear-to-row as females in detasselling blocks whereas the opposite population is used as males to produce test crosses.

Season two: 100 to 200 test crosses (seeds from each female row constitute one entry) half-sib (A × B) and the same number of test crosses half-sib (B × A) are evaluated in replicated yield trials (2 or 3 replications and 3 or 4 locations). Identify the best test crosses half-sib (A × B) and half-sib (B × A).

Season three: Remanent seeds of half-sib families (A) corresponding to the best test crosses half-sib A × B are planted for recombination. The diallel system of recombination can be used but a simpler procedure is the use of an isolated block in which the selected progenies are used as females (detasseled) and the bulk as males. The same procedure is used for the half-sib (B × A). At harvest, select 100-200 open-pollinated ears, getting about the same number from each female row.

FIRST RECURRENT CYCLE (C_1)

First cycle of recurrent selection in C_0 populations and additional cycles may be carried out in a manner similar to that described for the original cycle (C_0).

6. RRS Based on Half-sib Progenies of Prolific Plants

This is another modification of the original RRS. The main difference between this modification and RRS is that individuals within the selection units (test crosses) are related only as half-sibs and the recombination unit is a half-sib family instead of an S_1 family.

Method of RRS Based on Half-sib Progenies of Prolific Plants

ORIGINAL SELECTION CYCLE (C_0)

Season one: Population A is planted as female in an isolated detasselling block (Field 1), where population B is planted as male rows with a ratio 1:2,

1:3 or 2:4. Population B is planted as females in an isolated block (Field 2) where population A is used as males.

The first (upper) ears in each field are open pollinated. Protect the second (lower) ear shoots of prolific plants in female rows in both fields. Pollen is bulked, after passing through a sieve to eliminate anthers, and a pollen gun is used to pollinate the second ears in the opposite field. Alternate pollination can be used, i.e., from field 1 to field 2 in one day and vice versa the next day. Harvest both ears together with an appropriate identification. The upper ear is an interpopulation half-sib family, whereas the lower ear is an intrapopulation half-sib family.

Season two: 100 to 200 test crosses from each field, i.e. half-sibs (A × B) and half-sibs (B × A), are evaluated in replicated trials (2 or 3 replications, 3 or 4 locations). Identify the best test crosses A × B and B × A.

Season three: Remanent seeds of lower (hand-pollinated) ears of population A harvested in season 1, field 1 corresponding to the best test crosses A × B, are planted ear-to-row as females in a detasselling block, as in season 1. Male rows are a bulk of remanent half-sib seeds of population B that correspond to the selected test crosses B × A. Remanent half-sibs B related to the best test crosses B × A are also planted ear-to-row in another isolated field where selected half-sibs A are bulked and planted as males.

FIRST RECURRENT CYCLE (C_1)
Repeat procedure for C_1 cycle and additional cycles described for original cycle.

7. Reciprocal Full-sib Selection

The operational procedures of reciprocal full-sib selection are similar to those for RRS. The main difference is that full-sib progenies rather than half-sib families are evaluated.

Method of Reciprocal Full-sib Selection With use of Primarily One-ear Plants

ORIGINAL SELECTION CYCLE (C_0)
Season one: Self-pollinate S_0 plants in the two populations (C_0) under selection. Practise selection among S_0 plants at pollination and harvest. Note date of pollinations of S_0 plants.

Season two: Plant the S_1 progenies of two C_0 populations in alternate rows in the breeding and pest nurseries (in pairs according to the date of pollination of S_0 plants). Produce full-sib progenies on a portion of the S_1 plants.

Season three: Evaluate full-sib progenies in replicated yield trials (2-3 replications, 3 to 4 locations). S_2 progenies can also be grown in breeding nurseries for additional self-pollination and cross-pollinations.

Season four: Recombine remanent seed of 20 to 30 S_1 or S_2 progenies to produce C_1 cycle population.

First Recurrent Cycle (C_1)
Repeat procedure described for original cycle.

8. Method of Reciprocal Full-sib Selection with Prolific Plants

Original Selection Cycle (C_0)

'eason one : Produce S_1 progenies and full-sib crosses. Plant two populations (C_0) under selection in alternate rows in the breeding nursery. Self-pollinate one ear and cross the other ear. Practice selection among S_0 plants at pollination and harvest.

season two : Grow the full-sib crosses in replicated yield trials (2 or 3 replications, 3 to 4 locations). Pairs of S_1 progenies may also be grown in the breeding nursery for additional selection. Self-pollinate to produce S_2 progenies and cross-pollinate plants selfed to produce full-sib crosses between S_1 plants.

Season three : Recombine the selected S_1 progenies of the selected full-sib crosses (e.g. 20 to 30) to from C_1 syn_1 for each of the two populations.

First Recurrent Cycle (C_1)
Repeat procedure described for original cycle.

9. RRS With Use of Inbred Lines

Russel and Eberhart (1975) and Walejko and Russel (1977) suggested that RRS be conducted with use of inbred lines as testers. Choice of inbred lines to use as testers should be the same as choice of two populations to include for RRS, that is, tester lines should conform to the heterotic pattern expressed in hybrids. Instead of crossing plants from population A with plants from population B, cross plants from population A with an inbred tester extracted from population B. The reverse situation would be used for population B.

RRS with use of an inbred line was designed for improvement of lines for one or both sides of the hybrid pedigree. It is a useful method for supplementing the pedigree selection programme for improvement of inbred lines.

17

Germplasm Composites and Synthetic Varieties

GERMPLASM COMPOSITES

The term *germplasm composite* is commonly used to designate a broad group of materials mixed together in many different ways, and includes breeding materials put together on the basis of desirable characters, such as yield potential, maturity, disease resistance, etc., followed by random mating. A germplasm composite of corn involving interracial crosses is usually made by combining two or more genetically diverse but elite races of corn followed by random mating. Only those races that show significant heterotic response and high yields in F_1 combinations with little decline in F_2 performance are used to build the germplasm composite.

A germplasm composite is thus a genetically complex and extremely broad-base population that constitutes a reservoir of superior genes (Dhawan, 1965). Enough experimental evidence has established the preponderance of additive gene action in respect to yield and other desirable traits in germplasm composites. That is why we find minimal yield decline in F_2 and continued improvement in subsequent generations. Diverse but elite genetic material in germplasm composites makes them reservoirs of diverse useful genes and extremely broad-base populations. Yield and other characters are raised by repeated cycles of selection. A number of recurrent selection procedures may be adopted for further genetic improvement.

Application and Uses of Germplasm Composites

1. The outstanding germplasm composites can be recommended for commercial cultivation over a wide range of environments.
2. They may serve as a source population for extraction of superior inbred lines.

SYNTHETIC VARIETIES

The term *synthetic variety* is used to designate a variety that has been synthesized by hybridizing in all combinations a group of genotypes, be it inbred lines, clones, mass-selected populations or any other materials that are selected on the basis of general combining ability. Only such genotypes are included which combine well with each other in all combinations. Synthetic varieties are maintained by open pollination in isolated fields. The difference between open-pollinated and synthetic varieties are given in Table 17.1.

Table 17.1. Difference between Synthetics and Open-pollinated Varieties

Synthetics	Open-pollinated varieties
Selection of genotypes is based on combining ability.	(a) In mass selected varieties genotypes are bulked without previous testing of progeny performance or in hybrid combinations.
	b) In line-bred varieties progenies from superior lines are composited on the basis of lines tested individually.

Application and Uses of Synthetic Varieties

1. Synthetic varieties offer a good opportunity for controlled utilization of an appreciable amount of heterosis. They are thus the best available alternative in all such crops wherein large-scale commercial hybrid seed production is not feasible for various reasons.
2. Owing to greater genetic variability they are more flexible in adaptation over a wide range of changing environmental conditions.
3. Farmers can save their own seed.
4. Like germplasm composites, they may also be used for extraction of superior inbred lines for hybrids.

Genetic Basis of Synthetic Varieties

Selection of constituent inbred lines (or other materials) for synthetics is based on general combining ability. In terms of gene action, simple additive gene action predominates. The method of combining inbred lines into synthetics on the basis of a general combining ability test provides the means for increasing the frequency of desired yield genes with continued improvement in yield.

In terms of number of selfed generations, selfing is limited to one generation only. The reason is that any high combining S_0 plant contains many favourable genes and gene combinations. If further inbreeding is done these favourable genes or gene combinations may be lost, as specific individuals within segregating populations are selected to propagate the line. And hence the purpose of selfing is limited to control the parentage, while the selected S_0 plants are subjected to tests for general combining ability. Selfed seed from the selected S_0 plants is combined to constitute the synthetic population.

In terms of adaptation over a wide range of environments the genetic diversity of material(s) used in constituting a synthetic is important. The material(s) from a single source or closely related sources cannot be expected to show a wider adaptability than the source material itself.

Prediction of Performance of a Synthetic Variety

The expected F_2 performance of a synthetic variety is calculated by Sewall Wright's (1922) formula.

$$\hat{F}_2 = \overline{F}_1 - \frac{\left(\overline{F}_1 - \overline{P}\right)}{n}$$

where $\hat{F}_2$ is the estimated performance of the F_2 generation, $\overline{F}_1$ is the mean performance of all single crosses among n lines. $\overline{P}$ is the mean performance of parental lines and n is the number of lines involved.

Factors Affecting the Performance of Synthetic Varieties

1. *Number of parental lines* : Performance of synthetic varieties is affected by the number of parental lines involved in constituting the synthetic. Since synthetic varieties are maintained by open pollination it is important that F_2 yields be high. Theoretically, genetic equilibrium is reached in the F_2 generation. Therefore the F_3 generation should produce the same yield as the F_2 generation. This has been experimentally proven. As a matter of fact yields obtained in F_3 and F_4 generations may be slightly higher, provided the number of lines involved is not very small. Therefore number of lines selected should be kept at optimum, without undue sacrifice in high combining ability. When the number of lines is relatively small the F_1 yields are high but the decline in F_2 yields is more; when the number of lines is relatively high both the F_1 yields and the F_2 decline are low. To balance the two, that is, F_1 yields and F_2 decline, optimum number of lines are used. The optimum number of lines for constituting synthetic varieties usually varies from 5-6 best combining lines.

2. *Mean yield of parental lines* : Synthetic varieties made from a combination of relatively high-yielding parents (short-term inbreds) are expected to yield more compared to synthetic varieties from long-term inbreds which are low yielding and less vigorous. It is for this reason that precluding inbreeding altogether would be desirable. Use of non-inbred lines or lines developed by limited inbreeding (S_0 or S_1 lines) is therefore emphasized.

3. *Mean yield of all F_1s among the parental lines* : The mean yield of all F_1s should be sufficiently high so that even after some decline in the F_2 generation of the constituted synthetic, the yield level remains comparable to or better over the check variety or hybrid and significantly better over standard open-pollinated varieties and synthetic varieties.

Methods of Developing Synthetic Varieties

A. Corn

1. From Long-term Inbreds
Season one : Make all possible single crosses among selected inbred lines representing rather wide genetic diversity.

Season two : Conduct replicated yield trials over environments (3-4 or more locations). Select best inbred lines on the basis of performance in all single-cross combinations with each other. Mix an equal quantity of seeds of each of the single-cross combinations among selected inbred lines to constitute the synthetic (syn_0).

Season three : Grow population in an isolated plot (0.2 ha). Select desirable plants. Save about 100 ears without making close selection for ear or plant type. Bulk the seed to constitute Syn_1 population for next year's increase and yield trials.

Seasons four to six (Syn_1 to Syn_3) generation : (1) Conduct replicated yield trials including check hybrids and varieties. Determine the superiority of newly constituted experimental synthetic over checks and proceed for release etc., if found superior over checks. (2) Grow next generations of constituted (Syn_1) synthetic in isolation.

Repeat the process for another two years.

2. Development of Synthetics from Short-term Inbreds (S_0 or S_1)

	S_0	S_1
First season	Self approx. 200 S_0 plants in each selected open-pollinated variety(ies) and outcross them with the parent variety. At harvest discard any undesirable plant.	Self approx. 250-500 S_0 plants in selected variety(ies). At harvest discard any undesirable plant.
Second season	(1) Grow test cross progenies in replicated yield trials over environments. (2) Select topcrosses on the basis of their performance. (3) Composite 50 seeds from each of the S_0 plants of the selected topcross to constitute Syn_0 population.	Grow S_1 rows (25-30 plants) in a topcross block. Use parent variety as a tester parent.
Third season	Grow Syn_0 in an isolated block and further handle the material in the same way as described earlier in development of synthetic varieties from long-term inbreds.	(1) Grow topcross progenies in replicated yield trials over environments. (2) Select topcrosses on the basis of their performance. (3) Composite 50 seeds each of the component S_1 lines to constitute Syn_0.

Fourth	Grow Syn_0 in an
season	isolated block and
onward	handle the material
	in the same way as
	described earlier in
	the development of
	synthetic varieties
	from long-term inbreds.

Further, improvement in the performance of synthetic varieties may be effected through any of the recurrent selection schemes.

B. Forage Crops

The principles applicable to corn also apply to forage species. The important differences vis-a-vis corn pertain to ability to produce inbred and crossed seed, and methods of evaluating test cross progenies. These aspects are discussed below.

TESTS FOR COMBINING ABILITY IN FORAGE CROPS

The controlled pollination required for inbreeding and making crosses in most forage species is rather a difficult task and hence in these species it is necessary to rely on natural crossing to obtain the progenies required for tests of combining ability. The following four types of tests of combining ability can be made on progenies obtained through natural crossing in forage crops (Allard, 1960).

1. *Polycross test*: The polycross method is a good and widely accepted method for measuring general combining ability. In this method selected clones are grown in an isolated block so as to prevent contamination from other sources. The clones are replicated many times (10 times or more) in the block so as to promote random mating *inter se*. The seeds are formed as a result of natural crossing. The progeny of each line thus obtained is called a polycross. Polycross data is considered very valuable in identifying plants that can make a worthwhile contribution to the yielding ability of synthetic varieties. If a large number of clones is included in a polycross test, the average genetic composition of the male parentage is not expected to be much different from the male parentage in open-pollinated or top-cross tests. If, however, only a few highly selected clones are included in a polycross block, the performance of test cross progenies might be very different from open-pollinated or topcross progenies. The polycross test combines economy of effort with estimates of combining ability that are more relevant to performance in synthetic varieties.

2. *Topcross test*: In this method selected clones are planted alternately with a single tester variety. The test cross seed thus consists partly of topcrosses to the variety and partly of intercrosses with other selected clones. The proportion of topcross seed can be increased by increasing the number of plants of the tester variety sown in the topcross block relative to the number of plants of the selected clones. This test is a measure of general combining ability.

3. *Single-cross test*: In a single-cross test every selected clone is crossed with a number of other selected clones, usually by growing together in isolation each pair of clones to be crossed or by enclosing an inflorescence from one clone in a bag with an inflorescence of the other clone. Single crosses measure the combining ability of particular pairs of clones. The average combining ability of any single clone can be calculated as the mean performance of that clone in its crosses with other clones.

Due to practical limitations the use of single crosses is restricted to the final stages of a breeding programme when it might be desirable to determine the combinability *inter se* of a few strains of high prepotency.

4. *Open-pollinated progeny test*: In this case progenies are derived from seed produced on selected plants. The seeds are formed as a result of natural crossing with the other plants in the nursery. This is a test for measuring general combining ability.

Development of Synthetic Varieties of Forage Crops from Clones

The principal objective is to establish a variety with a sufficiently wide range of genotypes to maintain vigour, and yet approach homozygosity in respect of the traits under selection. For example, all the plants entering into a synthetic variety of some pasture grass might vary markedly in genotypes for plant growth, and yet be reasonably pure for resistance to a specific disease(s).

Synthetics are constituted by allowing random inter-pollination among clones selected on the basis of their combining ability. Equal quantities of seeds obtained from each clone are mixed and grown in open-pollination in an isolated plot. Open-pollinated seed is used in subsequent multiplication. Synthetics are constituted regularly from the original clones.

PROCEDURE FOR DEVELOPING A SYNTHETIC VARIETY FROM CLONES

First year: Assemble desirable plants in sufficient number from various sources to form a source nursery. The range of variability should be sufficiently large. Select 200-400 superior plants on a phenotypic basis for desirable characters, such as growth habit, vigour, head emergence, disease resistance, winter hardiness, etc. as per objectives of the breeding.

Selected plants may also be inbred[1] for one or more generations to fix desirable characteristics, if deemed desirable.

Second year: Establish clonal lines of selected plants through asexual propagation. Screen the clonal lines for desirable characteristics as mentioned above in the previous year. Clonal lines may also be planted in disease

---

[1] The controlled pollination required for inbreeding is rather a difficult task. When practised, the degree of decline in S_1 progeny performance reflects the amount of additive variance. Obviously, the progenies showing less decline are selected as parents.

nurseries and/or subjected to adversities such as severe clippings, cold tests, etc. to identify clones with superior qualities.

When deemed desirable, and already not done so, inbreeding followed by selection within clones may be practised now to fix desirable characters in the homozygous form. Select 25-50 superior clones for further evaluation and testing.

Third year: Evaluate the combining ability of the clones selected in the previous year. The polycross test is the most widely used method for determining general combining ability in clonal lines. Bulk the seeds obtained from each plant belonging to a clone for use next year.

Fourth year : Bulk seed obtained from each clone grown in the polycross nursery in the previous year is planted for progeny test for evaluation of yield and other desirable characteristics.

Clones with superior combining ability and possessing desirable characteristics as determined from the progeny test conducted in the previous year, are chosen to produce the synthetic. The number of clones selected to constitute a synthetic may vary from 4-10. Equal quantities of seeds of selected clones are mixed to constitute the Syn_0 population.

Fifth year: Grow the Syn_0 population in an isolated plot to grow the Syn_1 generation for next year's increase and yield trials. A recurrent selection cycle may also be started.

Sixth to eighth year: Conduct replicated yield trials at several locations including check varieties to establish the superiority of newly constituted synthetic. Record data in respect of yield and other desirable characters as mentioned earlier. Process for release and make further increases of seed, if found consistently superior over check varieties.

18

ASEXUALLY PROPAGATED CROPS

GROUPS OF ASEXUALLY PROPAGATED CROPS

Asexually propagated crops can be classified into the following four groups:

1. *Normal seedset but propagated vegetatively*: The crop species included in this group flower, set fruits and produce seeds like sexually propagated crops. However, they are propagated vegetatively.

2. *Poor seedset and propagated vegetatively*: The crop species included in this group flower and set seeds under certain environmental conditions.

3. *Apomicts*: The crop species included in this group produce seeds asexually (agamospergony). The resultant plants are like the female parent.

4. *Flowerless or sterile*: The crop species included in this group do not flower and/or set seeds. Propagation is only through vegetative parts of the plant.

METHODS OF ASEXUAL (VEGETATIVE) PROPAGATION

The vegetative propagation methods of some important crops are given below:

1. *Stem cutting*: Sugar-cane, sweet potato, pepper, etc.
2. *Suckers*: Banana, pineapple, chrysanthemum
3. *Stolons*: Mentha
4. *Tuber, rhizomes, corms and bulbs*: Potato, ginger, canna, yam and garlic
5. *Division of crown*: Fodder grasses
6. *Grafts and buds*: Fruit trees and flower plants
7. *Layering*: Rose

BREEDING METHODS

The basic principles applicable to sexually propagated crops are equally well applicable to asexually propagated crops. There are, however, some important differences which necessitate differential treatment, namely:

1) the extent of genetic variability, especially in crop species which are

totally sterile, is rather limited to that already existing or that which may arise through natural or induced mutations, and

2) once a desirable plant has been obtained, that is selected, regardless of heterozygosity, chromosome imbalances, sterility problems, etc., the same can be further propagated and multiplied through vegetative propagation for commercial use without genetic change. This is an important advantage to breeders of asexually propagated crops, in sharp contrast to what a plant breeder handling a sexually propagated crop is required to do.

The following breeding methods are employed for varietal improvement of asexually propagated crops :
a) clonal selection,
b) hybridization followed by clonal selection,
c) mutation breeding.

The methods of breeding apomictic grass species are discussed in Chapter 19 and mutation breeding in Chapter 22.

Clonal Selection

Clone

A clone is a genetically uniform material derived from a single individual, propagated exclusively by vegetative means. The differences between a pure line, inbred line and a clone are given in Table 18.1.

Table 18.1. Differences between pure line, inbred line and a clone

	Pure line	Inbred line	Clone
1. Propagation	Sexual (self-pollinated)	Sexual (cross-pollinated)	Asexual
2. Phenotype	Homogeneous	Homogeneous	Homogeneous
3. Genotypic constitution	Homozygous	Homozygous	Highly heterozygous
4. Pedigree	Progeny of a single self-fertilized plant	Selfed progeny of a selected open-pollinated plant	Progeny of a single selected plant
5. Utilization	As an improved variety	Used in hybrid varieties and/or synthetics	As an improved variety

Sources of Clonal Variation

The heritable variation in clones may occur through mutations in any of the following ways.

a) **Spontaneous mutations :** Clones are highly heterozygous and often complex hybrids. This seems to increase the chances of change within a somatic cell due to mutations that can occur as follows :

   i) Changes resulting from chemical alterations of the chromosome material (DNA) at specific locations on the chromosome (point mutations).

   ii) Gross structural changes in the chromosomes (deletions, duplications, inversions and translocations), or as a result of aneuploidy or polyploidy; or the plastids that are independently involved in determining plant characteristics. Variegated plants have a tendency to produce both normal and defective plastids.

Any point mutation or a chromosomal change *per se* is a relatively rare event. However, since vegetative growth of clones involves billions of cell divisions the chances are not all that rare. If such changes do occur within a vegetative cell (somatic cell), and are followed by mitotic divisions, this may lead to permanent changes in the clone provided the subsequent daughter cells occupy a substantial portion of the growing point. This gives rise to chimeras[1] and bud sports, which are readily discernible. Buds or scions taken from this altered branch would then constitute a new clone which can be asexually propagated.

b) **Induced mutations :** Treating the growing points with a mutagenic agent, such as colchicine, bombarding seeds or scions meant for grafting with X-rays or radioactive emissions or other radiomimetic chemicals may induce mutations and chromosomal changes. Polyploidy may give rise to 'giant bud sports'. Success in reproducing chimeras depends, however on their structure and stability. Many of them may revert to plants showing one or more types of tissue from which they are formed.

Types of Chimeras

Sectorial, when the growing shoot is composed of two genetically different tissues situated side by side, occupying distinct sectors of the stem.

Periclinal, when the tissues of one genetic composition occur as a relatively thin skin (circumference composed of one or several outer layers) over a genetically different core.

Mericlinal, when the outer layer of genetically different tissue does not extend completely around the core, occupying only a segment of the circumference.

Graft chimera, one established artificially by grafting.

Bud sports, that is, a branch which shows changes in one or more inheritable characters.

Virus-Like Genetic Disorders

Many abnormalities observed in clones resemble known viral diseases but are considered to be genetic disorders, although their basic nature is not

---

[1] A chimera is a plant, or part of a plant composed of two or more genetically distinct tissues growing adjacent to each other as parts of the composite plant.

understood. Such disorders are non-infectious and may develop either progressively or sporadically from a particular clone. Such disorders are controlled by continuously observing the budded progeny and eliminating undesirable clones through selection.

Viral Effects in Clones

Clones may become infected with one or more viruses, probably over a period of time. Viral infection adversely influences the growth, appearance and production not only of the infected plant, but also of plants propagated from it permanently. Usually, plant viruses produce distinct symptoms at low temperatures which are not detectable at higher temperatures. Multiple infections of several virus components may be involved in many viral diseases. For example, latent viruses, when present individually in a clone, may cause slight symptoms or none at all, but will produce a severe reaction when present in combinations of two or three. This necessitates identification of the presence of a virus in the clones and exclusion from those meant for propagation, because viral infections can spread from diseased to healthy plants through insect vectors, notably aphids and leafhoppers. Even a clone carrying a single latent virus may subsequently be infected in the field by other viruses and if that happens viral damage may be severe.

A viral infection in a clone is detected by
 i) visually recognizable symptoms, and
 ii) indexing, which makes use of several serological and histological techniques that are now available for determining the presence of latent viruses.

Goals of Clonal Selection

There are two important goals of clonal selection:
1) maintenance of disease-free and true-to-type clones, and
2) development of improved varieties (clones).

Methods of Clonal Selection

1. Maintenance of Pathogen-Free and True-to-Type Clones

Clones do deteriorate or degenerate with continued production because of infection by pathogens, namely fungi, bacteria, nematodes and viruses, as well as from genetic modifications. The concept of pathogen-free clonal stock has gained currency in recent years; it involves elimination of pathogens[2] through the following discrete steps.

A. Initial selection

Initial selection of plant or plants (source material) is made on the basis of uniformity and conformity to variety characteristics (true-to-type) from good propagating material (clones). Individual plants are carefully examined for

---

[2] A pathogen-free stock differs from a disease-free stock in that the pathogen is eliminated while in a disease free stock the disease, though controlled, may still be present in latent form.

genetic disorders, bud sports and symptoms of virus or other diseases. The selected plants are correctly identified (labelled). The clones from selected plants are indexed for viruses and other pathogens, using preselected serological, histological and visual inspection methods. When the test is negative, that means there is no evidence of infection, the clone is selected as a suitable initial propagation source. Indexing, however, does not prove that the clone is pathogen or virus free, since the tests used might not be sufficiently sensitive to demonstrate the presence of all pathogens. When the test is positive, i.e., clean clone is not found, it becomes imperative to eliminate the virus or pathogen. This can be done in one of the following ways:

i) *Selection of uninfected parts*: Pathogens, such as *Verticilium* and *Phytophthora* can be avoided by taking cuttings from growing plants only from tips not touching ground.

ii) *Meristem tip culture*: The terminal growing points are sometimes free of virus, even if the rest of the plant is infected. Excision and aseptic culture (tissue culture) of this small segment may produce a virus-free source. Examples : carnation, chrysanthemums, strawberry and orchids. Tissue-culture techniques are discussed in Chapter 28.

iii) *Short-duration heat treatment*: This procedure is widely used to free the material from fungi, bacteria and nematodes. The material is subjected to temperatures ranging from 43.5°C to 57°C from 1/2 hour to 4 hours (varies according to crop) either by soaking in hot water or exposure to hot air or to aerated steam. Hardening-off vegetative material or reducing the moisture content prior to heat treatment is usually desirable.

iv) *Long-duration slow-intensity heat treatment*: This procedure is used to free many kinds of plants from viral diseases. The procedure consists of growing plants in containers until they are well established and then transferring them to a controlled-temperature (37°C to 38°C) chamber for two to four weeks. Buds may be taken from the treated plants and inserted into virus-free rootstocks, or cuttings may be taken and rooted. Alternatively, a bud from an infected plant may be inserted into a disease-free rootstock prior to heat treatment. A combination of heat treatment and meristem shoot-tip culture may be more successful.

v) *Chemical treatment*: Externally carried pathogens may be eradicated through the use of appropriate chemical agents.

vi) *Growing seedlings*: Barring some exceptions, most viruses are not transmitted through seeds. Growing apomictic seedlings (in species in which these occur) provides the means to preserve the clone and eliminate the virus at the same time. This procedure has been utilized in *Citrus* to obtain nucellar seedlings that become the basis for new virus-free strains of old varieties that have been severely affected with serious viruses.

B. Maintenance of stock

The main features are described below.

  i) *Isolation*: Isolation of selected clones is necessary to prevent recontamination with diseases that are transmitted through clones This can be done either by growing them in a separate isolated block or in containers placed in screen cages or greenhouses. The important consideration is to keep them well protected from insects which act as vectors for viral diseases.

 ii) *Sanitation*: Total sanitation, including that of equipment, soil mixtures and growing environment helps considerably in preventing recontamination of clonal stock during maintenance.

iii) *Inspections and roguing*: Clonal stocks under maintenance propagation need to be inspected at periodic intervals, more so, at critical stages of crop growth (the growth stages of a plant when it may express disease symptoms), in conjunction with indexing. Culturing may have to be resorted to again, if the tests for indexed diseases (viruses and other pathogens) are found positive. Any clone showing disease symptoms is forthwith rogued out and is not reused for propagation of maintenance stock.

Variations in the procedure outlined above are possible, depending on needs. Subsequent propagations of the maintenance stock may involve only visual inspections or simple tests compared to more elaborate serological tests, depending on the need to do so.

C. Development of improved varieties

Clonal selection is an easy and quick method of developing improved varieties of asexually propagated crops since propagation is asexual the genotypic constitution of a clone is not altered. For good results availability of a large number of clones (clonal material) obtained from varied sources is necessary. The generalized procedure for developing improved varieties through clonal selection is outlined below.

PROCEDURE FOR DEVELOPING A VARIETY THROUGH CLONAL SELECTION

First year: Grow available clonal population(s). Throughout the season, more particularly at critical stages of crop growth, when viral and other diseases express their symptoms, a thorough examination is necessary. Use of disease nurseries to identify susceptible clones using special techniques is of great help in the selection of clones. Select plants that are evaluated to be the most desirable from the standpoint of plant type, freedom from genetic disorders and resistance to diseases, and label them.

Harvest each selected plant separately and record the yield and quality of the produce. Reject those found to be unacceptable due either to poor yield or quality. Clones that are diseased, malformed or affected by any disorders are rejected forthwith. The selected clones are retained and properly labelled.

Second year: Grow progeny of each clone selected in the first year and evaluate them in a similar manner. If the material was not grown in the disease nursery in the first year, it is time to do so. Select a few of the best clonal progenies.

Third year: Conduct preliminary replicated yield trials including check varieties. Record the yield, disease incidence and quality characteristics. Retain the few best ones.

Fourth to seventh year: Conduct multilocation trials to establish superiority of selected clones. Data is recorded as done in the third year.

Eighth year: Increase the best clones through vegetative propagation. Process their release as a commercial variety.

Limitations
1. Selection within a clone is not usually effective until a good amount of variability exists in the population to make it worthwhile.
2. The multiplication ratio is low.

Hybridization Followed by Clonal Selection

There are many asexually propagated crop species which may be sexually propagated with or without adoption of special techniques with varying degrees of success. This provides an added opportunity to a plant breeder to transfer the desired characteristics in a variety through hybridization, including other species and genera. The improved varieties can then be further propagated vegetatively.

Goals of Hybridization

There are two major goals in sexual reproduction and hybridization of asexually propagated crops.

1. *To create genetic variability*: The asexually propagated crops are highly heterozygous. Self-fertilization and hybridization between clones lead to considerable genetic variability in the resultant clonal populations in which clonal selection may be practised. Since clonal populations are highly heterozygous segregation occurs within the F_1 generation itself. Each F_1 plant is thus a potential source for a clone. The F_2 generation is usually not raised.

2. *Transfer of specific characters*: Transfer of disease resistance or any other desirable character from related species and genera (interspecific and intergeneric hybridization) in asexually propagated crops has long been practised.

Genetic Basis of Hybridization Followed by Clonal Selection

Inbreeding is usually accompanied by loss of vigour and fertility, but this is not always the case. In most instances the clones selected in inbred populations for specific traits, for example, sugar content in sugar-cane, are used as parents for hybridization. The purpose of inbreeding is limited to concentrating desirable polygenes or increasing homozygosity of simple inherited characters. The polyploid nature of most asexually propagated crops makes the procedure more difficult.

The success likely to be achieved by hybridization depends on the choice of parents. The breeding value of parents is assessed by the performance of hybrid progeny. If a cross is found to have desirable seedling in the hybrid progeny it may be repeated when a particular clone is found to contribute desirable characteristics in many crosses. This would mean it possesses good general combining ability and therefore may be used in a large number of crosses.

Wide-crosses are frequently attempted in such crop species. F_1 plants, however, are less desirable from an agronomic viewpoint due to inheritance of undesirable genes from the wild species. This problem is overcome through repeated backcrosses. More than one parent may be used as the recurrent parent in successive backcrosses to avoid any reduction in vigour. Genetic constitution, regardless of chromosomal imbalances or heterozygosity, is easily maintained through clonal propagation. Viral diseases, as mentioned before, are usually not transmitted through seeds.

Procedure for Developing Varieties by Hybridization Followed by Clonal Selection

For the sake of better understanding, the general procedures applied in sugar-cane breeding and potato breeding are described below.

A. Sugar-cane[3]

First year: Plant selected clones of parents chosen for hybridization. Several types of crosses, namely biparental crosses, area crosses or melting-pot crosses are made in sugar-cane. Any one type or more types of crosses can be made, depending on need. Crossing techniques have already been dealt with in Chapter 7. Make several crosses and harvest the seeds.

Depending on whether the female parent is self-sterile, the hybrid and/or both the selfed and crossed seed will be obtained from biparental crosses. Area crosses will yield only hybrids, since only the male sterile or self-incompatible females are used in this method. Melting-pot crosses are actually polycrosses.

Second year: About 50,000 seedlings are selected on the basis of vigour, height, tillering, freedom from mosaic and other visible characteristics and six- to twelve-week-old seedlings are transplanted in the field nursery at 25×25 cm spacing. A group of 3-15 seedlings may be set in one bunch, at 25 cm (bunch to bunch) and 1-2 metres between rows. At about harvest time select approximately 4000 superior canes on the basis of vigour, height, thickness, leaf width, freedom from diseases and sugar percentage.

---

[3] Facilities and methods required for successful hybridization program are determined largely by the latitude where the work is undertaken. In more tropical climates between approximately 20° North and South latitudes, temperature is generally favourable for flowering and pollen fertility. At high latitudes, low temperature during the fall and winter adversely affects pollen fertility and depresses flowering. Photoperiod houses may be required to induce flowering. Heated green houses are required to protect parents (to be used in hybridization) from low temperature.

Third year: First clonal nursery is grown. Each clone selected in the previous year is planted in a separate row. The row length usually depends on the sets available for planting. If sufficient sets are available clones may be simultaneously planted in a separate disease nursery for screening the material with respect to important diseases. Check varieties are also included at regular intervals for comparison. Record data, namely vigour, tillering, height, disease resistance, sugar percentage, as well as flowering behaviour. Select approximately 400 clones possessing high yield, higher sugar percentage and disease resistance.

Fourth year: Second clonal nursery is grown. If seed supplies permit a replicated trial, including check varieties for comparison, is also conducted. Data is recorded in a manner similar to the previous year.

Fifth year: Multiply selected clones for yield trials. Data is recorded as in previous years.

Sixth to tenth year: Conduct replicated yield trials at several locations. Data is recorded as in previous years. Any clone found to be inferior is discarded. Identify superior clones for release and multiplication.

B. Potato[4]

First season: Tubers of the chosen parents are placed on a brick and covered with sand and peat. The roots grow over the bricks and when they have penetrated the soil on which the brick is lying the covering sand is washed away. Removal of tips of stolons, which would otherwise produce new tubers, is then possible. This promotes vigorous stem growth and enhances flowering. Make the crosses between the chosen parents. Interspecific crosses often involve use of specific techniques.

Second season: Sow F_1 seeds in wooden boxes. Sowing time should be so adjusted that during the growing period of the crop aphid population is low. This is to minimize the possibility of viral infection. Transplant approximately 500 vigorous F_1 seedlings per cross to flower pots. Artificial infection may be used at this stage for screening the material for resistance to *Phytophthora* and potato virus YN. Harvest the tubers from each plant separately and label them.

Third season: Grow individual F_1 rows of each cross in a block. Each row represents the clonal increase from a single F_1 plant. Rapid growth and early tuber formation are two very important characters which allow early harvest (or haulm cutting) and thus of value in the prevention of transmission of viral infection through aphids. Selection against undesired plant types and diseases

---

[4] Clones of temperate zone *tuberosum* will generally bloom 60 days after planting in the field. Pollination of flowers on the plant in the field is generally unsuccessful for producing seeds. This difficulty is overcome by using cut stems. The technique involves cutting a stem that contains a large inflorescence with several unopened buds. Any opened flowers are removed. The cut stem is immediately placed in tap water and shifted to controlled environment conditions (the sunlight should be available for 16 hours or provide 20 k lux of artificial light, the temperature should be 19°C and relative humidity should be high during the period of pollination).

is practised during the growing season. At harvest the tubers from each plant are judged for yield, uniformity and number of tubers, tuber shape, depth of eyes and length of stolons. The number of plants retained varies, depending on the cross. Usually 10% plants are retained (4-12 tubers of each plant). After harvest one tuber from the retained selections is used for screening against cyst nematodes of pathotype A. The clones found susceptible at this stage are eliminated.

Fourth season: Plant tubers from each selected clone in single rows. Clones are spaced one metre apart. The standard checks are also planted to facilitate comparison between clones. Selection in the field starts at the early growth stage. All entries with undesirable plant types, poor soil coverage, viral infection, too high susceptibility to *Phytophthora infestans* and *Alternaria* are eliminated. When aphids begin to appear with consequent danger of viral infection, the haulms are destroyed. At harvest good clones are selected on the basis of characters already mentioned above in the previous year. Clones found susceptible to cyst nematodes of pathotype A are discarded.

Fifth season: Selected clones are planted for further evaluation and increase as in the previous year. If the seed supplies are sufficient, replicated yield trials at several locations may also start. To ensure that healthy plant material will be available, haulms of the two rows meant for increase per plot are removed before the appearance of aphids and the seed potatoes are left in the field until the whole trial is harvested. Observations on disease resistance until maturity and yield are recorded for the remaining rows in a plot. Tubers from these rows are further used for determination of cooking and processing quality and resistance to *Synchytrium inocobioticum* and nematode resistance.

Detection of Viral Infection
Selection for virus-free material is also practised, usually under greenhouse conditions. For this purpose, sprouting is induced in approximately 150 tubers of each selected clone. Each tuber is indexed for virus X, S, M, YN and A. The corresponding tubers are destroyed if found virus infected. Only the healthy tubers of a clone are used in subsequent yield trials and for maintenance of stock (stock multiplication). The method of tuber indexing is discussed in detail in Chapter 30.

Sixth to tenth season: After healthy seed tubers of the selected clones have been multiplied in the required quantities, replicated yield trials are conducted at several locations to establish superiority in different production areas. Data is recorded in respect of characters, such as growth, earliness, yield and susceptibility to diseases already mentioned earlier.

Clones which have consistently performed well are further multiplied and recommended for release as new varieties.

Model Envisaged for Selective Mating in Tetraploid Potato
Selective mating in tetraploid potato may be achieved in the following four steps (Fig. 18.1) :
1) *in vitro* production of haploids;
2) clonal propagation of haploids for evaluation of genotypes for disease resistance, yielding ability and quality;

3) assessment of combining ability of haploids by test crosses;
4) fusion of somatic cells from haploids selected for good agricultural characters and combining ability.

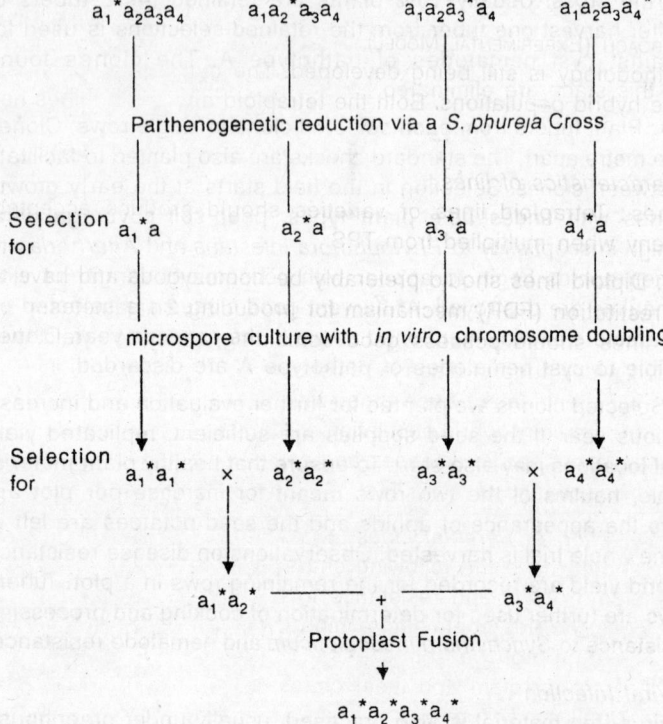

a = potato genome, a* indicates favourable genes

Fig.18.1. Breeding scheme for tetraploid potato (Wenzel et al., 1979) combining tissue culture techniques and conventional steps.

Steps one and two can be carried out with relative ease, but lack of fertility in most haploids is a big obstacle in the assessment of combining ability, and fusion of somatic cells is not yet a routine method.

Breeding Tetraploid Potato at Diploid Level
Steps (2) and (3) mentioned above basically constitute a selection programme at the diploid level.

Breeding For True Potato Seeds (TPS)

Advantages of TPS
Provided the breeding results TPS varieties (hybrid varieties) comparable to or better in yield, quality, uniformity and resistance to diseases, TPS have several advantages over tuber-propagated varieties :

1) TPS do not carry soil-borne or viral diseases, apart from new infections.
2) Use of TPS in potato cultivation is more economical besides saving considerable food (tuber) used as planting material.
3) Storage and transportation are safer and economical.

Breeding Approach (Experimental Model)

Breeding methodology is still being developed. The current approach is to breed suitable hybrid populations. Both the tetraploid and diploid lines have to be bred.

Desirable characteristics of lines

Tetraploid lines: Tetraploid lines or varieties should produce acceptably uniform progeny when multiplied from TPS.

Diploid lines: Diploid lines should preferably be homozygous and have the first division restitution (FDR) mechanism for producing $2n$ gametes.

Selected lines should possess good combining ability for yield, wide adaptability and wide range of disease resistance. The two parents involved in a cross should complement each other with regard to biotic and abiotic factors so that the resultant hybrid population or the greater part of it combines parental resistance to specific diseases and preferably shows transgression for any polygene resistance.

Criterion for Selection of Parental Lines

Parental lines should have the following characteristics
1) good yielding ability and resistance to major potato diseases;
2) ability to set berries containing a high number of bold seeds;
3) the resultant crop grown from seeds (TPS) should be early maturing, homogeneous for plant and tuber characteristics, and possess high yield potential as well; and
4) the lines should be able to transmit the above-mentioned traits to the hybrid progeny.

Studies have revealed manifestation of positive heterosis involving *S. tuberosum* as female parent and *S. andigena* as male parent. The final selection of a cross combination depends on two factors however, namely field performance of the hybrid family and the quantity of hybrid TPS produced per unit area for a unit amount of input.

Maintenance and multiplication of parental lines

For maximization of TPS production, and also to prevent TPS transmitted diseases, namely PSTVd, APPLv, APMY and PVT, it is imperative that stocks of parental lines be maintained pathogen free. Parental clones are therefore maintained in healthy conditions through well-established standardized techniques (tuber indexing).

Breeding scheme

With the desirable tetraploid and diploid lines (as described above) available as starting material, either a double or three-way cross may be produced as shown in Figure 18.2.

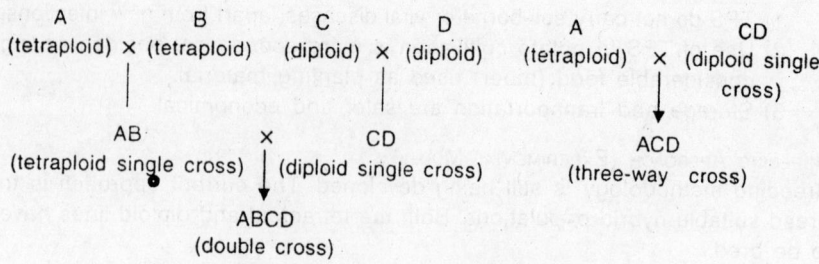

Fig. 18.2. Method of producing double and three-way cross.

It is also possible to streamline hybrid seed production by incorporating cytoplasmic male sterility into one diploid parent. *Solanum verrucosum* is a good source of cytoplasmic male sterility. Hybrid seed can then be harvested from the male sterile single cross in a seed production field with a tetraploid parent as pollen parent.

19

Apomictic Grasses

Over 100 species of perennial grasses are apomictic. Varietal improvement in these crop species is done through ecotype selection and through hybridization followed by selection.

ECOTYPE SELECTION

Grasses display considerable variation in populations adapted to various natural or man-made environments (ecotypes). Ecotypes usually show variation both at the individual plant level (being cross-pollinated and highly heterozygous) and at the population level. As the greater part of this variation is additive, phenotypic selection can be effectively practised.

Generalized Procedure for Ecotype Selection

The various steps involved in selection are :

1) Seeds or vegetative material collected from as many ecotypes as possible are screened in comparative trials. Promising introductions may be multiplied directly for commercial use.

2) Phenotypic selection in early years is done mainly for characteristics with high broad-sense heritability from plants grown at wide spacing. Data such as head emergence, growth habit, disease resistance and winter hardiness, etc. is recorded. The clones are then subjected to increasing selection intensity in successive years. Success depends how accurately plants with a superior breeding value are recognized from their phenotypic expression.

3) Selected individual single plants are evaluated in a dense sward of either another genus, or in monogenotypic swards, separated from each other by a barrier (a different type of grass grown in-between two swards) to prevent invasion by one another to ensure equal competition.

4) Phenotypic selection is carried out in two different ways.
   i) *Negative selection*: This consists of removing off-types in space-planted isolated fields, before flowering and allowing remaining plants to intercross for seed production.
   ii) *Positive selection* : This consists of selection for further evaluation in clonal rows or for intercrossing with similar plants to produce seeds.

HYBRIDIZATION FOLLOWED BY SELECTION

a. Obligate Apomicts

Obligate apomicts reproduce only through apomixis. Therefore, in a grass species, until exceptional sexual types are discovered hybridization is not feasible. Fortunately, a few sexual types have been discovered in some grass species of economic importance. A few examples of breeding schemes are mentioned below.

i) *Cenchrus ciliaris*: The sexual plant was discovered in 1958, which enabled starting a breeding scheme. The essential features of this scheme are shown in Figure 19.1. With the sexual S_2 and F_2 plants the scheme can be repeated.

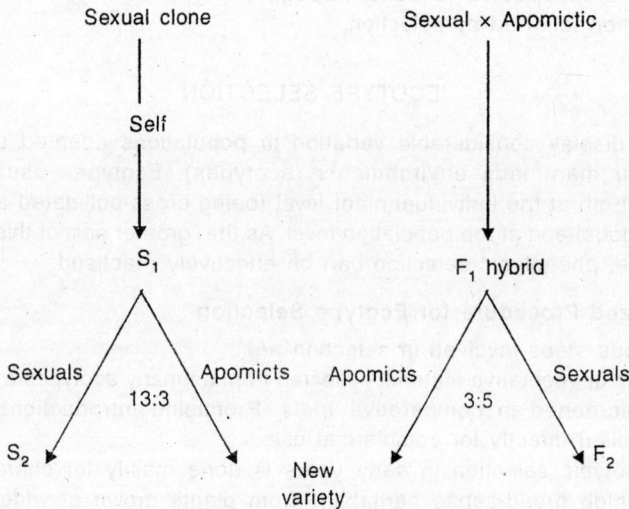

Fig.19.1. Breeding scheme for *Cenchrus ciliaris* (Taliaferro and Bashaw, 1966).

Apomictic hybrids are increased immediately by clonal propagation and seed, and evaluated. Inferior types are eliminated and seeds of promising ones used for extensive field tests.

ii) *Paspalum notaum*: Burton and Forbes (1961) doubled the chromosome number of sexual diploid accessions by colchicine treatment of seeds and crossed these new tetraploids with apomictic tetraploids. Both obligate apomictic and completely sexual F_1 hybrids were recovered. Heterosis was maintained in the progenies of apomictic hybrids.

b. Facultative Apomicts

The distinction between obligate and facultative apomicts is rather vague in many instances. A facultative apomict with a low degree of sexuality approaches the obligate type. In some species, for example *Panicum maximum*, both obligate and facultative apomicts and sexuals have been found. A few examples of breeding schemes are described below.

i) *Panicum maximum* : Savidan (1982) obtained sexual tetraploids by treating sexual diploids with colchicine. A breeding scheme involving recurrent crosses, sexuals x apomict at the tetraploid level enriched the sexual gene pool (Fig. 19.2).

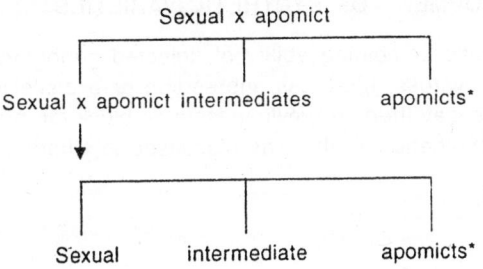

Fig. 19.2. Apomictic breeding scheme.
*Refers to plants to be evaluated as potential varieties.

The usually high degree of recovery of apomixis and the low degree of sexuality in the apomictic plants have greatly helped the breeding programme in this species.

ii) *Poa pratensis* : Breeding of *Poa pratensis*, a facultative pseudogamous species, is rather complicated. Within the species nearly 100% apomict to completely sexual genotypes can be found. A high level of apomixis (90-95%) is common. Average sexuality varies from 5-10%. The genetic background of apomixis has not been elucidated. Many loci with recessive gene action have been reported.

Generalized Breeding Procedure

(a) *Induction of sexuality* (Funk and Han, 1967): Male and female plants of good potential are raised in the greenhouse at staggered intervals to ensure overlapping flowering periods. A 20-hour daylength, and a diurnal 16/8 hour temperature (16°C/10°C) is maintained until flowering starts. After flowering has started the temperature is raised to 18°C/15°C.

Just prior to anthesis 2-5 panicles of the female parent are bagged together with 15-20 panicles of the pollen parent, the latter being arranged above the former. The pollination bag is connected with a device that shakes the plants every 20 minutes, day and night, freeing pollen for pollination. The pollination bags are removed after a fortnight, and three weeks later seed is harvested from the female parent.

(b) *Seedlings are raised and transplanted in the field* : Only those that deviate[1] phenotypically from the female parent are selected[2] for clonal evaluation and

---

[1] The percentage of aberrant progeny depends entirely on the female parent. Large varietal differences occur. The choice of male parent has little influence.

[2] The percentage of plants selected depends on both parents.

subsequently for turf performance. A high degree of apomixis and good seed-yielding ability determine the ultimate selection of hybrid plants.

(c) Remanent seed is used for assessment of the degree of apomictic reproduction.

Interspecific hybridization offers many possibilities, namely the introduction of an entirely sexual phase, and the possibility of introgression.

DEVELOPMENT OF SYNTHETIC VARIETIES

The general and specific combining ability of selected genotypes may be determined through polycross, top-cross, inbreeding or a diallel cross and the selected genotypes may then be grown in isolated block for intercrossing to produce seeds of synthetic varieties, as discussed in Chapter 17.

20

Corn Hybrids

What is a commercial hybrid (hybrid variety)?
A commercial hybrid is the first generation progeny from a cross involving selected inbred lines, that are genetically dissimilar.

TYPES OF HYBRIDS

Single-cross hybrid: A single-cross hybrid is a cross (F_1) between a pair of two unrelated inbred lines (A x B) selected on the basis of their combining ability and resultant yields in cross combinations.

Single-cross hybrids are usually higher yielders than any other type of commercial hybrid. They are genetically homogeneous and phenotypically exhibit maximum uniformity of appearance, maturity and other characters. Expression of heterosis is maximum. The genetic base of a single cross being narrow, these crosses are more vulnerable to disease epidemics, if the host-pathogen relationship within each field area is developed. Their adaptation is specific and usually covers a narrow range of environments. Nevertheless, through extensive field testing and by adopting appropriate strategy single-cross hybrids with high yields and adaptability can be identified.

For the reasons of significantly higher yields and/or uniformity in expression of desired traits, single-cross hybrids are more commonly used today despite low seed yields (F_1 seed yields) (commercial hybrid seed) compared to other types of commercial hybrids.

Modified single-cross hybrid: The modified single-cross hybrid is a cross between a single cross (involving two closely related inbreds, A x A') used as female parent for hybrid seed production and the other unrelated selected inbred line. These two closely related inbreds have common ancestry and the difference is in the degree of ancestry, that is, the proportion of common parent, for example inbred H 84 (B 37 x GE 440) Ht Ht) and H 93 (B 37 x GE 440) B 37^4 Ht Ht) contains parent B 37 but in different proportions. The objective of modified single-cross hybrids is to obtain higher F_1 seed yields and reduced seed costs.

Three-way cross hybrid: A three-way cross (A x B) x C is produced by crossing a single cross, involving two unrelated inbred lines (A x B) and using it as a seed parent, with a third unrelated inbred line (C). Seeds of three-way crosses are less expensive to produce than that of single crosses. They tend to be uniform and slightly higher in yield than double-cross hybrids.

Modified three-way cross : The modified three-way cross involves crossing a single cross between two unrelated inbred lines (A x B), with another single cross between two closely related inbred lines (C x C')

Double-cross hybrid : A double-cross hybrid is produced by crossing two single crosses, both involving unrelated inbred lines. (A x B) (C x D). Double crosses are the most widely used type of hybrid maize. They are slightly more variable in plant characters than single crosses or three-way crosses which may be an advantage when the crop is grown under relatively adverse conditions.

Double topcross hybrid : A double topcross hybrid is produced by crossing a single cross (between two unrelated inbreds) and using it as a seed parent, with a selected open-pollinated variety (A x B) Va.

PRINCIPLES OF DEVELOPING CORN HYBRIDS

The principal objective of developing hybrid varieties is to exploit hybrid vigour (heterosis) which expresses itself primarily in the F_1 plants raised from seed. The various steps involved in the development of hybrid varieties are:
1) development of inbred lines,
2) assessing hybrid combinations, and
3) assessing the double-cross performance.

1. Development of Inbred Lines

Inbreeding

An inbred line is a relatively homozygous pure breeding strain developed by controlled self-fertilization, that is, inbreeding and selection. The main objective of inbreeding is to obtain pure breeding lines (inbred lines) that are homozygous, uniform, superior in combining ability and performance in cross combinations besides possessing other desirable characters, such as resistance to diseases and pests. Controlled self-fertilization may be through selfing[1] alone or through a combination of selfing and sib pollination[2], or through sib pollination alone. Both the latter methods are less restrictive forms of inbreeding, which is preferred in many instances for maintaining more vigorous inbred lines. The only disadvantage is delayed fixation of deleterious genes compared to the selfing method. Self-fertilization is three times faster in effecting requisite homozygosity in inbred lines. Usually 5-7 generations of selfing are considered sufficient.

---

[1] Selfing means transfer of pollen of a plant to the stigma of flowers on the same plant.
[2] Sib-pollination means transfer of pollen of a plant to the stigma of flowers of sister plants.

Effects of Inbreeding

The obvious effect of homozygosity brought about by inbreeding are:

1. Marked decrease in vigour and productivity of the inbred lines derived by repeated self-fertilization compared to original materials from which the lines have been derived. Theoretically, about one-half of the total decrease in productivity will occur in the first generation of selfing. On average, about 97% of reduction in vigour occurs in the first five generations of selfing.
2. Traits become fixed as the plants approach uniformity within any progeny or line.
3. Differences among lines are increased, while variability within lines is decreased.
4. Inbreeding brings high lights and makes possible the elimination of deleterious or inferior recessive characters.

Source of Parental Material for Inbreds

The usual source materials are open-pollinated varieties, germplasm composites, synthetics, populations derived from recurrent selection and even hybrids. Inbred lines derived from the various unrelated sources are more likely to yield better results in hybrid combinations because of greater genetic diversity compared to those derived from closely related materials. It is therefore appropriate to use genetically diverse and unrelated elite germplasm or breeding materials for the extraction of inbred lines.

Selection of Inbred Lines

Selection of inbred lines is based on visual inspection and performance in hybrid combinations. The various aspects of evaluating performance in hybrid combinations are separately discussed in this chapter. The performance of inbred lines in cross combinations and exhibition of other desired characters is influenced by several factors. Important among these are *per se* yield of inbreds, vigour of inbreds and yield of F_1 as well as of F_1 crosses.

Vigorous inbred lines tend to produce the better crosses. From the standpoint of practical production they are certainly much more desirable. Through careful selection from among a large number of progenies it is possible to select reasonably vigorous inbreds with desirable phenotypes. The least desirable lines should be discarded early in the inbreeding period. The plants should be selected on the basis of desired traits, such as vigour, yield, standability, maturity, ear height, grain quality, resistance to diseases and pests, least number of barren plants, freedom from any kind of abnormality, etc. The effectiveness of visual selection, however, depends on the heritability of the trait(s). The final selection of an inbred line is based on the combining ability and productivity in cross-combinations.

Generalized Procedure to Develop Inbred Lines of Corn

First year: Self-pollinate desirable plants (S_0) in open-pollinated variety or any source material(s). The better ears from selected plants should be saved.

Second year:

i) Grow 25-30 plants in a row from each self-pollinated ear. Self-pollinate five to eight desirable plants in each' row. Practise selection within and between progenies. Ears are saved from the better plants in each selected row.

ii) The selected S_1 plants may be crossed with an adjacent tester for preliminary evaluation the following year.

Third year:

i) Grow ear-to-row progenies of selfed ears from the second year.

ii) Self-pollinate and save the selfed ears from desirable S_2 plants.

iii) The selected S_2 plants are crossed with a tester for evaluation in the following year.

iv) Evaluate test crosses made in the second year in replicated plots. The performance of a test cross is compared with one another and with the tester. Inbred lines involved in poor performing test crosses are discarded.

Fourth year:

i) Inbred lines selected on the basis of test cross performance are planted ear-to-row.

ii) Self-pollinate and save the selfed ears from desirable S_3 plants and make test crosses with the tester as in the previous year.

iii) Test crosses made in the third year are evaluated in the performance trials.

The above procedure is continued for three or more generations until each inbred line is relatively homozygous.

FURTHER IMPROVEMENT OF INBRED LINES

Selection in large populations, backcrossing and convergent improvement are some of the methods that may be employed for improvement of established inbred lines. Each of these procedures has certain advantages and disadvantages. Variations and combinations of some of these methods are desirable.

1. Pedigree selection

Pedigree selection is used in the recycling of lines that have known strengths and weaknesses for specific traits. Pedigree selection could be practised with progenies developed in open-pollinated varieties, germplasm composites, synthetics, backcross populations, mixtures of germplasm and F_2 populations as well. The selection unit is a combination of progeny row and individual plants within a progeny row. Selfed seeds of selected plants within selected progeny rows are planted ear-to-row the following season. Progeny rows usually are non-replicated but because of controlled pollination selection is effective for traits of relatively high heritability. For traits such as yield, replicated yield tests become necessary at some stage of pedigree selection.

2. Backcross selection

The principal objectives of backcross selection are to transfer a specific trait from a donor parent to an elite inbred line and incorporation of exotic germplasm. Since the traits transferred by backcrossing usually can be classified into discrete classes, the selection unit is a combination of individual plants within non-replicated progeny rows.

3. Convergent improvement

Convergent improvement involves the reciprocal addition to each of two inbred lines, the dominant favourable genes lacking in one parent but present in the other. The value of the method, which is equivalent to double backcrossing, is that it furnishes a plan for the improvement of each of two inbred lines that combine well in a single cross without modifying the yielding ability of the single cross. For complex traits such as yield the method is less efficient.

4. Single-hill method

Single-hill selection is similar to the standard method of developing inbred lines, except that only one hill of three or four plants is grown from each ear. This procedure limits selection primarily to differences between families. This method has not been used so far to any great extent.

5. Gamete selection

Gamete selection is another application of the principle of early testing. The method should increase the frequency of the exceptional genotypes from the varietal population. Stadler (1944) theorized that if the frequency of superior zygotes was p^2, the frequency of superior gametes was p, because the theoretical frequencies of superior gametes are greater than for superior zygotes (if $p = 0.5$, $p^2 = 0.25$). Therefore, gamete sampling should be more efficient than zygote sampling. This procedure is described in detail below by an illustration of improvement of inbred WF 9, a component of hybrid US 13 (WF 9 × 38-1) × (Hy × L 317) (Jugenheimer, 1958).

First year: Cross WF 9 with bulk sample of pollen grains from selected plants of an open-pollinated variety or other materials.

Second year: Self-selected F_1 plants. Cross these same plants with Hy × L 317. Cross the original WF 9 with Hy × L 317 to use as a standard check.

Third year: Test the crosses produced in the second year. Select the better performing lines for continued inbreeding. Self-pollinate selected plants of the S_2 inbreds or store them pending evaluation of the test crosses.

Fourth year: Self-pollinate selected lines. Retest the crosses produced in the second or third year.

Fifth year: Continue self-pollination of selected lines until homozygous.

Sixth year: Substitute new WF 9^r in original hybrid and use commercially as (WF 9^1 × 38-11) × (Hy × L 317).

The underlying assumption is that any test cross that exceeds the elite line × tester combination presumably receives a superior gamete from the source population. Although the test cross performance identifies the superior gamete, the same cannot be fixed in homozygous inbred lines *per se*. Selection units are individual plants from the source population but include progeny row and individual plants within the selfed progenies of the F_1 plants. This method has generally not been used.

6. Recurrent selection

The various methods of recurrent selection discussed in Chapter 16 may be suitably employed to improve and extract new inbred lines.

7. Doubling of monoploids

Lines produced by doubling monoploids are homozygous at all loci. The process greatly reduces the time needed to produce inbreds and selection for highly desirable genetic combinations can be made more efficiently on the gametic basis. Also, individuals possessing deleterious genes are eliminated since the monoploid sporophyte is subjected to intense natural selection. Successful production of homozygous diploids from monoploids, is tedious and complicated, however and depends on production and recognition of monoploids, and deriving homozygous, diploid progeny from the isolated monoploids. So far the method has not been practically implemented by corn breeders.

2. Assessing Hybrid Combinations

a. Preliminary Evaluation

Final evaluation of inbred lines can best be determined by hybrid performance. According to Jugenheimer (1958) in earlier years of corn-breeding programmes, inbred lines were not usually evaluated in hybrid combinations until they had been inbred for several generations. Many of the inbreds were divided into groups of about 10 lines. These were combined into all possible single crosses, which were then tested in field trials. The lines were saved or discarded on the basis of the mean performance of the crosses. Since only 10 inbred lines were evaluated at a time, the need arose for a simpler, more rapid and less expensive method.

Topcross performance : The inbred ∨ variety cross is known as topcross. It has been well established on the basis of a large number of studies that use of topcrosses provides an efficient method for preliminary evaluation of inbred lines. They are especially useful for determining general combining ability (*gca*) of a large number of lines. Topcross seed of corn can be produced in an isolated field. The inbreds are used as female parents and detasselled, and the open-pollinated variety is used as male parent (tester). Fifty per cent or more of the inbred lines may be discarded on the basis of a preliminary test.

Testing inbred lines in three-way crosses is an efficient method when a desirable single-cross seed parent is available. The procedure permits the testing of a large number of inbred lines at one time and eliminates the need for preliminary evaluation of inbred lines in topcrosses. This method, however, provides information primarily on specific combining ability.

CHOICE OF TESTER PARENT

The choice of appropriate tester for raising test cross progenies is important in the evaluation of breeding value of genotypes (plants). It is generally based on the objectives of the improvement and the gene action involved. A good tester correctly classifies the relative performance of genotypes and discriminates efficiently among genotypes under test. Hallauer (1975) pointed out that a suitable tester should include simplicity in use besides providing information that correctly classifies the relative merit of lines and maximizes genetic gains. For improvement of random mating populations the best tester is the one that maximizes the expected mean yield of the population.

In selection for general combining ability (*gca*) a broad-base heterogeneous population is used as a tester. It can be either the parental population or any unrelated broad-base heterogeneous population (synthetic or open-pollinated variety). In all instances genotypes are tested with a representative sample of genotypes in the tester, i.e., each plant in the base population is crossed to a random sample of gametes from the tester. Thus, each test cross is a type of half-sib family. General combining ability measures the average performance of an inbred line in a number of hybrid combinations and thus provides information on the relative usefulness of an inbred line. Also, this is indicative of the additive gene action.

In selection for specific combining ability (*sca*), a narrow genetic base (inbred line or single cross) is used as a tester. When the breeding objective is replacement of a line in a specific combination, *sca* is of prime importance and the most appropriate tester is the opposite inbred line parent of a single cross or the opposite single-cross parent of the double cross (Matzinger, 1953). Specific combining ability measures the performance of an inbred line in a single specific hybrid combination and is dependent on dominance, epistasis and genotype × environment interactions.

The expected genetic gains in a population have been given in Table 20.1. It is clear from the Table 20.1 that under *no dominance* there is no difference among testers; with *partial* to complete *dominance* ($d > 0$), the best tester is homozygous recessive; with *underdominance* the best tester is homozygous dominant; with *overdominance* the best tester is homozygous recessive.

Table 20.1. Expected genetic gain at one locus level in a population for three types of testers (Allison and Curnow, 1966)

Tester	Gene frequency	Expected change in population mean
Homozygous ($A_1 A_1$) dominant	$p = 1$	$\dfrac{i}{2\sigma}\left(\sigma^2 A - 4pq^2 \, \alpha \, d\right.$
Homozygous ($A_2 A_2$) recessive	$p = 0$	$\dfrac{i}{2\sigma}\left(\sigma^2 A + 4p^2 q \, \alpha \, d\right.$
Parental population	$p = S_0$	$\dfrac{i}{2\sigma}\left(\sigma^2 A\right.$

Where, σ_A^2 = additive genetic variance at the locus; $d = a + (q - p)d$ = average effect of the favourable allele.

Evidence presented by Rawlings and Thompson (1962), Comstock (1964), and Allison and Curnow (1966) leads to the conclusion that either an inbred line homozygous recessive or a population with low gene frequency at important loci is the most effective tester for discriminating among inbred lines in a hybrid breeding programme.

EARLY VERSUS LATE TESTING

There used to be considerable diversity of opinion as to the best time of assessing value of an inbred line in hybrid combinations, say through topcross performance. Many researchers believed that self-fertilization for three to five years was necessary before assessment of an inbred line in hybrid combinations. They advocated within progeny selection for general vigour, resistance to lodging, diseases and insects, and other desired characters. Their belief stemmed from the notion that the performance of inbred lines in crosses may change while they are becoming homozygous, and that the testing programme is much more expensive and laborious than developing inbred lines. Numerous studies, however, have established the usefulness of early testing. Early testing is based on two assumptions:

i) there are marked differences in combining ability among the plants of a population selected for inbreeding; and

ii) a selected sample based on tests of combining ability of S_0 plants provides a better sample for further inbreeding and selection than does a more nearly random sample drawn from the same population on the basis of visual selection alone.

A few breeders attempt to compromise between extremely early testing (from S_0) and extremely late testing (from S_3, S_4). They limit evaluation of new lines to visual selection during the first year (S_0) or up to the second year (S_1) of inbreeding. Promising S_2 inbred lines are crossed with a suitable tester and evaluated in performance trials. The better performing inbreds may be re-evaluated in the S_3 or S_4 generations. Testing of siblings, traceable to S_1, provides an opportunity for detecting significant segregation within families.

b. Hybrid Performance

After determining the more desirable inbreds by testing them for general combining ability, it is necessary to test their value in single crosses and, or in three-way or double crosses, if these are to be released as commercial hybrids. It is indeed the performance of selected inbred lines when combined into specific hybrid combinations for commercial use that matters the most.

i) Single-cross performance: The number of single-cross combinations that can be had from n lines is $n \times (n-1)/2$. In the 10 top inbred lines that have been selected on the basis of topcross performance for making single crosses, the total number of single crosses would be $10 \times (10-1)/2$, that is, 45. Single crosses are pair crosses between the most desirable parental inbred lines. The system of making all possible single crosses is usually known as diallel crossing. The performance of single crosses is evaluated in replicated yield trials over environments (locations and years). The top performing single crosses are selected as new hybrid varieties, or as the potential parents for the production of three-way cross commercial hybrids or double-cross hybrids.

ii) Three-way cross performance : Three-way crosses are pair crosses between most desirable single crosses and desirable inbred lines. The inbred line used in making a three-way cross is an unrelated inbred line not included in the single cross as one of the parents. The performance of three-way crosses is evaluated in replicated yield trials over environments in a manner similar to that described earlier for single crosses. The top performing three-way combinations are selected as new hybrid varieties.

iii) Double-cross performance: Methods have been evolved that enable breeders to accurately predict the probable comparative performance of double crosses from single-cross data without making and testing literally thousands of undesirable crosses. Predicted hybrid combinations, however, should always be thoroughly tested under field conditions before being put into commercial production.

PREDICTION METHODS

The following four methods were used by Jenkins (1934) to predict double-cross performance. He also compared the relative efficiency of each of these methods.

A. The mean value of all possible six single crosses between four inbred lines (A, B, C, and D), namely, A x B, A x C, A x D, B x C, B x D and C x D.

B. The mean value of four non-parental single crosses of a double-cross hybrid (A x B) x (C x D), namely, A x C, A x D, B x C and B x D.

C. The mean value of all single crosses in which the parental inbred lines of the double-cross hybrid, namely A, B, C and D were a parent. For example, if 10 inbred lines were included in a diallel crossing programme each of the inbred line would be involved in 9 single crosses. Four inbred lines thus would be involved in a total of 30 single crosses in the diallel set.

D. The mean value of topcrosses for the four inbred lines involved in the double cross. This information obviously would come from topcross performance data available in respect of the lines involved.

Methods A, C and D assume additive gene action, that is, a gene contributed by any line would produce its characteristic effect regardless of the order of pairing. Method B permits the recognition of non-additive effects arising from dominance, epistasis, etc.

The effectiveness and relative efficiency of these four methods in predicting the performance of double crosses has been studied by many workers who obtained high correlation between the predicted and observed in fields. The r value for the four methods were 0.75, 0.76, 0.73 and 0.61 respectively. Method B has been accepted as a relatively reliable method and has been widely used.

3. Assessing Double-Cross Performance

Based on predicted double-cross hybrid combination performance, experimental double-cross hybrids are produced by crossing the involved single crosses. The experimental hybrids are then tested in replicated yield trials including standard checks over locations and years to establish the superiority of the experimental hybrid, if any. The superior hybrids are released for commercial cultivation.

21

Hybrid Varieties

The general principles of development of corn hybrids outlined in Chapter 20 are applicable to other crops as well. However, most of the important economic crops which exhibit heterosis worth commercial exploitation differ from corn in their floral morphology. Commercial F_1 hybrid seed production in these crops is not feasible unless mechanisms such as male sterility, self-incompatibility are available which make the F_1 hybrid seed production commercially feasible. Even in many of the monoecious crops it becomes imperative to develop lines having specific sex expression, for example gynoecious lines. In other instances, that is, in the absence of satisfactory male sterility or self-incompatibility mechanisms, hybrid seed production has to be done through hand pollination.

In this chapter we discuss the development of hybrid varieties based on the following systems which are in common use :

1) Cytoplasmic-genetic male sterility system
2) Self-incompatibility system

The various aspects related to development of hybrids using sex expression forms e.g. in cucumber are discussed along with seed production in Chapter 32.

CYTOPLASMIC-GENETIC MALE STERILITY SYSTEM

Types of Inbred Lines

The salient features of cytoplasmic-genetic male sterility were earlier discussed in Chapter 5. Three types of inbred lines must be developed and maintained when a cytoplasmic-genetic male sterility system is used to produce hybrids. These are:

    i) Male sterile line (A line): This line is used as female parent (seed parent) line of a hybrid.

    ii) Maintainer line (B line): This line is used as a male parent line (pollinator) for maintaining the A line.

    iii) Restorer line (R line): This line is used as male parent line (pollinator) in hybrid seed production. This is also known as the C line.

Development of New Male Sterile (A Line) and Maintainer (B Line)

The availability of male sterile line (donor line) is the necessary prerequisite for development of new A and B lines. To utilize the male sterile condition in hybrid development, the breeder is required to develop a pair(s) of breeding lines known as A and B line(s). The best way to develop a new A line is to make test crosses of selected plants from open-pollinated varieties with known A lines (used as donor line) to identify any *rf/rf* genotype within those populations. Here the known A line is used as cytoplasmic male sterile tester. Therefore by pairing up a number of selections from open-pollinated varieties or breeding lines with known A-line plants one can determine which pollinators were B lines by growing out the F_1 progenies. The F_1 progenies with 100 per cent male sterile plants indicate which selections were B lines. At the same time, that is, when the selected plants are paired with known A-line plants to produce test crosses, they are selfed as well, and the selfed seed from each plant is saved so that once the B lines are identified, this selfed seed could be used in a backcross programme to develop the A line to the point where it is identical in genotype to the B line selection. In this manner, once several A lines are developed they are tested with other inbred B lines or C lines (restorer line, R line) in hybrid combinations to determine, whether they will make superior hybrids.

Generalized Procedure for Developing New Hybrids
First year:
  i) Make test crosses by pairing sufficient number of selected plants from open-pollinated varieties or breeding material with the known A-line (cytoplasmic male sterile tester) plants, taking care to include selections from a range of acceptable sources to maintain diversity.
  ii) Also, self-pollinate[1] selected S_0 plants to produce selfed (S_1) seed.

Second year:
  i) Grow F_1 progenies of the test crosses made along with S_1 progenies of selections made in the first year.
  ii) Examine F_1 progeny lines very carefully at the flowering stage for male sterility. Select 100 per cent male sterile progenies and discard all other segregating or fertile progenies. The corresponding S_1 progeny is now identified as maintainer B line selection.
  iii) Make first backcross of 100 per cent male sterile F_1 line by pairing it with the companion S_1 line (B line selection). The seeds obtained from the male sterile F_1 plants will be $BC_1(S)$ seed (male sterile), and those obtained from selfing (selfed seed) from identified S_1 line will be S_2 seed.

Third year:
  i) Grow $BC_1(S)$ progeny lines alongside S_2 companion lines. If pollen fertile plants are observed, all the BC_1 progeny should be discarded

---

[1] Artificial self-pollination is not required in cleistogamous species.

and the companion S_2 plants isolated to produce S_3 lines that may then be evaluated as pollen parents. (Pollen fertile plants in BC_1 progeny indicate the presence of restorer genes in the recurrent line.)

ii) Examine and classify BC_1 lines as 100 per cent male sterile and segregating or fertile in order to identify maintainer line. Select sterile BC_1 (S) lines which resemble their companion S_2 line as nearly as possible. Most uniform S_2 lines may be massed rather than selfed in order to minimize inbreeding depression. This decision must be made on a line-by-line basis. Some lines may be massed while a few selected plants are selfed as well. The criterion in this case is to arrest the inbreeding process as soon as the desired level of uniformity is achieved. Massing at this stage can avoid the unacceptably low seed yields that result from inbreeding depression.

iii) Plants from selected BC_1(S) progeny are paired with 5-10 single S_2 plants of companion S_2 line to produce BC_2(S) seed and S_3 seed. S_3 seed is mixed together to constitute a mass S_3(M).

Fourth year:

i) Grow BC_2(S) alongside companion S_3 (M) lines. A few S_3 lines that show good tolerance to inbreeding should be selected for additional generations of selfing. (The inbred lines advanced to S_4 and beyond will provide parents for fertile x fertile crosses between lines of diverse origin. This type of recycling is a most productive source of improved inbred lines.)

ii) Pair 5-10 BC_2(S) plants (phenotypically similar to companion S_3 (M) with S_3 (M) plants to produce BC_3(S) and S_3(M_1) seed in a manner similar to that described above. Some single-cross hybrids (sterile hybrids, BC_2(S) lines) with similar but unrelated parents should be selected for testing as F_1 (sterile hybrid) seed parents to be used in three-way crosses.

iii) For a preliminary test of combining ability isolate an additional 10-20 S_3 (M_1) plants with one or more pollen-sterile (S) lines or F_1 hybrid seed parents (pollen parents) to produce experimental prototype hybrid combinations.

Fifth year:

i) Grow BC_3(S) alongside maintainer companion S_3(M_1) to produce BC_4(S) and S_3 (M_2) seed of the best inbred lines in a manner similar to previous year.

ii) Evaluate the prototype made in the previous year to identify the lines that show promise as parents.

iii) There may be an assortment of BC_3 lines originating about the same time, but from diverse origin. One or more of these male-sterile inbreds (BC_3) may be isolated with a proven pollinator to produce experimental F_1 hybrids. This provides for an early test of combining ability of new male sterile inbred lines.

iv) Make crosses of BC_3(S) plants with selected pollen parents to test new lines as potential seed parents.

Sixth year: Isolate maintainer with selected female parents to reproduce the best hybrids in preceding trials and to produce additional experimental combinations.

Seventh year onwards

   i) Test new hybrid combinations in observation plots, advance the best from preliminary trials to replicated trials at various locations.

   ii) Increase ($BC_5(S)$ and $S_3(M_3)$ of the maintainer) lines identified in replicated trials as potential hybrid parents.

   iii) Start pilot seed production of candidate hybrids to produce hybrid seed for commercial trials. Determine whether the seed yield is acceptable.

   iv) Distribute the seed for commercial trials in production areas using all standard production, harvesting, processing, packing and distribution procedures. Plant the parent lines in seed-producing area to provide hybrid seed for commercial trials and to evaluate seed-yield potential.

   v) Distribute seed for second year and conduct more extensive commercial trials in all producing areas where the candidate hybrid is likely to be used. Release a hybrid variety and its inbred parent components, if warranted by performance in trials.

SELF-INCOMPATIBILITY SYSTEM

The self-incompatibility character is used to enforce the cross-fertilization required in producing hybrid seed of cabbage, cauliflower, broccoli, Brussels sprouts and kale. Sibling plants of inbred lines that have been selected for homozygosity of an *S*-allele followed by selection for strong expression of this *S*-allele of self-incompatibility will not cross-fertilize each other. As a result there will be little selfed seed. (Dickson and Wallace, 1986).

When a homozygous *S*-allele line S_1S_1 is planted in rows alternating with another inbred with allele S_2S_2, the two are readily cross-fertilized. Fertilization will be by pollen carried form one inbred to the other by pollinating insects (mostly bees). The cross-compatibility between inbreds S_1S_1 and S_2S_2 assures the production of F_1 hybrid seed. Since both the inbreds are self-incompatible the seed produced on both of these lines is hybrid ($S_1 \times S_2$ or $S_2 \times S_1$) and possess similar characteristics. Thus, commercial seed of the same F_1 hybrid (single cross) can be harvested from both inbreds. The homozygous *S*-allele self-incompatible lines selected for opposite *S*-alleles from the same I_1 population are often used as two inbred parents of a single cross F_1. Top cross hybrids are produced by crossing a selected homozygous *S*-allele self-incompatible line (used as female parent) with a selected open-pollinated variety (used as male parent). A three-way-hybrid is produced by crossing a selected single cross F_1 (used as female parent) with another selected inbred line or open pollinated variety (used as male parent). Similarly, four-way-hybrids are produced by crossing two selected single cross F_1's (highly self-incompatible single crosses).

The breeding of hybrids using self-incompatibility thus requires development of inbred lines homozygous for the S-allele with strong expression of self-incompatibility.

Development of Homozygous S-Allelle Self-incompatible Inbred Lines

Incompatibility Specificities

Incompatibility specificities are controlled by one locus, called the S-gene. The number of S-alleles each giving one specificity needs to be identified.

GAMETOPHYTIC INCOMPATIBILITY

Self-incompatibility has gametophytic control when the haploid S-allele genotype of each pollen grain (male gamete) exactly indicates that gamete's expressed incompatibility specificity. In a heterozygous plant $(S_1 S_2)$ two types of gametes shall be formed and 50% pollen grains will have S_1 specificity and 50% S_2 specificity.

Whether the gametophytically controlled pollen grain will function compatibly or incompatibly in any given self- or cross-pollination depends on whether the same allele occurs in the female flower. If not, pollination will be compatible; if yes, regardless of whether the female plant is homozygous or heterozygous, pollination will be incompatible.

SPOROPHYTIC INCOMPATIBILITY

For sporophytic incompatibility one has to know whether the S-allele in the pollen grain also occurs in the female flower of this pollination. If not, pollination will be compatible; if yes, pollination may be compatible or incompatible depending on the following:

a) Whether the male plant is homozygous for one allele or heterozygous for two S-alleles. If heterozygous, which of the four levels of interaction between the two alleles, namely dominance $(S_1 < S_2)$, codominance, $(S_1 = S_2)$, mutual weakening (no action by either allele) or intermediate gradations (0-100% activity of each allele) characterises the incompatibility specificity of the pollen.

b) Whether the female plant is homozygous for one allele or heterozygous for two S-alleles? If heterozygous, which of the four levels of interaction (dominance, codominance, mutual weakening or intermediate activity) between the two S-alleles characterises the incompatibility specificity of the stigma.

Advance knowledge of whether a pollination will be compatible or incompatible requires knowing whether either or both plants are homozygous, and whether the two S-alleles of a heterozygous plant interact with dominance, codominance, mutual weakening or intermediate activities, and which of the alleles is dominant and which one is recessive.

Methods for Quantifying Self-incompatibility

The methods used for quantifying self-incompatibility are as follows. (Dickson and Wallace, 1986)

1) *Counting the number of seeds that develop to maturity after each specific self- or cross-pollination*: The principal disadvantages of this system are the time duration between pollination and maturity, and the number of seeds developing and ultimately reaching maturity may be reduced due to disease, water shortage, high temperature or other stresses. Thus seed counts at maturity often do not strictly reflect the intensity of expressed compatibility or incompatibility.

2) *Fluorescence test*: The ability of the fluorescence microscope to readily display those pollen tubes that have penetrated the style provides a direct measure of incompatibility that can be completed within 12-15 hours. It is adequate and convenient as well to pollinate on day 0. The pollinated flowers are then collected 16-30 hours later (day 1). On the same day, the excised ovaries are softened in 60% NaOH and placed in analine blue. At about 48 h (day 2) after pollination, the stigma and style are squashed on a microscope slide.

The aniline blue stain accumulates in the pollen tubes and fluoresces when irradiated with ultraviolet light. Therefore, with appropriate light filters under a fluorescence microscope, the pollen tubes are visible, whereas the background of the stylar tubes is largely unseen. Penetration of the style by none or few tubes indicates incompatibility, while penetration by many tubes indicates compatibility; penetration by intermediate numbers indicates intermediate strength of the expressed incompatibility/compatibility.

ADVANTAGES OF FLUORESCENCE TEST

The incompatibility data and attendant conclusions can be had within 2 days compared to the 60 days required for seed counts, or the 20-40 days if developing seeds are counted. Also, with the conclusions available additional pollinations can be specifically planned to verify a conclusion. This early acceptance facilitates moving on to work with other populations remaining flowers are sufficient to maximize implementation of the conclusions.

Generalized Procedure for Developing Homozygous S-Allele Genotypes (Inbred Lines)

Homozygosity for a single S-allele is an essential step in developing the inbreds to be used as parents in producing hybrid seeds. This is done in the following steps.

Step 1. (a) Select plants in open-pollinated varieties or hybrid populations on the basis of desirable characteristics. Individual plants selected in this manner are usually heterozygous for two S-alleles and represent random selection with respect to S-alleles present in the source population.

(b) Self each selected plant (I_0) by bud-pollination for maintenance and seed increase. Ensure that no outcrossing takes place during the bud-pollination. The seeds thus obtained are used for growing I_1 generation.

(c) Also self open flowers on each selected plant (I_0). Ensure that no outcrossing takes place during the self-pollination.

This is done to check seed set or pollen tube penetration from the open flowers. These two observations are used as a measure of intensity of self-incompatibility of the selected plant. If the selected plant is found to be compatible or weakly compatible all the resultant seeds from this plant, including those obtained from bud-pollination are discarded.

(d) Selfed-seed (bud-pollinated) of only those selected plants (I_0) are carried further which show strong expression of self-incompatibility.

Step 2. Grow the I_1 generation. This population contains three types of genotypes, namely, S_aS_a, S_aS_b and S_bS_b in 1:2:1 ratio. The question here is how to identify the genotype of an individual plant. This is done by making reciprocal crosses between the subsets of two of the I_1 sibling plants at a time.

An efficient procedure is to begin by reciprocally crossing a series of individual I_1 sibling plants. This procedure is continued until the three different genotypes becomes simultaneously evident. Thus the three possible genotypes will each be represented by one of the three I_1 plants.

A group of 11 plants from an I_1 population provides a 95% probability of having at least one plant each of the three genotypes, namely S_aS_a, S_aS_b and S_bS_b.

Reciprocal Crosses

It is expressed compatibility for one or both of the pairs of reciprocal crosses between two I_1 plants which is the first information that is required to identify the S-allele genotype of any and then all I_1 plants.

A reciprocal difference concomitantly indicates three facts:

(i) One of the I_1 plant (say Plant 1) is S_aS_a (homozygous recessive);

(ii) The plant 2, with its dominant S-allele phenotype is genotype $S_a < S_b$; and

(iii) The plant 2 is not S_bS_b.

A reciprocal compatibility demonstrates that one of the two I_1 plants (say Plant 1 or Plant 2) must be recessive genotype (S_aS_a) but cannot indicate which plant carries which of the two genotypes.

Differences in reciprocal pollinations and their interpretation

On the assumption[2] that each I_1 plant is self-incompatible, and assuming

---

[2] When the above assumption is not valid, the heterozygote will be of a type intermediate to the extreme types I, II, III, IV. Such intermediate types lack strong dominance or codominance in the heterozygote. Intermediate activities in the heterozygote by both S-alleles constitute mutual weakening. This weakening in the S-allele heterozygote is specific for given pairs of S-alleles. Such mutual weakening is indicated when many of the reciprocal crosses between I_1 plants give intermediate seedsets or penetrations of pollen grains into the stigmas that indicate neither definite incompatibility nor definite compatibility.

I_1 Populations involving S-allele interactions with mutual weakening of both S-allele activities in the heterozygote are discarded soon after it is determined to be so.

that each S-allele heterozygote will belong to either of the dominance or Codominance types, the differences from reciprocal crosses are interpreted as follows:

    (i) *Codominance × dominance and dominance × Codominance types of S-allele interaction. (Type II and Type III)*

In this type there is dominance of one of the two alleles of heterozygote in either the pollen or stigma, but not in both the organs.

(a) The observation of a reciprocal difference between any two plants of an I_1 population (Plant 1 x Plant 2 is incompatible, while Plant 2 x Plant 1 is compatible or vice versa) indicates that one of the two I_1 plants (Plant 1 or 2) is the genotype S_aS_a (recessive) while the other is a heterozygote S_aS_b.

(b) Make more reciprocal crosses between these plants to confirm the findings.

(c) Cross these two I_1 plants (Plant 1 and Plant 2) with additional I_1 plants. When it is found that one of these two plants (say Plant 1) is having reciprocal compatibility with a third plant (say Plant 3), it is positively established that Plant 1 is S_aS_a, since it was both reciprocally compatible with Plant 3. Plant 3 therefore must be S_bS_b genotype. Since it had a reciprocal difference in crosses with Plant 2, the Plant 2 therefore must be heterozygous S_aS_b.

To sum up

Plant 1 × plant 2	incompatible	$(S_aS_a × S_aS_b)$
Plant 2 × plant 1	compatible	$(S_aS_b × S_aS_a)$
Plant 1 × plant 3	compatible	$(S_aS_a × S_bS_b)$
Plant 3 × plant 1	compatible	$(S_bS_b × S_aS_a)$
Plant 2 × plant 3	incompatible	$(S_aS_b × S_bS_b)$
Plant 3 × plant 2	incompatible	$(S_bS_b × S_aS_b)$

with all three of the possible genotypes now identified the plant with S_aS_a genotype becomes the most efficient for identifying the genotypes of all the additional plants. It would behave as follows:

Known genotype S_aS_a	× unknown S_aS_a	Reciprocally incompatible
	× unknown S_aS_b	reciprocally different
	× unknown S_bS_b	reciprocally compatible.

(ii) *Dominance × Dominance type of S-allele interaction (Type I)*
In this type there is dominance for one of the two alleles of the heterozygote in both the pollen and stigma. This is therefore indicated when say plant 1 is reciprocally compatible with either plant 2 or plant 3 (say plant 2), that are reciprocally compatible with each other while the plant 1 is reciprocally incompatible with the Plant 3.

Plant 1 x plant 2	reciprocally compatible
Plant 2 x plant 3	reciprocally compatible
Plant 1 x plant 3	reciprocally incompatible.

In this situation one of the two (plant 2 or 3) is the homozygous recessive genotype S_aS_a and the other plant is either the heterozygous S_aS_b or homozygous dominant S_bS_b.

For further differentiation of the three genotypes it is necessary to grow I_2 generation.

In the I_2 generation all the plants having genotype S_aS_a or S_bS_b will breed true, while the heterozygote plant will segregate.

The true breeding dominant homozygous I_2 plants are identified by their incompatibility with about 3/4 of the plants in the I_2 population and compatibility with about 1/4 of the plants in the I_2 population (S_aS_a genotype). Similarly, the recessive homozygous (S_aS_a) will be indicated with compatibility about 1/4 of the plants in the I_2 population (S_bS_b genotype).

(iii) *Codominance x Codominance type of S-allele interaction (Type IV)*: Type IV has codominance in both pollen and stigma of the heterozygous S-allele genotype. Thus the heterozygote has strong activity by both alleles in both its stigma and its pollen. Type IV is therefore indicated when an I_1 plant is reciprocally incompatible with both of two I_1 plants that are reciprocally compatible with each other. The I_1 plant that is reciprocally incompatible with both reciprocally compatible plants is the heterozygous genotype. The two reciprocally compatible I_1 plants must be one plant of each of the two homozygous genotypes. These two plants can be arbitrarily assigned tentative S_aS_a and S_bS_b identities, because the alleles are codominant and there is no recessive versus dominant S-allele interaction. That is, both alleles are simultaneously expressed with near equal intensity; the genotypes of all of the other I_1 plants can now be designated after each plant of unknown genotype has been reciprocally pollinated to two of the three known I_1 genotypes.

Permanent S-allele identities

Most permanent S-allele identities (S_1, S_2... Sn) are assigned by the breeder. The National Vegetable Research Station at Wellesbourne, England, has a collection of S-alleles, which constitutes the internationally accepted nomenclature. A homozygous plant of tentative S_aS_a or S_bS_b genotype will be known to represent the breeder's S_3S_3 genotype when S_aS_a or S_bS_b is demonstrated to be reciprocally incompatible with the breeder's inbred of S_3 genotype. Similarly, it will be $S_{26}S_{26}$ of the international allele nomenclature when it is reciprocally incompatible with plants of known international S-allele genotype S_{26}. The S_aS_a and S_bS_b will be compatible with all plants having other S-allele genotypes.

Unknown alleles in plants of heterozygous S-allele genotype can usually be specifically identified by partial incompatibility with reciprocal crosses with the corresponding homozygous S-allele genotypes. Thus with dominance

202

the S-allele activity will be strongly expressed only in the stigma (type IV) or only in the pollen tube (type II) or strongly expressed in both (type I). Both alleles will be strongly expressed in both the stigma and pollen for type IV. Alternatively, with recessiveness the allele will be weakly expressed in the pollen but strongly in the stigma (type III), or weakly in the stigma but strongly in the pollen (type II), or weakly expressed in both stigma and pollen (type I). Because of these complexities, assignment of permanent S-allele designations is most easily done using plants known to be homozygous $S_a S_a$ or homozygous $S_b S_b$. (Dickson and Wallace, 1986).

22

Mutation Breeding

Mutation breeding refers to isolation and selection of desirable induced mutations in segregating generations that follow a mutagenic treatment of seeds or other plant parts. The selected mutants are used in breeding programmes and are also released for commercial cultivation depending on their usefulness.

APPLICATION AND USES OF MUTATION BREEDING

Mutation breeding is usually practised as a standard method of plant improvement in vegetatively propagated crops, such as ornamentals, fruit and forest species, etc. In ornamental species, where any novelty will be of commercial interest, mutation breeding is especially useful. Also, in situations wherein a plant breeder is confronted with little genotypic variation, it appears to be the only method available to him to bring about desired improvement.

In sexually propagated economically important agricultural crops the rate of success of mutation breeding in terms of efforts made and commercial varieties released, has been extremely low, however. Notwithstanding this extremely low rate of success, mutation breeding is still considered a useful complementary method to conventional breeding even in these crops.

MAIN FEATURES OF MUTATION BREEDING

1. Selection of Mutagen

The frequency and spectrum of mutations (gene mutations, chromosomal aberrations) depends on the choice of mutagen and the dose applied. Among physical mutagens, X-rays and gamma rays (sparsely ionizing radiations) are widely used. The ultraviolet light which has low penetration power may be effectively used with materials such as pollen. Densely ionizing radiations, such as thermal and fast neutrons, cause more chromosomal aberrations. Chemical mutagens produce a higher rate of gene mutations. However, these mutagens present particular problems, such as uncertain penetration to the relevant target cells, poor reproducibility, persistence of the mutagen or its metabolites in the treated material and finally the risk of safe handling.

Table 22.1 presents a list of mutagens together with doses used by plant breeders in some of the important crops.

Table 22.1. Mutagens and recommended doses for inducing mutations in some crops

Crop	Material	Recommend Mutagen	Dose
A. Sexually propagated crops			
Wheat	dry seeds	gamma rays	10–25 k rad
		fast neutrons	600–800 rad
		EMS	3.76%
	Pollen	gamma rays	0.75–3.00 k rad
Barley	dry seed	X-ray	10–22 k rad
	pollen	gamma rays	0.6–2.0 k rad
Oats	dry seeds	X-rays	12–24 k rad
Rice	dry seeds	X-rays	14–28 k rad
Corn	dry seeds	gamma rays	0.6–2.0 k rad
Pea	dry seeds	X-rays	10 k rad
		DES	2 per cent
	pollen	gamma rays	600 rad
Chick-pea	dry seeds	X-rays	10–16 k rad
Groundnut	dry seeds	gamma rays	20–30 k rad
Soybean	dry seeds	gamma rays	10–20 k rad
Tomato	dry seeds	EMS	0.8 per cent
Pepper	dry seeds	fast neutrons	2.4 k rad
		gamma rays	14–22 k rad
B. Asexually propagated crops			
Potato	tubers	EMS	100–500 ppm
Sugar-cane	single-budded sets	gamma rays	2–6 k rad
Sweet potato	cuttings	gamma rays	20 k rad
		EI	0.5 per cent
Cassava	dry seeds	gamma rays	5–20 k rad
	stem	gamma rays	< 1.5 k rad
	pollen	gamma rays	2–6 k rad
Coffee	dry seeds	gamma rays	2.5–20 k rad
Apple	one-year-old dormant shoots	gamma rays	2.5 k rad
Citrus	Dry seeds	gamma rays	1.0–2.5 k rad
	buds	gamma rays	5 k rad
Grape	dormant buds	gamma rays	2.5–3.5 k rad
		EMS	0.15–0.20% k
Dahlia	dormant tubers	X-rays	1–4 k rad
Chrysanthemum	rooted cutting	X-rays	1200 rad
Carnation	nodal stems	X-rays	1.5–2.0 k rad

2. Starting Material

Choice of Genotype

The choice of genotype for mutagenic treatment should be based on familiarity with the available germplasm. The best varieties are usually, selected. In some instances F_1 is used. The obvious advantage of using F_1 is that two different genomes are exposed to a mutagen at the same time, which may increase gene recombination and thus reveal a wider spectrum of gene mutations.

Choice of Plant Material

In seed-propagated plants, the physical or chemical treatment of dry seeds is the most convenient and practical. The difficulty, however, is that a seed is a multicellular organized structure, in which a mutation induced in a single cell gives rise to a chimeric plant (M_1) which will have to face both diplontic[1] and haplontic[2] selection pressure to be included in tissues forming seed on M_1 plants. In the case of vegetatively propagated crops the material for treatment should come from a disease-free plant. The vegetative propagates, such as bud woods, scions, cuttings, tubers, bulbs, or various explants of plant organs (shoot tips, meristems, epidermis, ovaries, nucellar tissues, etc.) are subjected to mutagenic treatment. These organs consist of many cells and after mutagenic treatment give rise to a chimeric structure consisting initially of genetically different cells in the form of a sector within a histogenic layer (mericlinal chimera). A mericlinal chimera occurring in the M_1V_1 generation may convert into a periclinal chimera with a uniformly mutated cell layer. To favour this development and eventually obtain non-chimeric mutant plants, the following established techniques may be applied:

i) Pruning or cutting back the primary shoot and using the axillary buds in propagation.

ii) Repeated budding of those buds coming from preformed primordia and located generally in the basal and middle part of the M_1V_1 shoot.

iii) In vitro multiplication of uniformly mutated plants through propagation of axillary buds

iv) Adventitious bud technique. It is assumed that adventitious buds develop from a single cell and hence an adventitious bud developed on a mutagenized leaf, stem or root cutting is expected to give rise to a non-chimeric homohistont mutant. However, this technique is not applicable to all plant species.

---

[1] Diplontic selection refers to competition between mutated and normal tissue during the vegetative stage or during ontogenic differentiation of reproductive organs. When the mutant sector is large enough, or is able to compete, only then can it enter into the formation of reproductive parts.

[2] Haplontic selection refers to similar competition between mutated and normal pollen during fertilization of the ovule. When the mutant pollen is able to compete, i.e., is able to fertilize the ovule, only then can the mutant be recovered in M_2 or later generations.

3. Nature of Induced Mutations

The evidence for mutagenic specificity is very limited and control over the mutation spectrum is very limited. Some specificity of action is shown by different mutagens in the ratio of point mutations to chromosomal aberrations. A tendency to affect particular heterochromatic regions is typical of certain chemical mutagens. In higher plants different loci may respond differently to mutagens. Thus one particular mutagen tends to produce only part of the entire spectrum of possible genetic changes, and the type it produces is a reflection of its energy potential.

Densely ionizing radiations (neutrons) largely induce chromosomal alterations (segmental rearrangements, losses and so on) and a few point mutations. Some of these point mutations may be produced by direct mechanical action, while others may be caused by secondary and indirect action of chemicals produced by the action of ionization of the protoplasm. The more sparsely ionizing radiations (X-rays, gamma rays) induce a larger number of point mutations. UV light has even lesser tendency to induce chromosomal alterations and a still higher proportion of point mutations. The chemical mutagens, namely mustard gas and its derivatives produce some chromosomal alterations and some point mutations. The mutagenic nucleosides and other chemical mutagens with special effects on nucleic acids appear to produce mutations by special chemical affinities. Other factors, namely metabolism of the treated plant, pretreatments (if any), or change in mineral nutrition may also influence or alter the genetic response to a particular mutagen.

4. Frequency of Desirable Mutations

Deleterious mutants outnumber desirable mutants by a factor of several hundred to one. The recovery of desirable mutants at an economic cost is an important consideration. It is greatly influenced by the manner in which treated material is handled and the method of selection applied.

GENERALIZED PROCEDURE OF MUTATION BREEDING

A. Sexually Propagated Crops (Seed-propagated Crops)

Step 1. Mutagenic Treatment

Seeds, shoot tips, pro-embryos, gametes, zygotes, single cells (*in vitro* propagation) of selected genotypes are treated with mutagens, such as, gamma rays, X-rays, fast neutrons, thermal neutrons or chemical mutagens. The choice of mutagen and the dose depends on the kind of material to be treated and the availability of mutagen. (see Table 22.1 for recommended mutagens and recommended dose for mutagenic treatment.)

Step 2. Grow M_1 Generation

Treated material is grown in isolation or bagged[3] along with control. There are many common effects observable in the M_1 generation. Look for chimeric plants and non-chimeric heterozygous plants. Depending on the objective of breeding, decide carefully whether to carry all M_1 harvested seeds, or a single seed per plant, or one or more reproductive organs (spike, fruit, pod capsule, etc.) per M_1 plant and whether the M_2 generation is to be grown in bulk or in individual progenies. Harvest and keep the material accordingly.

Step 3. Grow M_2 Generation

i) Grow the material raised in M_1 generation. For diploid species, M_2 progenies of 20 plants should be sufficient to obtain a certain number of segregants homozygous for a mutant trait. For polyploids, and in general for quantitatively inherited characters, it is advantageous to allow more recombination to take place by delaying selection to the M_3 or M_4 generation.

ii) In diclinous species such as corn, the chimeric structure of the M_1 plant does not allow self-pollination within a mutated sector and therefore no segregation of homozygous mutants in the M_2 is observed. A practical procedure here would be an interpollination of M_1 plants, which would bring all induced mutations into heterozygous M_2 plants. An alternative would be the mutagen treatment of pollen and pollination of non-treated mother plants, which would lead to non-chimeric M_1 plants heterozygous for induced mutations. These plants should then be self-pollinated as usual to obtain homozygous mutants. In cross-pollinated species that are self-incompatible mutation breeding should involve some form of recurrent selection.

iii) Look for possible segregants, identify induced mutations and harvest seeds from mutated plants.

Step 4. Grow M_3 Generation

Grow the material harvested in M_2 generation. Look for segregants. Verify mutants selected in M_2. For indirect use in a breeding programme crosses of desirable mutants with other breeding material may be made.

Step 5. Grow M_4 Generation

i) Selected mutants are included in a station trial for preliminary evaluation of agronomic value of selected mutants.

ii) Evaluate the genetic stability of the selected mutants.

iii) For indirect use in a breeding programme crosses of desirable mutants with other breeding material may be made.

---

[3] M_1 plants show some degree of sterility. This increases the potential outcrossing even in species classified as cleistogamous. Therefore, particular measures, such as spatial isolation or bagging are necessary in the M_1 generation.

Steps 6 to 9. Grow M_5 to M_8 Generation

Evaluate the selected stable mutants recombinant lines in multilocation trials. On the basis of performance, selected mutants may be directly released as new improved varieties and or indirectly used for incorporating the desirable trait they carry into other breeding material of promise.

Step 10. Official Testing and Release of Mutant Variety(ies)

B. Asexually Propagated Crops

Step 1. Mutagenic Treatment

Plant organs, such as shoot meristems, bulbs, tubers, cuttings, etc. of selected clones are treated with mutagens, such as X-rays or gamma rays, or with chemical mutagens. The choice and dose of mutagen depend on the kind of material to be treated and the availability of the mutagen. (See Table 22.1 for recommended dose for mutagenic treatment.)

Step 2. M_1V_1 Generation

Look for chimeric development from apical and axillary bud meristems (mericlinal, sectorial). Cut back the M_1V_1 shoot, bud grafting, etc.

Step 3. M_1V_2 generation

Grow M_1V_1 material. Mutant selection usually starts in M_1V_2 generation. Identify periclinal or uniformly mutated scion, branch, tree, etc. Isolate induced somatic mutations and establish clones. Cut back non-mutant shoots from chimeric plants.

Step 4. M_1V_3 Generation

Grow M_1V_2 material. Verify genetic uniformity within a mutant clone. Further isolate somatic mutations. Vegetatively propagate mutant plants. Preliminary evaluation of mutants may also be done at this stage.

Steps 5 to 10. M_1V_4 to M_1V_9 Generations

In the M_1V_4 and subsequent generations the stability and uniformity of the clone is assessed. Only uniform and stable clones should be planted for agronomic evaluation.

a) The performance of mutant clones is evaluated. The improved mutants are released for commercial cultivation.

b) During this period evaluation of sexual transmission of the mutations is also done for indirect use of mutants in a breeding programme.

23

Polyploid Breeding

Polyploid breeding refers to induced chromosome manipulation. Success is wholly dependent on the control of chromosome pairing and recombination in polyploids and their hybrid derivatives. Breeding strategies for transferring genes across ploidy levels depend on their origin. From a practical viewpoint of plant breeding we shall therefore discuss the various breeding approaches separately for autopolyploids, allopolyploids and then the gene transfers involving chromosomes or chromosomal segments.

AUTOPOLYPLOIDS

Autopolyploidy originates from doubling the chromosome number of a diploid species, or a hybrid between races of the same species, resulting in two pairs of homologous chromosomes.

Types of Autopolyploids

Monoploid, that is, only a single set of chromosomes is present. A haploid (*n*).

Triploid, that is, three full sets of chromosomes are present (3*n*).

Tetraploid, that is, four full sets of chromosomes are present (4*n*).

Similar terminology is applied to many other combinations.

General Features of Autopolyploids

Autopolyploidy is generally associated with larger vegetative parts and reduced fertility (lower seedset). *Gigas effects* are mainly attributable to increased cell size. There are, however, limits to the *gigas effect* and beyond a certain point ploidy may have an adverse effect. The varieties or strains within the same species possessing the same number of chromosomes and similar sized chromosomes may differ in their response to tetraploidy in respect of vegetative growth and seed fertility.

Autotetraploids are usually slower in all phases of growth and take more time to germinate, flower and maturity. This is mainly attributed to lower rate of metabolic activities, reduction in rate of cell division, amount of growth hormones, lower transpiration rate, etc. A different photoperiod may be necessary to improve seed setting.

Genetics of Autopolyploids

The genetics of autotetraploids is rather complicated primarily due to seed sterility and their cytological behaviour. Segregation ratios vary according to chromosome pairing (random chromosome association at meiosis and random assortment of chromatids), degree of quadrivalent formation, distance of the gene from the centromere, and nature and number of chiasmata. Also, the different genes may reach their separation point of phenotypic expression at different levels. In other words, genes may differ in the degree of cumulative action. Genes which are completely dominant over their partners in equal doses, may exhibit incomplete dominance in hybrids where three doses of recessive are present to one of dominant. A disturbance of genic balance may also occur due to cumulative action of some genes and non-cumulative action of others, which may cause differential reaction. Autopolyploids of self-incompatible diploids do not exhibit self-incompatibility. Lewis (1949) explained that this may be either due to the interaction between two S-alleles in the diploid pollen resulting in weakening of the inhibitory stimulus, or due to the behaviour of one of the alleles as completely dominant over the other, with the result that pollen containing these two factors behaves as if it were homogenic to the dominant allele and grows normally in a style containing the other allele.

Seed Sterility

Seed sterility may be due to following causes :

i) *Cytological irregularities* : Seed sterility or lower seedset can be traced back to non-viable, unbalanced gametes resulting from the disjunction of multivalents formed at meiosis in almost all tetraploids. Other meiotic abnormalities, namely lagging of chromosomes at anaphase I and II and spindle abnormalities may also result in unbalanced and less viable gamete formation.

ii) *Genetic causes* : Seed sterility may be due to genetically controlled physiological factors of an unknown nature. The differential reaction of genes may upset the delicate genic balance leading to seed sterility in tetraploids.

iii) *Physiological disturbances* : Physiological disturbances brought about by the change in cell surface-cell volume relationship may be responsible for lower seedset. There may be reduction in number of flowers produced in the whole season, number of pollen grains produced per anther, and increase in the percentage of malformed anthers, flower shed, etc. This may be due to low nutrient supply to reproductive tissues.

iv) *Fertilization of ovules* : Rajan and Ahuja (1956) suggested that seed sterility in low-fertility tetraploids is due to lack of fertilization in normal embryo sacs. Pre- and post- fertilization abnormalities thus also contribute to seed sterility.

METHOD OF BREEDING INDUCED AUTOTETRAPLOIDS

Factors to Consider

Levan (cited by Dewey, 1980) concluded that a crop most amenable to improvement through chromosome doubling should have the following features:

1) low chromosome number,
2) be harvested for vegetative growth, and
3) be cross-pollinated,

Dewey added:

4) have perennial habit, and
5) have ability to reproduce vegetatively.

Selection of Material for Chromosome Doubling

A sufficient number of tetraploid lines should be synthesized from diverse diploid lines to insure allelic diversity in the breeding programme. Single and double crosses of the diverse 4 x lines are required to produce varieties with maximum hybrid vigour.

Colchicine Treatment and Its Mode of Action

Colchicine treatment for inducing autopolyploidy, i.e., doubling the chromosome numbers, is by far the most effective treatment. Colchicine is water soluble and produces a high proportion of polyploid cells at concentrations that are non-toxic to a wide variety of plant species. When colchicine is applied to plants in a lanolin paste, dripped onto a cotton pad held against meristematic tissue or applied to plants in several other ways, spindle fibres fail to form in many cells. The chromosomes do not line up on the equatorial plate and divide without moving to poles. The duplicated chromosomes then go through a regular telophase, and a membrane forms around the nucleus with a doubled chromosome number.

Table 23.1 gives the details of colchicine treatment applied to some of the important crops.

Proper aftercare of colchicine-treated material and isolation of the crop to avoid genetic contamination are the basic requirements. A selection strategy for each crop is imperative if plants with high fertility and ploidy level are desired.

Selection Strategies

Direct selection for ploidy level based on chromosome counts is rather laborious. The possibilities of indirect selection, such as stomatal length in C_0 and C_1 generations that could serve as a reliable criterion for selection at the ploidy level therefore need to be explored. In vegetatively propagated crops heterozygosity of elite genotypes can be fixed instantly but in seed crops strategies based on maximizing heterozygosity within populations have to be implemented. Aneuploidy in advance generations is another important factor to consider in devising breeding strategies for induced autopolyploids.

Table 23.1. Colchicine treatment for production of induced autopolyploids

Crop	Plant material	Colchicine Conc. (%)	Duration of treatment (h)
Pearl millet	seedlings	0.1 to 1.0	The aqueous solution is forced into the tissue under reduced pressure; roots are least brought into contact with the colchicine solution
Cotton	Shoot tips of young plants (4-5 leaves)	0.08	at intervals for 12 hours
Chic-pea	germinating seeds (radicles just emerging)	0.25	30 minutes
Chilli	seeds	0.1	8 days
Sesame	Vegetative buds	0.4	Sprayed twice daily for 3 alternate days
Egyptian Clover	4-day-old shoots	0.1	8 hours
Melilotus sp.	4-day-old seedlings	0.05	4 hours
Tobacco	growing vegetative buds	0.4	Wetted twice a day for 10-15 minutes, for one week

Schwanitz (1951) has emphasized that selection for smaller cell size and increased nuclear size and volume should be practised for improved fertility of tetraploid strains. Selection for 5-6 generations is usually sufficient to achieve the desired type with tetraploidy and high fertility (seed-set).

Parthasarthy and Rajan (1953) used mass pedigree selection for improvement of fertility in tetraploid Brassica campestris var. toria. Plants were selected on the basis of certain previously determined minima for certain characteristics contributing to yield and mixed seeds grown from them as an 'elite population' in the next generation with the necessary spatial isolation to avoid genetic contamination. Weak segregants were eliminated before flowering. Generation after generation selection was practised as in the previous year and seeds of selected plants bulked and sown as the 'elite population' the next year. They were thus able to obtain in three or four generations, a tetraploid population with a fertility distribution curve very nearly equal to that of a diploid variety.

Handling of Breeding Material

The other aspects of handling the material are similar to any other breeding procedures discussed in earlier chapters.

Application and Uses of Autopolyploids

1. *Triploid sugar-beets :* Triploid sugar-beets are prevalent in Europe, but their superiority over diploids is not universal. Triploidy appears to be the optimum levels of polyploidy. Triploid roots are longer compared to diploid varieties and also yield more sugar per unit.

2. *Rye (Secale cereale)* : Among grain crops, rye is the only crop to be successfully developed as an autotetraploid. Tetraploid rye has larger kernels, superior ability to emerge under adverse conditions and higher protein content.

3. *Seedless watermelons (triploids)* : The yield of triploids and consumer acceptance has been good in Japan and the USA. However, it is necessary to interplant diploid, plants among triploids as pollinators in a commercial field for fruit setting (1 diploid: 5 triploids).

4. *Tetraploid grapes* : Tetraploids have large berry size and fewer seeds per berry.

5. *Forage crops* : Swedish tetraploid strains of alsike and red clover have given higher hay yields than corresponding diploids.

6. *Ornamentals* : Induced autopolyploidy has been most successful in ornamentals, for example snapdragons. In such crops novelty itself is a virtue. Autopolyploids may have bigger flower size, longer blooming period and relatively longer lasting flowers.

The outlook for utilization of autopolyploidy in crop species where vegetative parts are important is better, for example, in medicinal plants.

ALLOPOLYPLOIDS

Allopolyploidy usually involves hybridization between diverse species followed by doubling of the chromosome number and thus combines two genomes. These are also known as amphidiploids. The primary objective of allopolyploid (amphidiploid) breeding is to combine complementary characters of two different species into a new, synthesized species. The progenitor species may be closely related or show a high level of divergence. The net result of interspecific hybridization and polyploidization is that the genomes of distinct species are combined in the allopolyploid.

Fertility and stability of induced amphidiploids depends on the extent of chromosome homology between the genomes of constituent species. When the two genomes are highly divergent, as indicated by little pairing in the undoubled hybrid, induced amphidiploids are often fertile and reasonably stable, because the chromosomes tend to form bivalents fairly regularly at meiosis, and disjunction is often normal. When the genomes are partly divergent, as judged from considerable pairing in the undoubled hybrid, fertility and stability are often low in induced amphidiploids, because of irregular pairing among incompletely homologous chromosomes leading to unequal partition of the chromatids to gametes. Besides cytological behaviour, the physiological disturbances associated with genetically controlled imbalance may also be involved.

Uses and Application of Allopolyploid Breeding

Limited success has been achieved so far. Only triticale (wheat-rye amphidiploid) and a tetraploid between Italian rye grass and perennial rye-grass have been grown in farming systems to a rather limited extent. Poor success in allopolyploid breeding is mainly due to plant's inability to stabilize meiotic behaviour to a level comparable to that found in natural allopolyploids. Irregular meiosis arises mainly as a result of homoeologous chromosome pairing leading to the formation of unbalanced gametes. The major issue for success in induced allopolyploids therefore, is regular diploid-like chromosome pairing. If regular meiosis could be achieved it could be successfully translated into developing varieties of commercial interest.

Gene Transfer Involving Chromosomes or Chromosomal Segments Using Aneuploidy

Induced allopolyploids (synthesized allopolyploids), however, provide an effective bridge for controlled introgression of desirable gene(s) into a cultivated species. The objective in such gene transfers is to introduce the smallest possible segment of the chromosome of the alien genome into a cultivated species, so that the resultant genotype is agronomically more acceptable. In other instances wherein the possibilities of recombination between the alien genome and cultivated species are limited, the whole chromosome of the alien genome carrying the desired gene(s) is introduced into the cultivated species either through addition or substitution.

Types of Aneuploids

Aneuploidy involves an unusual number of repetitions (plus or minus) of one or a few particular chromosomes in the normal set of somatic chromosomes for species. Many aneuploid combinations are possible.

Nullisomic : both members of one pair of chromosomes are missing from the normal set of somatic chromosomes for a species.

Monosomic : one chromosome is missing from the normal set of somatic chromosomes for a species.

Double monosomic : two non-homologous chromosomes are missing from the normal set of somatic chromosomes for a species.

Trisomic : two sets of chromosomes present, plus a single extra chromosome, that is, one particular chromosome is present in triplicate.

Double trisomic : two sets of chromosomes, plus extra non-homologous chromosomes, one each of two different pairs, that is, two particular chromosomes are present in triplicate.

Tetrasomic : when two extra chromosomes of one particular chromosome are present.

Monosomic-trisomic : one chromosome of a particular pair of the complement is missing, and an extra chromosome of another pair of the complement is present.

Chromosome Addition

Production of chromosome addition lines : The synthesized allopolyploid (F_1) is propagated by back-crossing to cultivated species to develop single chromosome addition lines. F_1 is used as the seed parent and the cultivated species is used as the pollen parent. This is because male gametogenesis is more easily disturbed by chromosomal or genic disharmonies than in the female gametophyte. Further backcrosses invariably produce single chromosome addition lines in the absence of recombination. The line carrying the desired chromosome is screened out and selected for further use.

Chromosome addition lines may not be sufficiently stable, however. Instability is correlated with the failure of the pair of alien chromosomes to synapse in a small proportion of pollen mother cells. The unpaired chromosomes usually fail to become incorporated into the second telophase nuclei and form micronuclei and tetrads. This leads to the formation of haploid gametes which have a selective advantage over the normal $n + 1$ gametes pr·duced by the addition lines, and monosomic addition lines ($2n + 1$) appear in the progeny. On selfing these monosomic addition lines produce up to 90% euploid progeny that do not express the transferred character. This causes variation that may not be acceptable for release as a commercial variety.

Chromosome Substitution

An alternative to adding an alien chromosome is to substitute a chromosome of the recipient species (cultivated species) with an alien chromosome from donor species. Availability of a series of aneuploid lines and chromosome addition lines is necessary to produce specific substitution lines.

Cytogenetics of Aneuploids

Usually two members of a pair of homologous chromosomes regularly segregate during meiosis to give a haploid set of chromosomes in each gamete. Sporadic failures of a chromosome to pair initially, and with a random passage to one or the other of two poles may occur, or passing of both chromosomes into the same nucleus, imbalanced gametes (gametes deviating from the normal haploid complement) arise. Such gametes on union with normal gametes give rise to aneuploids, namely nullisomics ($2n - 2$), monosomics ($2n - 1$), double monosomics ($2n - 1 - 1$), trisomics ($2n + 1$), double trisomics ($2n + 1 + 1$). Monoploids (haploids) and triploids are other sources of aneuploids. Trisomics can more readily be obtained by selfing triploid plants, or by crossing a diploid with a triploid. The irregular distribution of chromosomes in a triploid is such that imbalanced gametes are frequently formed. Secondary trisomics arise from primary trisomics. Tertiary trisomics have an extra chromosome which is made up of parts of two non-homologous chromosomes.

Monosomics produce two types of gametes, n and $n - 1$. The odd chromosome passes at random to either pole in meiosis, but frequently it will lag at anaphase and not be included in either daughter nucleus. For this reason gametes with $n - 1$ chromosomes are more frequent than gametes with n chromosomes. This bias towards $n - 1$ gametes is usually not reflected as strikingly in the zygotic chromosome numbers, because gametes with $n - 1$ chromosomes often do not function, and also the zygotes with $2n - 2$ chromosomes (nullisomics) may not be viable. Thus most of the progeny of monosomics are either normal diploids or monosomics. Trisomics are relatively more stable compared to monosomics. Tetrasomics $(2n + 2)$ often behave more regularly than the aneuploids with an odd number of chromosomes. Nevertheless, quadrivalents may not always be formed or disjunction may not always be regular and hence these are also not genetically stable.

Production of specific chromosome substitution lines

1) Specific chromosome substitution lines are developed by crossing the monosomic line $(2n - 1)$ with a disomic addition line $(2n + 2)$ (Fig. 23.1). The monosomic line produces two types of gametes, i.e. with n and $n - 1$ chromosome numbers. The disomic addition line produces only one type of gametes $(n + A)$. Here, A refers to the added chromosome in the chromosome addition line. F_1 resulting from the union of $(n - 1)$ gamete of the monosomic line with the gamete $(n + A)$ from the chromosome addition line, is selected and again crossed with the disomic addition line. F_1 produces 4 different types of gametes, namely $(n - 1)$, (n), $(n - 1 + A)$ and $n + A$. The hybrid resulting from the union of gamete with $(n - 1 + A)$ and the gamete $(n + A)$ from the disomic addition line is selected. This plant is selfed and used as a disomic substitution line.

2) Intervarietal chromosome substitution lines may be produced by using the monosomic series through a backcross programme. After 6-8 backcrosses the genotypic background of the variety (selected for substitution) is adequately recovered. Comparisons between the recurrent variety and the substitution line developed will provide a precise assay of the genes located on the substituted chromosome.

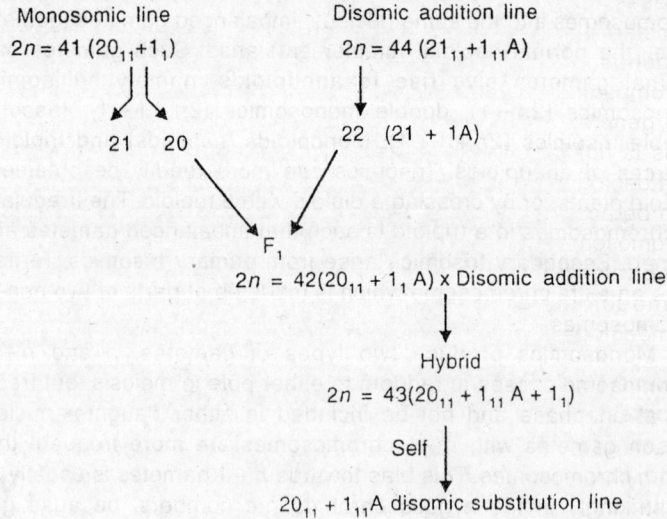

Fig. 23.1. Crossing scheme for development of chromosome substitution lines in wheat (from Thomas, 1993)

Practical Application of Chromosome Addition/Substitution

Whole-chromosome substitutions or additions have usually been found to be unsatisfactory from a practical viewpoint, apparently because so much genetic material is introduced from the donor species that undesirable characters are brought along with the wanted characteristics. There is only one disomic substitution line known as *weique* (Wienhues, 1965) which proved to be of commercial value and is being grown as a commercial crop. In this variety a pair of *Agropyron intermedium* chromosomes carrying black rust resistance replaced a pair of wheat chromosomes.

Mesbah et al. (1997) used sets of *Beta procumbens* (2n = 18) and *B. patellaris* (2n = 36) derived monosomic additions (2n = 19) in *B. vulgaris* subsp. *vulgaris* (cultivated beet, 2n = 18) for chromosome localisation of resistance genes of beet cyst nematode (*Heterodera schachtii*), *Cercospora beticola* (a foliar disease of beet) and *Polymyxa betae* (a soil borne fungal vector for beet necrotic yellow vein virus causing rhizomania). They reported that genes conferring full resistance to *H. schachtii* in *B. patellaris* are located on chromosome 1.1 and the other tested chromosomes of *B. patellaris* are not involved in the expression of resistance. High levels of resistance to *C. beticola* were observed in *B. procumbens* and *B. patellaris* was not found in any of the monosomic additions tested. The *B. patellaris* dervied monosomic additions showed that families of group 4.1 had strong partial resistance to *P. betae*, while the addition from group 8.1 appeared to be completely resistant to the pathogen.

Transfer of a Segment of a Chromosome

Where genetic recombination is limited, for example in several interspecific crosses, the target gene(s) may be introgressed into cultivated species through transfer involving small segments of the alien chromosomes carrying the target gene(s). When the alien chromosome is homologous with one or more chromosomes of the cultivated species, chromosome pairing is normal and the gene(s) may be transferred using any conventional breeding procedure. This is not the case, however when the chromosomes of alien species are homoeologous, and the pairing does not take place. In such situations pairing can be accomplished through manipulation of gene controlling pairing of the chromosomes. For example, in wheat the pH gene on chromosome 5B controls the bivalent pairing. Manipulation of the 5B system relaxes the restriction on homoeologous chromosome pairing and allows pairing between homoeologous chromosomes.

GENETIC MANIPULATION OF pH LOCUS

Suppression of pairing between homoeologous chromosomes is reportedly due to a recessive gene pH on chromosome 5B. Introduction of dominant pH gene from *Aegilops speltoides* to the wheat complement enabled pairing between both homologous and homoeologous chromosomes. Consequently, the recombination between the alien chromosome carrying the desirable disease resistance and homoeologous wheat chromosome took place. The first successful transfer of yellow rust resistance from *Aegilops comosa*

($n = 14$) to wheat ($n = 42$) based on modifying the 5B system was in the development of wheat variety *Compair* by Riley, Chapman and Johnson (1968). By repeated backcrossing of the hybrid wheat variety *Chinese spring* x *Aegilops comosa* using Chinese spring as the recurrent parent, they isolated a monosomic line which contained a full complement of Chinese spring chromosomes and one *Ag. comosa* chromosome. However, the transfer of *Aegilops comosa* genes to wheat could not be achieved, because the Chinese spring complement carries the pH locus which suppresses pairing and chiasma formation between homoeologous chromosomes. To overcome this problem they crossed monosomic alien addition line ($2n = 43$) with *Aegilops speltoides*. In the hybrids meiotic pairing took place between homoeologous chromosomes of *wheat* and *Aegilops speltoides*. These 29 chromosome hybrids were backcrossed to wheat variety Chinese spring for three generations and each time selection for disease resistance was done. A resistant derivative having 21 bivalents was finally obtained in the backcross progenies. This plant was heterozygous. Selfing produced homozygous stock, named *Compair*. Compair when crossed with other commercial varieties gave 21 bivalents in the hybrids and thus could be used in transferring resistance to other wheat varieties.

Sears (1977) described the method of induction of homoeologous chromosome pairing through the deletion of 5B and as a means for transferring genes for leaf rust resistance from *Agropyron elongatum*. Using substitution lines of chromosomes 3 *Ag* and 7 *Ag* lines Sears was able to establish hybrids that were heterozygous for the *Ag* chromosomes nullisomics, for 5B and trisomics for 5D. The extra dose of 5D compensates for the absence of 5B. There was evidence of homoeologous chromosome pairing and a range of transfers were identified which included variable length of 3 *Ag* and 7 *Ag*. The smaller the segment of *Ag* chromosome, the greater the utility in breeding since this reduces the chances of introducing deleterious genes with rust resistance. Repeating another cycle of induced homoeologous pairing of chromosomes resulted in more acceptable transfers by reducing the length of the segment of *Ag* chromosome through further recombination.

IRRADIATION INDUCED TRANSFERS

A mutation at the pH locus (which increases homoeologous chromosome pairing) reduces the effort required. Either the F_1 hybrids between wheat and the alien spp. or the monosomic/disomic alien addition line may be subjected to irradiation or chemical mutagenesis with the hope that mutation may be induced at the pH locus on 5B chromosome. Reports of such mutations in F_1 hybrids between wheat and alien spp. are available. Such a mutation will allow pairing between wheat chromosomes and the alien chromosomes. These hybrids showing homoeologous pairing may now be subjected to a back-crossing programme, selecting for disease resistance each time.

Sears (1956) transferred a segment from *Aegilops umbellulata* chromosome to wheat chromosome 6B for incorporation of leaf rust resistance in wheat. Knott (1961) produced a number of translocations, carrying stem rust resistance through the irradiation technique. However, only those translocations transmitted normally through the egg and the pollen which were located on wheat chromosome 6A. Later, it was established that the *Agropyron* chromosome carrying stem rust resistance belonged to the homoeologous group 6. In the case of 'translocation-4' (Agatha) obtained by Sharma and Knott (1966), brown rust resistance was transferred on the wheat chromosome 7D. It was later established that the *Agropyron elongatum* chromosome involved in this case belonged to the homoeologous group 7. However, in the case of *Transec* derived by Driscoll and Jensen (1964), transfer of a segment of rye chromosome belonging to homoeologous group 2 (2R) was brought about on wheat chromosome (4A) belonging to the homoeologous group 4.

In the reported cases of mutations induced in the F_1 hybrids, mutation at the 'pH' locus could not be recovered and therefore, such mutation could not be utilized in subsequent programmes. This mutation, now available, can be utilized for promoting chromosome pairing between wheat and alien chromosomes. Such a technique will eliminate the necessity of use of *Aegilops speltoides* or the 5B deficient lines.

FUTURE PROSPECTS OF POLYPLOID BREEDING

Failure to achieve a regularity of meiotic behaviour in induced autopolyploids and allopolyploids comparable to natural polyploid species sets the practical limitations in their successful development as crops. The approaches to gene transfer as discussed in this chapter may be overtaken by developments in the field of genetic engineering in years to come. Nevertheless, induced polyploidy will remain an important tool in the hands of a plant breeder for introgression of alien genes across species barriers in several ways.

24

Disease Resistance Breeding

The principles underlying breeding for resistance to diseases are much the same as for other characters, except for one important difference, that is, the diseases caused by pathogens are the product of interaction between two genetic systems, that of the host and that of the parasite. Both systems are capable of variation and evolution and affect the interaction between host and pathogen. Here the breeder is required to deal with the heritable characters of both the host plant and the pathogens invading it.

DISEASE RESISTANCE

Disease resistance of a crop variety (host genotype) refers to its inherent ability to rapidly express effective defence reaction which prevents the pathogen from penetrating the plant and/or halts microbial attack. The degree of damage, such as foliage damage caused by leaf spots, holes, necrosis, chlorosis, mosaic, transformation or destruction of grain (grain smuts, ergots), malformation of plant parts, and dropping of foliage, blossom, fruits, etc. is much less in resistant varieties compared to other varieties of the same crop under similar cultural conditions. Varieties which express severe disease symptoms are called susceptible varieties. Disease susceptibility results from the plant's failure to perceive and respond to an attempted infection or may be due to the ability of a pathogen to avoid or disarm plant defence mechanisms.

Other features of disease resistance are:
1) resistance genes are often effective against only part of the population of a pathogen species;
2) expression of disease is influenced by the environment, sometimes profoundly;
3) often resistance to a particular pathogen sp. is ineffective against other pathogen spp. of the same host.

HOST-PATHOGEN INTERACTION

From the viewpoint of plant breeding it is important to understand host-pathogen interaction.

1. Pathogen Aspect

Pathogenicity, that is, the ability of a pathogen type (race or pathotype) to cause disease symptoms is known as *virulence*. Races or pathotypes that fail to cause disease symptoms or attack a given genotype of host are known as *avirulent*.[1] These two terms correspond to susceptibility and resistance of the host genotype in a system involving *vertical resistance* (oligogenic-resistance). Similarly, the *aggressiveness* of a race or pathotype and *non-aggressiveness* corresponds to susceptibility and resistance of the host genotype in a system involving *horizontal resistance* (polygenic or minor genes resistance).

In breeding for disease resistance the virulence of the pathogen is overcome with the incorporation of a corresponding resistance gene in the host. However, when a new variety of host(s) with vertical resistance is introduced, the pathogen adapts or tends to adapt itself to the host, and all too often matches the resistance with newly accumulated virulence, that is, with new races of the pathogen. These new races arise through sexual recombinations and a variety of parasexual and asexual processes, and through mutations. The pathogen mutates from avirulence to virulence with ease for some resistance genes and with difficulty for others. Virulence, however, has its own structure, which is only indirectly influenced by the host. According to Vanderplank (1982) when the virulence in the pathogen is identified by the corresponding resistance gene in the host, virulence for some resistance genes strongly dissociates and for other resistance genes strongly associates in a way that seems to have little to do with pathogen/host adaptation. This enables classification of resistance genes in the host according to how the corresponding virulences in the pathogen behave. This behaviour enables a plant breeder to control disease by grouping in the host the best combination of resistance genes.

2. Plant Aspect

For a disease to occur, the genes of the pathogen must match those of the host. There are a number of ways in which a plant escapes, avoids or resists pathogens.

   i) *Preformed defence mechanisms* : Morphological characters, such as permeability and penetration barriers of the plant outer layers of epidermis or rhizodermis with the cuticle and its cutin or suberin layers, separation of the inner tissue from the environment (suberin, cork, lignin, callose layers), contribute to the defence mechanism, that is, resistance of the host to a pathogen. Highly bioactive secondary plant products, such as secondary metabolites (phenols, flavonoids, alkaloids, glycosides, etc.), also confer a certain degree of basic resistance on the plant because of their pronounced antimicrobial activity.

   ii) *Infection-induced defence mechanisms* : This involves the principle of

---

[1] In bacteria and viruses the virulent strain causes severe damage, while the avirulent strain causes much less damage.

differential gene activation and *de novo* synthesis of enzymes at different levels of plant metabolism. The reaction, such as hypersensitive response, biochemical reactions, hydrolases (chitinases, B-1,3 glucanases), synthesis of pathogenesis-related proteins, inhibitors of fungal polygalacturonases and other hydrolases, peroxidases and polyphenoloxidases, phytoalexins; re-enforcement of cell walls/isolation of lesions by hydroxyproline-rich glycoproteins, phenolic polymers, lignin, suberin and callose, are characteristic of a given plant species, a particular organ and most importantly which defence mechanisms are preferentially expressed in a certain type of interaction. Induction of active antimicrobial defence reactions results from perception by the plant cells or signal molecules (elicitors) of biotic or abiotic origin. Examples of biotic elicitors are the polysaccharides, oligosaccharide fragments, proteins, glycoproteins and fatty acids, and that of abiotic elicitors are UV light, heavy metal ions, detergents, xenobiochemicals, freezing or heating of plant cells. Abiotic elicitors exert various forms of stress and act in a manner not yet well known.

Genetically determined properties of the host and the pathogen decide the compatibility or incompatibility of an interaction. These properties can be studied when both the host and the pathogen are thoroughly investigated and the chemical molecules exchanged during the interaction are identified to explain the genes and their products which are activated during the interaction.

Genetics of Host-pathogen Relationships

1. Gene-for-Gene Relationship

The ability to show genetic differences in host lines with respect to reaction to a pathogen species is dependent on the genetic constitution of the host. The basic pattern that has emerged from studies of genetic variability in both host and pathogen is the pattern called a gene-for-gene relationship by Flor (1955). In this system a host plant will not be resistant to a pathogen unless the pathogen has the corresponding gene for avirulence. A pathogen will not be avirulent unless the host has the corresponding gene for resistance. It is genes of this pattern that plant breeders have used extensively over the years in breeding disease-resistant varieties. Resistance to a given pathogen is usually controlled by one to few major genes (oligogenic). It is possible to identify individual genes for resistance and to develop procedures for monitoring and modifying individual genes, and to select for individual plants or sectors of plants with altered phenotypes.

2. Independent Gene Action

There is considerable evidence that some genetic variability affecting host-parasite interactions does not follow the gene-for-gene pattern. When host varieties are tested against pathogen genotypes no interaction is observed and there is no race specificity. The observed variation in resistance of varieties and variation in pathogenicity are independent of each other.

The resistance therefore is non-race-specific. Usually it may involve many genes (polygenes). This does not necessarily mean the absence for gene-for-gene action, however. Vanderplank (1982) argued that most of what we have been calling polygenic resistance is for practical purposes oligogenic, the difference being the additive variance and the dominance variance.

Thus, the host plant resistance and pathogenicity of a pathogen may each be governed by one or a few genes or polygenically. It is through a combination of type of gene action and number of genes involved either race-specific (vertical) or non-race specific (horizontal) host-pathogen relationships arise.

VERTICAL RESISTANCE (PERPENDICULAR RESISTANCE, PHYSIOLOGICAL RESISTANCE)
Vertical resistance is race specific or differential resistance. It is usually governed by major gene(s) (oligogenes).

HORIZONTAL RESISTANCE
Horizontal resistance is non-race specific or uniform resistance (no differential interaction). Many minor genes may be involved. Terms such as, field resistance, generalized resistance, adult-plant resistance, etc., have often been synonymously used for horizontal resistance.

Method of Determination of Host-pathogen Relationship

Parlevliet and Zadoks (1977) suggested that whether one deals with vertical or horizontal or both, types of resistance can be determined by evaluating the degree of resistance of a number of host-genotypes (clones, varieties) for a number of pathogen-genotypes (isolates, races). Determination is based as follows:

(1) When all the non-environmental variation (genetic variation) can be explained by differences between varieties and differences between pathogen isolates, the effect being additive, it is taken as horizontal resistance and horizontal pathogenicity.

(2) When all the non-environmental variation (genetic variation) is caused solely by the interaction between varieties and isolates of pathogen, and the main effects (varieties or pathogen isolates) are nil, it is taken as vertical resistance and vertical pathogenicity.

(3) When both the main effects and interaction occur, it is taken as two-dimensional resistance. The consequence of vertical resistance is that the ranking of varieties according to disease severity may depend upon the pathogen isolates used for testing. In the case of horizontal resistance the ranking of varieties is independent of pathogen isolates.

Nelson (1975) suggested that each gene has a vertical and horizontal component. The vertical genes are considered to contribute something in the way of resisting the pathogen at a point beyond hypersensitivity. The net result of a number of these genes within a single host genotype appears to offer collective resistance against colonization. For example, five genes able to react vertically to some races of a pathogen but not to others may collectively react in a horizontal way to the other races.

Epidemiology of Resistance to Disease

The variable resistance constantly observed is almost incomplete resistance. Vertical resistance to avirulent phenotypes is usually complete because plant breeders select plants for complete resistance which give a resistance reaction of type 0 or 1 when challenged by relevant races of a specific pathogen. Complete horizontal resistance often takes the form of *population immunity* (Vanderplank, 1975). This happens when the parent/progeny ratio does not exceed 1.

Usually, vertical resistance delays the onset of an epidemic, while horizontal resistance slows the epidemic down. Ontological and seasonal changes can cause horizontal resistance to be manifest as a postponement of adult-plant susceptibility.

Disease Reaction

Disease reaction is usually classified on the basis of degree of damage (level of infection) caused by the pathogen on a score basis. The following categories are usually recognized.

- Immune: no infection takes place.
- Highly resistant: varieties exhibit minor damage.
- Moderately resistant: in between high and low resistance.
- Low resistance: relatively less damage compared to average damage to crop.
- Susceptible: crop damage is more than average damage to the crop.

Factors Affecting the Expression of Pathogen

A more susceptible host, a more aggressive pathogen and an environment more favourable to disease—all increase disease. This is known as the disease triangle, all of whose sides are equal, as all the three factors are equally important. Changes from susceptible to less susceptible host, or from aggressive to less aggressive races of the pathogen, or from favourable to less favourable environments for disease—all have the same effects.

BREEDING APPROACHES FOR DISEASE RESISTANCE

The various approaches for breeding disease-resistant varieties are discussed below:

1. Avoidance (Disease-escaping Varieties)

This type of resistance is due to specific plant characteristics, for example closed flowering, earliness, etc. that reduce the chances of contact between the pathogen and the host plant tissues. Advantage is often taken by growing early varieties with vigorous early growth to escape infection or damage caused by *Tilletia* (wheat), seedling blight (wheat), *Hevea mildew* (rubber), and closed flowering to escape loose smut infection (barley).

It may be noted, however, that there is no inherent resistance in disease-escaping varieties. When conditions such as rainfall, low temperatures, fertilization etc. are conducive to prolonging vegetative growth, varieties may suffer heavy loss on the appearance of disease. Therefore, an element of risk is always present.

2. Tolerance

Many crop varieties are known to be tolerant, that is, they are able to endure severe disease attack and give satisfactory returns to the growers. Tolerance does not reduce levels of infection. The extent of damage is reduced, however, due to factors such as compensatory growth. The exact mechanism of tolerance has yet to be explained. Vanderplank (1963) argued that tolerant varieties do not control the initial inoculum, infection rate and infection time and hence tolerance is not a desirable characteristic.

3. Replanting of Discarded Varieties

This approach is based on the assumption that in the absence of suitable hosts the pathogen involved earlier is materially reduced in numbers. The principles governing the biological equilibrium between the host and the pathogen are the same. Therefore, rotation of varieties would help to restrict the build-up of a pathogen in such quantities over a period of time to create an epidemic. It is understood that where varieties are replaced more frequently, with each rapid increase in number of new races due to new host varieties, there is a corresponding decrease in old races. After 5-10 years of widespread cultivation of a new host variety , it may be possible that formerly well-known races will become scarce. The approach may be worth considering provided in the intervening time newly introduced varieties are not superior in yield or other agronomic traits and that the old varieties are not found to be susceptible to new races which have appeared during the intervening period.

4. Cultivar Diversification

Diversification of varieties of a crop with resistance genes and alleles against a specific disease may help in devising schemes to restrict the spread of pathogens. Farmers may be advised to choose a range of varieties with different specific resistances.

5. True Resistance (Incorporation of Disease-resistance Genes in Agronomically Superior Varieties)

This type of disease resistance is specific. The resistance mechanism reduces the rate of colonization or growth of the pathogen and becomes operative with contact between the host tissue and the pathogen. There is thus an intimate contact between the two and the expression is the result of their mutual interaction.

The prerequisites to this system are as follows.

a) Finding the Source(s) of Resistance

A close collaboration between the plant breeder and a pathologist is necessary to identify source(s) of resistance to diseases and various races of diseases in diverse germplasm. Genetic variation in crop species and their wild relatives holds the key to the successful breeding of improved crop cultivars with durable resistance to disease. Similarly, adequate race-survey would enable identification of various races which are of immediate concern. The source(s) of resistance available in diverse germplasm then could be found through screening, using artificial inoculation with individual races and the appearance of disease symptoms. Resistance is determined by comparing the plant damage with that of resistant and susceptible check varieties. Development and standardization of efficient disease screening techniques is a basic requirement for identification of resistance source(s).

b) Inheritance of Resistance Genes

Once the source(s) of disease resistance genes have been identified, study of inheritance is necessary to evolve an appropriate breeding strategy. A great deal of work has been done in this regard. The work done in relation to important diseases of some of the economic crops is summarised in Table 24.1.

c) Characterizing Mechanisms of Disease Resistance

In breeding disease-resistant varieties it is equally important to study plant traits conferring resistance and their influence on disease expression and the mechanism involved in reducing the disease damage.

BREEDING TECHNIQUES

Once the source(s) of the disease-resistance gene(s), mechanism of disease resistance involved and its genetics are known, the resistance genes can be incorporated through hybridization followed by selection. The various methods, namely, pedigree method, backcross breeding, mutation breeding and recurrent selection discussed in earlier chapters may be suitably employed. Use of artificial epiphytotics during the breeding cycle is imperative for identification of resistant types. The permanence and stability of resistance is determined through evaluation of advanced genotypes in diverse cultural conditions.

1. Variety Mixtures

The use of cultivar mixtures may considerably restrict the spread of a disease. The principle behind use of variety mixtures is that the spread of a race affecting a particular genotype will be drastically reduced due to other genotypes which will act as physical barriers to spore movement as well and consequently the rate of spread of the disease. The components of cultivar mixtures can be varied according to current needs. For example, if races

Table 24.1. Inheritance of disease resistance

| Crop (1) | Diseases | |
	Oligogenic (2)	Polygenic (3)
Field Crops		
Wheat	Stem rusts, covered smut, hill bunt, common powdery mildew, leaf blight, common root rot, leaf blotch	Glume blotch, common bunt
Barley	Stem rust, leaf rust, blight, powdery mildew, stripe rust, loose smut	—
Oats	Crown rust	—
Rice	Grassy stunt virus, bacterial leaf blight, tungro virus	—
Corn	Blight, northern leaf blight, southern rust, maize dwarf mosaic virus	Downy mildew, brown stripe, southern corn leaf blight, stalk rots, smut, ear rot
Sorghum	Sorghum rust	Charcoal rot, grain mould, *Fusarium* grain mould,
Pearl millet	Rust	Downy mildew
Chick-pea	Blight, wilt	—
Peas	Blight, pea mosaic virus, powdery mildew	—
Lentils	Pea mosaic virus	—
Cowpea	Chlorotic mottle virus, cowpea mosaic virus, cucumber mosaic virus, *Cercospora* leaf spot, bacterial leaf blight, bacterial pustules, black-eye cowpea mosaic	—
Dolichos bean	Bacterial leaf spot	—
Broad bean	—	Chocolate spot
Snap bean	—	White mould, *Rhizoctonia* root rot, *Pythium* seed rot, damping off, halo blight
French bean	Angular leaf spot	—
Soybean	Mosaic virus, *Phytophthora* rot, rust, stem canker	—
Sunflower	Wilt	—
Safflower	Root rot, lettuce mosaic virus	—
Flax	Powdery mildew	—
Alfalfa	Wilt, *Phytophthora*	—
Italian rye-grass	Crown rust	Barley yellow dwarf, *Rynchosporium orthosporum*

Contd.

Table 24.1. Contd.

(1)	(2)	(3)
Cotton	Bacterial blight	—
Vegetables		
Tomato	Stem rot, foot and stem rot, early blight	—
Okra	Yellow vein mosaic virus	—
Cucumber	Angular leaf spot	—
Cauliflower	—	Black rot
Cabbage	Black rot	—
Lettuce	Downy mildew	—
Celery	*Fusarium* yellows	—
Grape vines	*Anthracnose*	—

adapted to one component of the mixture become especially prevalent, then that component may be replaced by a variety of different specific resistance. Also, as new varieties became available, they can be incorporated into the mixtures. A major advantage to the farmer is that the system costs little to implement in terms of seed costs and management. The use of variety mixtures has not, however, been adopted on a large scale because of preference for single varieties.

2. Multiline Varieties

Concept of multiline variety

Jensen (1952) outlined, the concept of multiline variety as follows:

A multiline variety is a blend of multiple pure lines, each of which is of a different genotype. Each line is chosen for uniformity of a different genotype. Each line is chosen for uniformity of appearance, particularly height and maturity, resistance to diseases, and other characteristics essential for a basic desirable agronomic type. Each line used should contribute additional desirable genetic factors without detracting from the uniformity of the composite. Through a knowledge of the characteristics, of the individual component lines, and with performance data gathered in the usual way, the plant breeder would be able to blend compatible lines in the proper proportion. He would retain the individual lines and could effect withdrawals or make additions of promising new selection in later years. The seed blends would proceed through established certified seed channels to the grower. A multiline variety thus would change little in outward appearance with the passing of time but might change in genotypic composition.

The principal feature of multiline varieties is disease protection. The principle sought to be exploited is the putting up of genetical barriers to the pathogen by diversifying the genotypes. The new races will not have a large continuous favourable substratum of a particular genotype to build up to epidemic proportions. On theoretical grounds this system would have longer variety life, greater stability of production and greater protection against diseases because no races are necessarily excluded from reproducing on it. As a result, the sudden shifts in racial proportions that repeatedly have caused pure lines to become susceptible should no longer occur in a multiline variety.

Method of Developing Multiline Varieties
The following steps are involved in the development of multiline varieties.

1. DEVELOPMENT OF COMPONENT LINES
The diverse source(s) of resistance are incorporated in the base line (agronomically superior line) by backcrossing that leads to production of near isogenic lines, or restricted backcrossing, or by double, or complex crosses that lead to production of genetically dissimilar lines carrying different resistance genes which resemble each other for the relevant agronomic characters. The important and necessary requirement is that recovery of the phenotype of the base line should be very high.

2. SCREENING AND GROUPING OF COMPONENT LINES
(a) Component lines are thoroughly screened for disease resistance at several locations. The resistant lines showing close similarity with the base line in respect of characters such as early growth vigour, height, plant type, maturity, ear size, shape and grain colour, texture, size, etc. are selected. Simultaneous selection for these several characters may be done by independent culling.
(b) Selected component lines are evaluated at several locations and those showing yielding ability equal to the base line are sorted out. Grain quality tests are also done.
(c) Sorted-out component lines are evaluated for their competing ability and ecological nicking ability at several locations.
(d) Finally selected component lines are composited to form a multiline.

3. EVALUATION OF MULTILINES
The constituted multilines are evaluated in multilocation tests including base variety and the promising ones are identified for release.

4. SEED MULTIPLICATION
The breeder seed of each component line is produced separately. The next generations, however, may be multiplied from the mixture. In order to keep a check on the challenge posed by possible shifts in the race structure it is necessary to watch resistance in the component lines through artificial epiphytotics or at 'hot spots'. Any component line found susceptible should be replaced.

Modifications in Breeding Methods for Incorporating Disease Resistance

1. Pyramiding of Resistance Genes (Multiple Resistance)

This concept is based on the assumption that simultaneous mutation in the pathogen at more than one locus is much rarer than mutation at one locus. Thus, the greater the number of introduced diverse genes in a variety governing resistance to various races of a pathogen present in an area, the longer the longevity of its resistance. A variety having such a combined resistance would offer more than one physiological barrier against the pathogen, and also prevent step-wise development of races virulent to varieties possessing different but single genes for resistance.

The breeding objective, i.e., simultaneous introduction of diverse resistant genes into a single variety, is achieved through a programme involving backcrossing.

2. Chromosome and Genome Substitutions

Classic examples of achieving disease resistance through chromosome substitution have been provided by Sears (1956) who transferred leaf rust resistance from *Aegilops umbellulata* into a wheat variety. The technique developed by Sears involving monosomic analysis may be used for transferring chromosomes carrying resistance genes.

Application and Uses of Disease Resistance

Production of disease-resistant plants is the most desirable method of achieving control of pathogens with the minimum of effort and expenses for growers and in an environmentally benign manner. It is invariably one of the principal objectives in any plant-breeding activity. Examples of spectacular success are too many to be listed here. Practically, in every crop of economic importance, wherever a source(s) of resistance to a disease of economic importance has been identified, plant breeders have not lagged behind to incorporate the resistance gene in agronomically superior varieties. Development of disease-resistant varieties is perhaps the most outstanding contribution by plant breeders in achieving global food security.

25

Insect Resistance Breeding

The principles and techniques used in breeding for disease resistance are, in substance, also applicable to insect resistance.

INSECT RESISTANCE

Insect resistance in crop varieties refers to their inherent ability to combat specific insect pests and to achieve better performance over other varieties of the same crop at the same levels of insect populations. Insect resistance thus means a relatively lesser degree of damage, such as chewing and cutting of the various plant parts, including underground parts, flowers and fruits; sucking plant sap and fruit juice; boring tunnels in stem and fruits and spread of fungal, bacterial and viral diseases of plants caused by insect vectors. Resistant varieties function in many different ways to reduce the effects of insect attack. In some instances, they may involve new chemicals (allemones) or increased levels of existing ones, and in others may be based on reduced levels of chemicals (kairomones). Also, in many instances physical factors may be involved.

INSECT AND HOST-PLANT RELATIONSHIP

A good insight into the basic relationship between plant and insects, usually referred to as insect-plant interactions, is necessary in breeding insect-resistant varieties.

a) *Insect aspect*: The insect aspect includes knowledge of the general habitat, host plant(s) and sufficiency of the plant as a host.

b) *Plant aspect*: Both morphological and physiological characteristics of the plant elicit given insect responses.

    i) *Morphological characteristics*: Plant morphological features, such as variations in foliage size, shape, colour and presence or absence of glandular secretions may determine the degree of acceptance or utilization by insects, in other words ultimate crop damage. Pubescence and tissue toughness may also affect insect mobility and feeding.

    ii) *Physiological characteristics*: These include chemicals that are the products of plant metabolism. The chemicals produced by primary metabolites may be feeding stimulants, nutrients or toxicants to an insect. The chemicals produced by secondary metabolites are thought to have arisen as mechanisms for chemical defence against plant eating. They may be stored in any convenient place in the plant structure and often are exuded from the outer layer of plant tissues. These may be sensed by insects and function as token stimuli. A token stimulus elicits a response initially and afterwards has no effect. Examples are pheromones, which promote communication between members of the same species of insects; allemones, which are defensive chemicals such as repellents, oviposition and feeding deterrents, and toxicants which produce negative responses in insects and thereby reduce the chances of contact and plant damage; and kairomones, which include attractants, arrestants, excitants and stimulants, which promote host finding, oviposition and feeding.

c) *Host-plant selection :* Host-plant selection by insects may involve both primary and secondary metabolites. Host-plant odour or taste for insects comes from nutrients and odd compounds that are intertwined as complex sensorial. These are interpreted by the insect's central nervous system to determine whether a given plant is a host. The genetically programmed 'correct' signal supports a lasting host association.

Mechanisms of Insect Resistance

Insect resistance in cultivars may be due to the following mechanisms (Painter, 1951).

1. Non-preference (Antixenosis)

This type refers to plant characteristics that lead insects away, that is, the plant is unattractive to an insect for feeding, oviposition or shelter. Non-preference may be due to allelochemics or morphological features of the plant. Use of some non-preference characteristics may be limited by the given cultural environment. Non-preference may break down in the absence of alternative hosts. Because of the widespread practice of monocropping and the breakdown of resistance, allelochemic non-preference is not of primary importance in plant breeding. On the other hand, morphological non-preference that impairs feeding behaviour is very important as a first defence against many insects. Non-preference is important under circumstances wherein, brief infestations cause severe plant damage, for example, stem borers in rice, and in transmission of viral diseases.

2. Antibiosis

This is the most widely sought objective of plant breeders. Both insect and plant factors are involved in this mechanism. This mechanism impairs an insect's metabolic processes and often involves consumption of plant metabolites. Examples are gossypol and related compounds in cotton; steroidal glycosides in potato, saponins in alfalfa and cyclic hydroxamic acids (DIMBOA) in corn.

Quantity and quality of primary metabolites may also be important in conferring antibiosis. Particularly significant in this regard are imbalances of sugars and amino acids that result in nutritional deficiencies for insects feeding on the plant. For example, pea varieties with low amino acid levels and increased sugar content show resistance to the pea aphid; rice varieties deficient in asparagine reduce fecundity in the brown plant hopper.

3. Tolerance

Unlike non-preference and antibiosis, only a plant response is involved in tolerance. Tolerant plants can withstand a heavy insect attack. The ability of tolerant plants to survive heavy insect infestation for a longer period permits a longer exposure of the insects to their natural enemies. It is attributable to plant vigour, compensatory growth in individual plants and/or the plant population, wound healing, mechanical support in tissues and organs, and changes in photosynthate partitioning. This type of insect resistance is particularly valuable in integrated pest control.

Apparent Resistance (Ecological Resistance, pseudoresistance)

Temporary host-plant resistance characteristics that rely heavily on environmental conditions come under apparent resistance. The varieties involved are potentially susceptible. This type of resistance is important in integrated pest control but needs to be carefully synchronized with prevailing environmental conditions for effectiveness. Three types of apparent resistance are recognized.

a) *Host evasion*: In this type the plant passes through a susceptible stage rather quickly or at a time such that its exposure to potentially injurious insects is reduced. Examples are use of early maturing varieties.

b) *Induced resistance*: This is a form of temporary resistance derived from plant condition or environment. Factors such as soil fertilization and soil moisture are involved. Recently, the role of phenolic compounds (phytoalexins) in inducing plant resistance has been highlighted. These compounds are produced by plants when they become diseased or are attacked by insects and enable them once fed upon, to resist further damage by the pests.

c) *Host escape*: This category refers to the plants which have escaped in an otherwise susceptible population of plants. The reasons for escapes are rarely understood.

Genetics of Insect Resistance

Effectiveness and stability of resistant varieties are determined by the plant genes that confer resistance and insect genes that enable plant resistance to be overcome.

(i) Gene-for-Gene relationship

Many pest populations have individuals with *virulent genes* which allow the pest species to overcome resistance and once more attack a plant. In such cases, one or more virulent genes may be present which enable an individual pest to overcome the effects of one or more plant genes responsible for resistance. This principle is called the gene-for-gene relationship. In the gene-for-gene relationship, plant varieties are resistant because they have a resistant allele at a gene locus that corresponds to an avirulent allele at an equivalent locus in the insect. Even though a plant variety is effectively resistant against most insects in the population, an occasional insect may have a virulent allele instead of the normally avirulent allele. This circumstance enables a virulent individual to attack an otherwise resistant plant and, over a period of time, the virulent genotype can replace the avirulent genotype. Eventually, the effectiveness of the resistant cultivar will decrease. Different populations of an insect species that vary in virulence to a variety are referred to as biotypes. Some insect species are known to have several biotypes.

(ii) Vertical and horizontal types of insect resistance
Vertical resistance refers to varieties with resistance limited to one or a few pest genotypes. Horizontal resistance, on the other hand, refers to varieties with resistance against a broad range of pest genotypes. Improvement in crops using vertical resistance is achieved by identifying plant genes that confer resistance and their incorporation into varieties. Improvement in crops with horizontal resistance is a building process based on step-wise accumulation of genes with favourable additive effects.

Insect Resistance-mode of Inheritance

1. *Oligogenic resistance* : This involves major gene(s). Resistance is conferred by one or only a few genes. This type produces vertical resistance against insects and may be inherited through dominant or recessive genes.
2. *Polygenic resistance* : Polygenic resistance is conferred by many genes, each contributing to the resistant effect. This type of resistance is usually complex. Horizontal resistance is usually polygenic.

Factors Affecting Expression of Insect Resistance

Resistance is primarily governed by genetics. However, physical and biotic elements of the environment often influence its expression. Abnormal deviations of the environmental factors may have pronounced effects on the performance of many resistant varieties.

a) Physical Factors
Temperature : Abnormally high or low temperatures for a period of time may cause loss of resistance.

Light intensity : Shade-induced loss of resistance has been found in several instances. With potatoes, shading was found to be associated with reduced levels of steroidal glycosides in leaves which are known to retard feeding and development of the beetle.

Soil fertility : Changes in soil nutrient levels also may mediate the expression of resistance in some plants, but little is known about the mechanism involved.

b) Biological Factors

Biotypes: New biotypes may appear that may lead to the breakdown of resistance. The time required for breakdown in resistance may be only a few years (8-10 years or earlier). It is important to distinguish among various forms of biotypes, namely, true resistance-breaking biotypes, unusually vigorous variants that have high reproductive potentials on all plant genotypes with no definite gene-to-gene relationship, different species, etc. in order to develop a proper counterresponse to the problem.

Plant age: Physiological responses in plants vary with age, and these can lead to changes in the expression of insect resistance.

Classifications of Insect Resistance

The following classes of insect resistance in relation to specific insects are recognized.

Immune: the varieties exhibit no insect damage (non-host).

Highly resistant: the varieties exhibit very little or minor damage.

Moderately resistant: intermediate between highly resistant and poorly resistant.

Poorly resistant: relatively less damage compared to average damage to the crop.

Susceptible: more insect damage to the crop compared to average damage to the crop.

METHODS OF DEVELOPMENT OF INSECT-RESISTANT VARIETIES

General Aspects

1. Source(s) of Insect Resistance

In breeding varieties for insect resistance the first requirement is to find source(s) of resistance to specific insect pests of economic importance. A close collaboration between entomologists and plant breeders is necessary to screen diverse germplasm for resistance to specific insect pests. Efficient methods of mass screening are necessary to achieve the desired objectives in the shortest possible time. Screening may be done in natural insect populations maintained in the field (endemic areas) that favour propagation of the insect species under consideration, or in artificially reared insect populations transferred to plants (in insect-tight cages) in the field or glasshouses to keep the insect pests in contact with the plants and to prevent infestation from other insects by natural means. Resistance is determined by comparing the plant damage with that of resistant and susceptible check varieties.

2. Inheritance of Resistance Genes

Once the source(s) of insect resistance has been identified, the next logical and important step is to determine the inheritance of the character in order to evolve suitable breeding strategy. The work done so far is rather limited and information regarding oligogenic/polygenic mode of inheritance has been collected only for some species of insects (Table 25.1).

Table 25.1. Inheritance of insect resistance

Crop	Insects	
	Oligogenic	Polygenic
Wheat	Green bug, stem sawfly, Hessian fly	Cereal leaf beetle, rice weevil
Barley	Corn leaf aphid, green bug	—
Oats	Green bug	Cereal leaf beetle
Rye	Green bug	—
Rice	White-backed planthopper, gall midge,	Stem-borer
Corn	European corn-borer,	Corn leaf aphid, corn ear worm
Sorghum	Green bug	Shoot fly
Pearl millet	Chinch bug	—
Cowpea	—	Weevil
Soybean	Cyst nematodes	—
Brassicas	—	Aphids
Cotton	Thrips, leafhoppers	—
Lucerne	Pea aphid	Spotted alfalfa aphid
Vegetables	Lettuce leaf aphid, melon aphid, red pumpkin beetle, fruit-fly	Striped cucumber beetle, squash bug

3. Characterizing Mechanisms of Insect Resistance

In breeding insect-resistant varieties it is equally important to study plant traits conferring resistance, their influence on pest performance and mechanisms of reducing insect damage.

Breeding Methods

Once the source(s) of insect resistance and its mechanism and genetics are properly understood, resistance genes can be incorporated into agronomically desirable varieties through hybridization followed by selection. The various breeding methods discussed in earlier chapters of this book may be suitably modified and employed for breeding insect-resistant varieties. It may be noted, however, that during the breeding cycle the material is exposed to insect populations so that resistant types can be distinguished from susceptible ones. Also, the permanence and stability of resistance needs to determined by evaluation of advanced genotypes in diverse cultural and environmental conditions, and its vulnerability by using infested versus non-infested comparisons. The influence of resistant traits on key predators and parasitoids of the cropping system should also be evaluated.

Application and Uses of Insect Resistance

Insect-resistant varieties provide inherent control of specific insect pests with no expense to the farmers and cause no environmental problems resulting from the use of chemical insecticides. In certain instances these constitute the only means of control. Use of insect-resistant varieties on an extensive scale by farmers constitutes the mainstay of the entire IPM programme being encouraged throughout the world. Resistant varieties may be used as a primary method of control or as an additional measure to control insect pests.

Successful Uses of Insect-resistant Cultivars

A large number of varieties resistant to specific insect pests have been released for commercial cultivation. However, the following stand out as having provided the most significant monetary gains to the users (Pedigo, 1991).

1. RESISTANCE TO HESSIAN FLY

Hessian fly-resistant varieties of wheat have reduced a very serious problem, with annual crop losses over $238 million to the status of minor pests. Resistance to the Hessian fly is mainly due to antibiosis, as larvae feeding on resistant plants die, while those that survive remain small in size.

2. RESISTANCE TO EUROPEAN CORN-BORER

Today, many commercial field corn hybrids possessing genes for resistance to the European corn-borer are grown, covering about one-third of the corn area in the USA, resulting in annual savings of about $150 million. Resistance results from the presence of the cyclic hydroxamic acid DIMBOA (2,4-dihydroxy-7-methoxy-1, 4-benzoxazine-3-one). DIMBOA levels are highest early in the growing season (mid-whorl growth stage), thus offering maximum resistance to first-generation corn-borers. DIMBOA levels decline as the season progresses and most varieties have very little resistance to second generation corn-borers.

3. RESISTANCE OF SPOTTED ALFALFA APHID

Use of resistant varieties has resulted in annual savings of about $60 million. Resistance is attributed to antibiosis caused by higher saponin content, particularly medicagenic acid.

4. RESISTANCE TO WHEAT STEM SAWFLY

The annual savings resulting from the use of resistant varieties are estimated to be over $4 million. Resistance is due to solid stems which results in the sawfly eggs being mechanically damaged and desiccated and restricts the movement of newly emerged larvae. Stem solidness is controlled by one or more dominant, recessive, or complementary genes, depending on the parents used, the crosses and the ploidy involved.

5. RESISTANCE TO THE GREEN BUG

Annual savings from the use of resistant varieties of wheat and barley are estimated to be over half a million dollars.

238

Resistance in barley is controlled by a single dominant gene.

The development and release of paddy varieties resistant to brown planthopper and gall midge are the other examples of successful breeding for insect resistance.

26

Abiotic Stresses

DROUGHT RESISTANCE

Drought implies a relatively prolonged dry period (absence of rains) resulting in moisture stress in the soil detrimental to crop growth, especially in rainfed agriculture. Severity of drought varies with timing (in relation to stages of crop growth) and intensity (duration of absence of rains). Other factors, namely soil characteristics (texture, depth, water retention and drainage characteristics) and agronomic practices are also associated and have a bearing on crop yields. Damage is more severe when the period of drought coincides with the critical stages of crop growth.

Concept of Drought Resistance

The ability of a crop variety to perform better over other varieties under specific drought conditions (environment) is known as drought resistance. It is linked with realized yields and potential yields achievable in an environment in the absence of drought conditions. Drought resistance is highly crop environment (area) specific. Yield stability may also be influenced by crop management practices, and/or physiological mechanism and not necessarily associated with the drought resistance ability of a genotype. In a drought-resistant variety, plant growth and development is well-matched to specific drought environment(s).

Characterization of Drought Environment

From a practical viewpoint of breeding for drought resistance it is necessary to characterize the drought environment in which crops/crop varieties are to be grown. This is necessary because selection of the appropriate genotype and its subsequent evaluation in yield trials need to be done under specified environment(s) for evolving drought-resistant varieties. It is also necessary due to the highly location-specific nature of drought resistance due to the several edaphic, biotic and agronomic factors involved.

Environment characterization should preferably be done on the basis of information collected through an organized field survey, and agronometeorological and soil characteristics, as well as water resources data available with official agencies. Remote sensing data, when available, should also be used in conjunction with all other available data to characterize crop growing environment(s).

Mechanism of Drought Resistance

Susceptibility to drought can occur during the early vegetative seedling stage, during the period of panicle development prior to flowering, and during the post flowering stage of grain development. Susceptibility during post-flowering stage is characterized by reduced seed size and grain yield, pre mature plant and leaf senescence and increased stalk lodging. Terminal post flowering drought results in an abbreviated period of grain development and therefore reduces seed size. Genotypes with a high rate and reduced duration of grain filling may be more tolerant under terminal post flowering conditions.

Identification of critical stages of crop growth, that is, the stages at which a crop is more severely affected by drought and the plant response under stress, and more particularly its response to stress, if any, is therefore important for understanding the mechanism of drought resistance and for evolving appropriate methodology for developing drought-resistant varieties. The usual mechanisms are as follows.

1) Escape : Early maturing crops/crop varieties often escape drought because they complete the critical stages of crop growth prior to drought conditions setting in. Early growth vigour may enable a variety to establish a good plant stand rather quickly while the moisture supply is good. Thus the crops/crop varieties involved escape the adverse effects of drought and perform relatively better. Many indeterminate crops respond to reirrigation by resuming their growth and perform still better.

2) Avoidance : Drought avoidance is an alternate mechanism by which plants can maintain positive tissue water relations even under limited soil moisture conditions. Mechanisms of drought avoidance typically involve water conservation at the whole plant level. Avoidance is accomplished by decreasing water loss from the shoot or by more efficiently extracting moisture from the soil. Many crop varieties/crops with deep as well as dense root system may be able to maintain minimal water uptake from soil to avoid internal stress, at least during the initial stages. High varietal resistance to water loss has also been observed in a few cases, for example, in rice the amount of epicuticular wax deposition is reportedly associated with water loss. Parker (1968) reported that some grasses of the Mediterranean region reduce transpiration rate by as much as 46 to 63%.

3) Tolerance : Drought tolerance is defined in a number of ways, namely, the performance *per se*, the stability of performance under drought and to focus on specific physiological or morphological traits that are believed to be associated with the expression of drought tolerance. Mechanisms that condition drought tolerance function at the tissue or cellular level. Under conditions of tissue desiccation, these mechanisms stabilize and protect cellular and metabolic integrity. Crop varieties may differ in their ability to survive under drought conditions. This has been demonstrated through tests, such as desiccation survival, heat tolerance and proline accumulation. The relationship between these tests and drought resistance under field conditions has not been established as yet, however.

4) Recovery: Droughts vary in duration, but when rainfall does commence the ability of a genotype (crop variety) to recover quickly and resume active growth is an important character. In rice it is reportedly associated with characters such as vegetative growth vigour, high tillering ability, shallow root system and rather long growth duration.

Assessment of Drought Resistance and the Plant Traits Associated with Drought Resistance

Asana (1957) concluded that drought resistance of an annual crop plant can at present be assessed for agronomic purposes only on the basis of yield. The argument is valid even today. Few of the many screening tests proposed have been adopted by breeders. Asana maintained that desiccation tests may be more useful in determining drought resistance of natural or pasture vegetation which may be subject to permanent wilting.

Several plant traits, such as dehydration avoidance[1] and dehydration tolerance[2] have been found to be positively associated with yield under stress across genotypes of wheat and barley. Leaf rolling, root system, pubescence of aerial organs, reflectance of incoming solar radiation, increased heat dissipation through decreased boundary layer resistance at the organ level (narrow leaves, awns), etc., are the traits that contribute to dehydration avoidance. In nature, a better balance is associated with a higher proportion of energy dissipated as latent heat and hence a lower canopy temperature. The fast crop-temperature assessment method is being used in wheat and maize breeding programmes (Blum, 1988; Bolanos and Edemeades, 1989) at a few research centres. Dehydration tolerance related to cellular and subcellular processes can be readily assessed by measurements of membrane stability with the electrolyte leakage test. It is difficult, however, to relate this type of test to plant production. Nevertheless, visual scores on morphological traits, such as leaf rolling, root habit, etc., and/or observations recorded through other methods, if any, in relation to the above-mentioned characters should invariably be used as an indirect measurement of drought resistance for practising selection in a breeding programme.

In sorghum 'staygreen' character is reportedly associated with post-flowering drought tolerance. Staygreen is characterized as resistance to premature leaf and stalk death induced by post-flowering drought. Resistance to premature leaf and stalk death is thought to increase the potential period of grain development and thereby stabilizing the expression of seed weight (Duncan et al., 1981). Sorghum lines with high levels of staygreen have been identified and are being used in some breeding programs.

---

[1] Dehydration avoidance is interpreted as the ability of a genotype to maintain higher leaf-water potential when grown under soil-water deficits.

[2] Dehydration tolerance is related to the ability of a genotype to sustain post-anthesis stress. This is due to low leaf water potential and encompass cytoplasmic tolerance.

Genetics of Plant Traits Associated with Drought Resistance

Genetic variation has been observed for a number of adaptive traits related to environmental stress. These include physiological traits, such as maintenance of relatively higher leaf-water potential under soil-water deficits, osmotic adjustment, tolerance to stress in plant or organ growth rate, plant recovery on rehydration, tolerance in the photosynthetic system or its components, tolerance in enzyme activities, tolerance in translocation, stability of cellular membranes; chemical traits, such as proline accumulation; epicuticular wax content; and morphological traits, such as root growth, leaf size, leaf area per plant, leaf orientation, tiller survival and organ pubescence.

In sorghum genetic studies of 'staygreen' have generally indicated a complex pattern of inheritance. Both dominant and recessive expression, strongly influenced by the environment have been reported. Tunistra et al. (1997) studied the inheritance of staygreen in a set of recombinant inbred lines of sorghum. Quantitative trait loci (QTL) identified 13 regions of the genome associated with one or more measures of post-flowering drought tolerance. Two QTL were identified with major effects on yield and staygreen under post-flowering drought. These loci were also associated with yield under fully irrigated conditions suggesting that these tolerance loci have pleiotropic effects on yield under non-drought conditions. QTL analysis indicated many loci that were associated with both rate and duration of grain development. High rate and short duration of grain development were generally associated with larger seed size, but only two of these loci were associated with differences in stability of performance under drought. It may be noted that associations between markers and QTL were somewhat variable across testing environments. This highlights the importance of multi-environment testing when evaluating drought tolerance.

Until such time as routine screening techniques are developed and the relationship between these adaptive traits and crop performance under stress is clearly established, incorporation of adaptive traits from one genotype to another shall remain elusive from the viewpoint of practical plant breeding. Selection for drought resistance will therefore continue to be primarily based on yield assessment under stress conditions as concluded by Asana (1957).

Procedure for Breeding Drought-resistant Varieties

The usual breeding strategies are adopted for developing drought-resistant varieties.

1. *Selection of drought-escaping varieties*: Selection of early maturing varieties that are likely to escape drought is the first line of defence against drought and the most widely used approach.

2. *Selection of genotypes under defined drought conditions*: This involves selection of varieties under specified drought environments and applying appropriate selection pressure.

3. Incorporation of relevant drought-resistance factors (traits) into agronomically superior varieties under optimum conditions to make them suitable for suboptimal conditions. Usual breeding methods are used for incorporating relevant genes.

SALT TOLERANCE

Salt-stress resistance denotes a plant's ability to prevent, reduce or overcome the possible injurious effects caused directly or indirectly by the presence of excessive soluble salts/toxic ions in its root zone. A 50% reduction in yield is taken as a measure of salt stress.

Characterization of Salt-affected Soils

Salt-affected soils is a general term applied to saline, sodic/alkaline soils and degraded alkaline or solodic soils. Salt status of a soil is determined on an extract drawn from the saturated soil paste on the basis of electrical conductivity (decisiemens of the saturation extract (1:2 soil/water suspension) (dSm^{-1}), exchangeable sodium percentage (ESP), sodium absorption ratio (SAR) and pH. There is an intimate relationship between ESP and pH of the saturated soil paste.

1. *Saline soils :* Soils having sufficient neutral soluble salts, such as chlorides, and sulphates of sodium calcium (Ca^{2+}), and magnesium (Mg^{2+}) are called saline soils. The pH is usually below 8.2 and ESP below 13. When the salinity level is low, the proportion of sodium is less and calcium and magnesium are more. In highly saline soils sodium predominates. The salinity status is based on dSm^{-1} as follows:

dSm^{-1}: 0-2, non-saline; 2-4, slightly saline; 4-5, moderately saline; 8-16, strongly saline; and more than 16 very strongly saline. In very strongly saline soils the dSm^{-1} in the surface layers may be exceptionally high.

Due to preponderance of neutral salts, the saline soils remain flocculated and possess good internal drainage. The texture of saline soils usually varies from loamy sand to loam and there is no hardpan (calcium carbonate nodules) in the subsoil. The groundwater table remains high for most of the year. Also, the groundwaters are highly saline.

2. *Sodic soils (alkaline soils)*: Soils having excessive exchangeable sodium (Na$_2$CO$_3$, NaHCO$_3$) are called sodic soils. The soil pH is rather high (up to 10.8) and ESP high (may be as high as 90%) throughout the soil profile. These soils have very low infiltration and hydraulic conductivity rates. The sodicity status is based on ESP as follows:

ESP: Up to 15, slight; 15-30, slight to moderate; 30-50, moderate to high; 50-70, high to very high; and more than 70 extremely high.
EC: Saturated salt extract measured as dSm^{-1} per metre.

These soils have developed under impeded drainage due to nearly impermeable, clay subsoils usually underlain by indurated hard pans and highly fluctuating groundwater table. The soil surface has extensive white salt-efflorescence during dry parts of the year, turning into pellets of black to brownish-grey flakes due to dispersion and dissolution of humic substances at high alkalinity levels.

3. *Degraded alkaline or solodic*: Rice cultivation on non-calcareous sodic soils for a long period has resulted in degradation or solodization of these soils due to excessive leaching, leading to dealkalization of surface horizon and lowering of pH to nearly 5.2. The degradation process leads to replacement of exchangeable Na^+ by exchangeable H^+ and Al^{+++} in the surface layer along with accumulation of crystalline, amorphous and water-soluble silica. Disruption of the clay complex is usually not conspicuous since exchangeable Al^{+++} is much less than exchangeable $3H^+$. The dark colour of the surface horizon is due to accumulation of organic matter under submerged conditions for longer durations. In subsurface horizons, the soil pH gradually tapers to neutrality with depth; the magnesium is very conspicuous however, which imparts poor physical properties and a prismatic structure to these soils, which break into medium blocks. The ESP is usually below 15. The groundwater table is within two metres even during the summer months and gleying in the subsurface is well marked.

Mechanism of Salt Stress and Plant Response

Stress factors, namely osmotic, ion toxicity, nutrient imbalance and pH alter the expression of several morphological, physiological and biochemical characteristics of plants. As the stress increases, plant growth is further restricted. In severe stress plants may die prematurely after germination or transplanting.

Seed germination is often hindered and or delayed. Seedlings often fail to survive and thus the plant stand may be thin initially. Plant growth is stunted and the number of tillers is relatively reduced. Other vegetative characters, such as leaf number, size, etc. may also be somewhat affected. Anthesis may be delayed and maturity hastened. The slow rate of starch accumulation leads to shrivelled grains. There is an overall reduction in yield. Realized yields are usually much less than potential yields under normal growing conditions.

1. *Soil-related stress mechanism*: Plant growth in saline soils is usually affected because of the osmotic effect in the soil solution. High salt concentration increases the potential forces that hold water in the soil and makes it more difficult for plant roots to extract soil moisture. During a dry period, salt in soil solutions may be so concentrated as to kill plants by sucking water from them (exosmosis). Salt in the soil solution forces a plant to exert more energy to absorb water and to exclude salt from metabolically active sites. As salinity increases, plant growth is further restricted. A saline soil should be kept wet to dilute the salt concentration so as to cause the least salt hindrance to the growing plants. Plant growth in sodic/alkaline soils is affected due to high ESP throughout the profile, very low infiltration and hydraulic conductivity rates. The exchangeable complex of alkaline soils is largely occupied by sodium ions which cause dispersion of soil due to the

breakdown of aggregates forming a dense surface crust which greatly hinders seedling emergence due to low permeability of the soil to water and air. Poor drainage in such soils is due to a high water table which further restricts plant's ability to absorb water and nutrients in required amounts. High pH results in reduced availability of some essential plant nutrients. Accumulation of certain elements in plant parts at toxic levels may result in plant injury or reduced growth and even death in extreme cases. Elements commonly toxic are sodium, molybdenum and boron. Selenium may also occur in toxic concentration. Plant growth in degraded alkaline or solodic soils is largely due to poor drainage. After making adequate provisions for dr ainage, crops such as rice and sugar-cane can thrive well.

2. *Plant responses to salt stress* : Crop species and varieties vary a great deal in their response to salt stress. Many naturally occurring plants in salt-affected soils have certain specialized structures, for example salt glands and salt hairs on their leaves. Detailed studies on salt glands in salt-tolerant plants, such as *Leptochloa fusca*, showed the presence of enlarged cells protruding above the epidermis of both abaxial and adaxial surfaces of leaves and also on the exposed side of the leaf sheath. These glands are associated with salt deposition (Na > K > Ca > Mg) on leaf surfaces. *Acanthus ilicifolius* and other crop species have salt glands on the adaxial leaf surface and studies have shown each gland to be surrounded by six collecting cells (salt-collecting cells). One of the most salt-tolerant plants, *Porteresia coarctata*, has unicellular salt hairs on the adaxial surface of the leaves. Analysis of its leaf washing showed that Na and Cl were predominantly excreted, followed by K, Mg and Ca. Some crop species have sunken stomata. Sensitive rice variety M 1-48 showed highly constricted vascular bundles of the stem and midrib of the third leaf at the tillering stage in response to salinity.

Plants subjected to salt stress face reduced availability of water and response to changes in the processes related to maintenance of a favourable water balance. Increase in salinity resulted in a decrease in transpiration in mustard, wheat and pearl millet, whereas leaf diffusive resistance (LDR) and leaf temperature increased. Higher LDR coupled with low transpiration might contribute to moisture conservation in plants under salt stress conditions.

Excessive salt in the root environment not only reduces availability of water to plants, but their excessive absorption of salt creates danger of ion toxicity and interference in the uptake of other essential nutrients. In general, with increase in salinity and sodicity, Na content increase in plants while K decreases. The antagonistic effect of both cation is well established. Tolerant varieties show a tendency to take up less Na while maintaining their K status.

Plants growing at sublethal levels of salt stress may often appear greener due to increase in chlorophyll. Accumulation of certain amino acids, sugars and other osmotically active organic substances in response to salt stress are indications of altered nitrogen and carbohydrate metabolism. Two-week-old wheat plants subjected to electrolyte concentration (EC) 22 doubled their amino acid content after 24 hours. Glutamine (1.6 x), phenylanine (3.2 x) and particularly proline (10.9 x) were the most prominent. There was significant decrease in alanine, aspartic acid and glutamic acid. The highest proline accumulation occurred in lamina followed by leaf sheath and roots. In barley moderately tolerant varieties accumulated more proline than sensitive ones.

In wheat, water-soluble proteins increased in leaves in response to salinity. In *Chloris gayana*, TCA and NaOH soluble proteins increased in response to salinity. Enzymes are also influenced by change in plant water status as well as ionic imbalance. Decrease in (a) amylase activity with increase in salinity was observed in wheat leaves after two weeks of exposure to EC 12, while activity of invertase and other enzymes of carbohydrate metabolism increased 60% in upper leaves. Nitrate reductase activity may also decrease with increase in stress level in many species. Tolerant varieties of pearl millet showed a tendency to maintain their nitrate-reductase activity. Polyphenol oxidase activity has been reported to be higher in sensitive varieties of wheat, barley and rice. Qualitative and quantitative changes in phenolic compounds at tillering and grain-filling stages may also occur.

Criteria for Salt Tolerance

Salt tolerance ratings are usually fixed on the basis of relative yields (50% reduction in economic yield compared to control) at defined stress levels (ESP levels) of salt stress.

1. *Seed germination and seedling survival*: To begin with, good field emergence (good seed germination followed by good seedling survival) under a defined stress condition is the prime requisite for salt tolerance. Crops and genotypes that even fail to establish themselves well under defined stress cannot be expected to do any better at a later stage.

2. *Grain yield*: Highly tolerant varieties are those that exhibit minimum reduction in relative economic yield with per unit increase in stress. The slope of regression of yield against stress gives a fairly reliable estimate of salt tolerance of a crop/genotype. This is by far the best index for identification and screening of salt-tolerant genotypes.

A number of other plant attributes, namely Na and K content in shoots/leaves, Na/K ratio, pH of the cell sap, proline content and enzyme response may also have some potential use. The only limitation to their practical use so far however, is that the differential genotypic response observed in various crops cannot always be explained on the basis of these data. Tables 26.1 and 26.2 give the relative crop tolerance to soil sodicity and salinity.

Procedure for Breeding Salt-tolerant Varieties

1. Intracrop Variability

Crop species differ considerably in respect of limits of salt tolerance (based 50 per cent reduction in crop yield) (Table 26.3).

Table 26.1. Relative crop tolerance to soil sodicity

ESP range			
> 50 tolerant	30-50 moderately tolerant	20-30 semi- tolerant	20 sensitive
Rice	Mustard & Rape-seed	Flax Groundnut	Peas Chickpea
Barley	Wheat Sunflower Sorghum Pearl millet	Garlic Onion Lentil	Soybean Safflower

Table 26.2. Crop groups based on response to salt stress (saline soils)

Highly tolerant	Medium tolerant	Medium sensitive	Highly sensitive
Barley	Spinach Sugar-cane	Radish Cowpea	Lentil
Cotton Sugar-beet Tobacco	Rice	Broad-bean Vetch Peas Cabbage	Chick-pea Beans
Safflower	Wheat Pearl millet Alfalfa Blue panic Para grass Sudan grass	Cauliflower Gourds Tomato Sweet potato Sorghum Millets Corn Clover	Carrot Onion Lemon Orange Grape Peach Plum Pear Apple

Table 26.3. Range of variability for salt tolerance

Salinity		Alkalinity	
EC_e dSm^{-1}	Crops	pH	Crops
4.0-11.0	Barley	9.2-10.2	Rice
4.0-10.0	Rice	9.2-9.8	Sugar-beet
4.2-10.0	Sugar-beet	8.8-9.3	Wheat
4.0-8.0	Pearl millet	9.0-9.3	Barley
4.0-7.0	Broad-bean		
		8.9-9.0	
			Sugar-cane
4.0-6.0	Flax, Safflower, Sunflower	8.7-8.9	Flax, Pearl Millet
		8.7-8.8	Safflower
		8.6-8.8	Cowpea
4.0-5.0	Cowpea, Chick-pea		Pea, Chic-pea Sunflower.

2. Breeding Approaches

The usual breeding procedures are the basic procedures that are also followed for breeding of salt-tolerant varieties using the following strategies.

i) Selection of genotypes under defined salt-stress conditions: This involves selection of genotypes under defined stress conditions, field and/or under simulated field conditions (using lysimeters). Initial screening of varieties at germination stage/seedling stage may be done in wooden trays.

ii) Incorporation of relevant salt-tolerance traits into agronomically superior varieties under optimum conditions to make them suitable for suboptimal conditions.

FLOOD RESISTANCE

Characterization of Environment

Deepwater rice has an agro-ecosystem which is much more complex than other cultural types. In this agro-ecosystem the rice crop is first required to withstand drought conditions, and subsequently after commencement of the rainy season (monsoon season) it is required to withstand rapidly surging floodwaters and grow under shallow flooding to a deeply flooded condition for the rest of the crop duration. The extent of flooding depends the amount of rainfall and its distribution. The flood-pattern thus varies a great deal. The soils may pose additional constraints, such as high or low pH, aluminium toxicity, phosphorus and micronutrient deficiency. The physiographic silting of wetlands is such that they all accumulate surface water, usually more than 10 cm for a substantial period or some of them even for the entire duration of the crop growth season. Often, more than one physical stress occurs at the same time or throughout the growing cycle, even though one stress may dominate. Rice crop failures or low production have been recurrent features of this ecosystem. The crop environment can be subgrouped as follows:

1. Stagnant-water Conditions

These are the low-lands in which the water level rises gradually with onset of the rainy season until the rains recede. It includes ponds, lakes and waterlogged areas. The water level recedes gradually and may last beyond December and January, that is during the entire period of crop growth.

1) Shallow-water rice (10-30 cm deepwater conditions)
2) Medium-deepwater rice (30-50 cm deepwater conditions)
3) Deepwater rice (50-100 cm deepwater conditions), and
4) Very deepwater rice (more than 100 cm deepwater conditions)

2. Flash-flood Conditions

These are the areas which are flooded with surging floodwaters from overflowing rivers following excessive rainfalls. Duration of flooding may be short or relatively long and frequency of floods may be once or twice to as many as five times during the entire crop season. The water level, duration and force of water current may differ each time The depth of water may even vary within a field depending on its slope. It is within this framework that the crop environment must be seen and understood.

Assessment of Traits Associated with Flood Resistance

Deepwater and 'floating' rice varieties possess certain unique characters not seen in other rice varieties. These are described below.

1) In deepwater rices submergence tolerance and in floating rice elongation ability is the prime need. Ability to rapidly elongate internodes is a trait which enables the rice to keep its head literally above the rapidly surging floodwaters. Deepwater rice varieties seem to be expanding so much energy in elongating their internodes that they are left with very little energy to produce grain. Consequently, their yields are rather low.

2) Ability to develop tillers from nodes, besides those arising from the base.

3) Plants should be photosensitive so that flowering commences from late October to the first week of November.

4) Plants should be able to withstand drought during the early stage of crop growth. Establishment of good plant stand to begin with is a good plant attribute.

5) The variety should be of medium plant height with strong culm with slight kneeing ability so as to withstand the pressure of water current during flash-floods.

6) The variety should be resistant to stem-borer and blast, especially the 'neck blast' disease.

Breeding Approaches

In recent years breeding of rice varieties for deepwater areas has received increasing attention as no other crop can successfully be raised in such situations. The major problems, however are to combine desirable characters such as, submergence tolerance. semi-tall height, photosensitivity, superior grain quality and resistance to insect pests in a single genotype. Conventional breeding procedures may be employed to breed improved varieties. Screening and evaluation would invariably need to be done under diverse stress conditions only to achieve the desired objective. The lack of target environment for testing remains a major constraint. The possible approaches are:

1) Screening and or development of varieties under defined stress conditions. This involves identification of suitable donor parents and incorporation of submergence tolerance and other desirable characters mentioned above.

2) The better performing germplasm (local and exotic collections) may be isolated and released. There is a need to select/evaluate the local germplasm for its wider adaptability.

HEAT RESISTANCE

The high temperatures during growth stage of the crop and more so at the time of anthesis adversely affect crop growth and reproductive processes leading to flower abscission, sterility and poor seedset. Asana and his group at I.A.R.I., New Delhi studied the relation between temperature from anthesis to maturity and grain development in two wheat varieties, namely, Pb C 281 and NP 720. They reported that for every 5°C rise in mean maximum temperature (between the limits 24.2°C to 32.3°C), the 1000-grain weight declined by 11.9 and 15.9% respectively. He later generalized that the reduction in grain size to be 16.4% for every 6°C rise in temperatures between the limits 24°C and 31°C. The rate of grain development showed that grain weight per ear consistently increased with an increase in temperature from 17 to 25°C, whereas it was significantly depressed at 28°C and 31°C, probably because of an increase in rate of respiration. It was also reported that although during the first fortnight after anthesis grain size increased with an increase in temperature from 27.9°C to 31.4°C in the final analysis grain weight per ear declined by 17.3% for every 5°C rise in mean maximum temperature (between 27.6°C to 32.3°C) during the whole growth period. The adverse effect of high temperatures is thus quite apparent (Asana, 1966).

Mechanism of High-temperature Effects on Crop Growth

Above-normal temperatures speed up the growth and development phases, usually negatively affecting crop establishment, leaf area growth and development, ground cover by the crop, radiation interception and photosynthesis per unit ground area. The yield sensitivity to high temperatures varies with growth stage.

Assessment of Characters Associated with Resistance to High Temperatures

Selection for resistance to high temperatures is related to the ability of a genotype(s) to maintain the duration of growth stages of the crop, for example duration between emergence to fifth-leaf stage on main shoot and up to anthesis, and from anthesis to maturity. Anthesis during the coolest part of the crop-growing season is the most desirable attribute. Other plant attributes, namely quick early growth to rapidly cover the ground, high leaf number, high transpirational cooling, glaucousness of plant parts and deep root system are other desirable characters. 1000-grain weight is a fairly stable character and considerable variability exists among wheat genotypes. At the cellular level, membrane stability at high temperatures has been reported to increase the heat tolerance of genotypes. The membrane stability test therefore may be useful in screening wheat genotypes for heat tolerance.

Breeding Approaches

Conventional breeding procedures are the basic procedures that may be used to breed heat-resistance genotypes. There are, however, two approaches:

1) Choosing a planting date that would produce anthesis during the coolest growing month leads to higher yields. In practical terms, the varieties developed for late sowing should possess quick early growth ability and somewhat shortened growth period from germination to anthesis, so that anthesis assuredly takes place latest by mid-February.

2) Incorporation of heat-resistance characters, such as cell membrane stability at high temperatures, higher 1000-seed weight and various other characters discussed earlier.

COLD TOLERANCE

Characterization of Low-temperature Environments

Two types of low-temperature environments are usually recognized depending on the type of injury they may cause to plants. These are chilling injury (0–10°C) and freezing injury (soil temperatures below 0°C). The combined effects of climate, soil, plant and cultural factors interact to determine the degree of injury incurred by a crop following the rigours of low-temperature winter regimes. Soil factors, such as compaction or water-holding capacity, may accentuate or diminish the influence of several climatic factors, including temperature and precipitation. Climatic conditions, such as amplitude of diurnal temperature fluctuations, or snow cover may mitigate soil factors. Cultural practices and previous climatic conditions may influence the plant's resistance which shows considerable diversity within a species, to the stresses imposed on it by the combined climatic and soil factors.

Effect of Low Temperature (Non-freezing Temperatures) on Plants

1) *Chilling injury*: Chilling injury refers to the physiological damage caused to plants by low temperatures. The degree of injury depends on the temperature, duration of exposure, the stage of development and genotype. The primary effect is expressed at the cellular level and, more precisely, on the membrane stability which decreases with modification of lipids and membrane-bound transport proteins. This leads to increase in solute leakage and alteration of chloroplast functioning due to the same effect on their membrane. Also, membrane alteration and reduction in metabolic activity at the root level reduce their hydraulic conductance, leading to a chilling-induced drought stress. Dysfunction of membranes at low temperatures is thus the primary cause of chilling injury.

At the crop level, injury is not usually visible during chilling but appears after an increase in temperature. Wilting and discoloration of the leaves are symptoms of chilling injury. With severe chilling, plants or plant parts are killed. Long-term exposure to chilling temperatures results in seed imbibition and mortality. Young seedlings may be killed by an exposure to 1°C or so for a weak or so because of ultrastructural changes in the meristematic cells of primary roots. The range of injuries includes a slight discoloration of roots and killing of the primary root.

(2) *Chlorosis*: In cool and bright weather conditions, seedlings are often partly chlorotic. Chlorosis occurs at a high light intensity in combination with temperatures of 10-15°C. Chlorotic plants may recover at higher temperatures but growth is somewhat inhibited in comparison with non-chlorotic plants.

(3) *Other adverse effects*: Seeds or seedlings may be killed by soil fungi, seedling malformations may occur as a result of the opening of coleoptiles below soil level and the seedling vigour may be reduced. Low temperatures during flowering may result in abnormal flowers or sterile flowers, and/or abortion of flowers.

Effect of Freezing Temperatures on Plants (Freezing Injury)

Freezing in plant tissues involves the redistribution of water with respect to both its physical state and its location. When intact plants, tissues, cells, or isolated organelles are subjected to temperatures that are decreasing below the freezing point of water, the water both in the cell and in the extracellular surroundings will initially supercool and then ice will be formed. The intracellular ice, created around small particles inside the cell, is responsible for cell dehydration, and later for cell membrane destruction. The extracellular ice, which produces a matrix around the plant cell, causes mechanical damage and the necrotic zones resulting from anoxia.

The most common symptoms at the plant level are wilting, death of organs (leaves and primary stems) or death of the plant. The visual manifestation of freezing injury is darkened, water-soaked, flaccid appearance immediately after thawing. The gross disruption of cellular architecture is evidenced by extremely leaky cellular membranes.

Mechanism of Resistance to Chilling and Freezing Injury

Plant resistance may take the form of certain physiological or morphological adaptations that allow the plant either to tolerate or to avoid imposed stresses. Tolerance or avoidance mechanisms may reside at either the whole plant, tissue or at cellular level.

Winter hardiness: Winter hardiness implies avoidance of or tolerance to all the cumulative effects of winter that a plant/genotype encounters, namely freezing, heaving, smothering, desiccation, diseases, etc. It is a composite of cold hardiness, desiccation resistance, ability to tolerate or resist frost, heaving, disease resistance and probably several other factors as well.

Cold hardiness: Cold hardiness refers to the ability of a plant/genotype to withstand low freezing temperatures.

Chilling Resistance

Chilling resistance during seed germination can be conferred by mechanisms which slow down seed imbibition. As a matter of fact, rapid imbibition at low temperatures leads to membrane destruction and subsequent loss of solutes at the expense of the seed and the advantage of soil pathogens. The most efficient characteristic in legumes seems to be the phenolic content of the seed-coat.

Various mechanisms at the cell membrane level have been proposed to explain increased membrane fluidity during growth. For example, the level of unsaturated fatty acids can reduce the temperature at which membrane fluidity- and thus function are reduced. At present, it is possible only to measure the change in membrane function associated with its structural alteration during chilling. The most common measurements are solute leakage (electrical conductivity test) and level of reduction of photosystem II (by fluorescence test). The drought stress induced by chilling can be limited by ABA production, leading to increased root hydraulic conductance at low temperatures. At flowering the chilling resistance of the pollen could be associated with proline accumulation leading to simultaneous increase in chilling resistance.

Freezing Resistance

With regard to position of stress these mechanisms concern the vegetative parts of the plant. There are three possible mechanisms:

a) *Freezing escape* : This involves managing cultural practices in such a way that the most susceptible crop stage does not coincide with the main freezing period and comes after the main freezing period is over. Another type of freezing escape can be obtained with a change in plant morphology. For example, in pea, genotypes with a prostrate growth habit with high branching rate and short internodes, giving a rosette-type plant during the winter, are more resistant to frost. Whether or not these morphological and phenological characteristics or escape are associated with physiological mechanisms of resistance is not known.

b) *Freezing avoidance* : Intracellular ice formation can be avoided by supercooling (due to lack of ice nucleation particles) and osmotic adjustment.

c) *Freezing tolerance* : This is the ability of the plant to withstand extracellular ice formation (by avoiding cell dehydration) or intracellular ice formation (with increased membrane stability). Osmotic adjustment is an interesting mechanism for retaining water in the cell but its effect is limited. Cytoplasmic water binding can account for much of the tolerance to frost. The presence of small leaflets is a characteristic reportedly associated with resistance to frost.

Freezing avoidance and tolerance mechanisms are included in the hardening process, which is a progressive adaptation of the plant to decreasing temperatures. It is a reversible process and its benefit can be lost if there is an increase in temperature. Several mechanisms, not necessarily specific to freezing stress, are involved in hardening. These are:

• Increased membrane fluidity.
• Osmotic adjustment, associated mainly with soluble sugars accumulated in the cell vacuole.
• Accumulation of more specific compounds, for example proline, ABA. Some phenolic compounds also play a role in frost tolerance.

The hardening process is associated with a switch of the plant metabolism towards slow growth and even dormancy. Meristems are more resistant than the other plant parts and can be the origin of new stems and roots, particularly if they are at the seedling level and protected by the soil.

Screening Methods for Winter Hardiness

Screening methods that allow for accurate assessment of winter hardiness potential are critical to breeding for winter hardiness.

1. Field Survival Test

Field survival is the most commonly used method of evaluating winter hardiness. The procedure consists of conducting trials over a large number of locations and assessing the percentage of plants that have survived the rigours of winter. It is considered the ultimate test of a variety's winter hardiness. However, field survival trials are often inconclusive either due to complete winter kill or lack of it. For this reason the breeder has to either wait for a winter of desired severity or conduct trials over several locations in the hope of encountering desired stress at any one or more locations. Even when the trials are conducted over a large number of locations one is not sure whether the desired stress conditions will be encountered. Moreover, even when differential winter killing is observed, it is often irregular, with the result that experimental error is usually high. Further detection of small, but important differences among genotypes or treatment is often difficult. Repeated field tests are therefore necessary.

Fowler and Gusta (1979) developed a field Survival Index (FSI) for measuring relative winter hardiness in cereals. The differences in FSI represent the average per cent difference expected in field survival (averaged over all trials). This is a modified procedure which reduces the experimental error associated with field trials, alleviates the difficulties arising from the absence of partial winter kill for a few varieties in each trial, and allows for pooling of the results from different trials. Only plots which have partial winter kill, that is, > 5 and survival < 95 per cent are used to develop a field survival rating. Differences in winter kill among varieties are determined for each replicate.

2. Controlled tests :

According to Brule-Bable and Fowler (1989) an ideal control test must be highly correlatable with field survival, simple, repeatable, rapid, non-destructive and require only a single plant for analysis. Fowler et al. (1981) found that crown and leaf water content of field acclimated plants and LT_{50} were the best predictors of field survival ability. This observation is in agreement with many other studies. The methods used by Brule-Bable and Fowler (1989) are described below.

LT_{50} AND TISSUE WATER CONTENT

The various steps involved for cereals were as follows.

    i) *Cold acclimation* : Seed of cultivars to be tested were surface sterilized with a 0.6% sodium hypochlorite solution for 10 min, rinsed thoroughly, subjected to imbibition at room temperature for 12-18 h, then placed on moist filter paper in petri dishes. The seeds were then germinated at a constant temperature (25°C) with a 16 h day/8 h night and kept under these conditions for 24-36 h. Germinated seedlings were transferred (embryo down) to plexiglass trays with holes backed by a 1.6 mm mesh screen and returned to germinating conditions until their roots were 1-2 cm long. They were then placed in hydroponic tanks filled with continuously aerated one-half strength modified Hoagland's solution. Temperatures were maintained at 25°C day and night with a 16 h day/8 h night for one day, then reduced to 22°C day/15°C night until the plants reached the 3-leaf stage. Daylength was then reduced to 12 h and the following temperature regime was followed: 5°C day/3°C night for 7 days; 3°C day/1°C night for 14 days; 2°C day/-2°C night for 14 days.

ii) LT_{50} determination : Plants were prepared and frozen by a method similar to that described by Fowler and Carles (1979) for the determination of LT_{50}. All plant parts 3.0-4.0 cm above and 0.5 cm below the crown tissue were removed. Five plant crowns from each cultivar were placed in labelled aluminium weighing dishes for each test temperature, covered with moist sand and set on an aluminium plate in a modified, controlled freezing unit (Gusta et al., 1978). Five test temperatures separated by 2°C intervals were selected for each cultivar. Twenty-four gauge copper-constant thermocouples were inserted in moist sand in the aluminium weighing dishes and placed at three locations within the low temperature chamber to monitor temperatures during freezing. Crown tissues were allowed to equilibrate at − 3°C for at least 12 h and the temperature was then lowered at a rate of 2°C per hour until a temperature of − 17°C was reached. At this point, almost all freezable plant water had frozen (Gusta et al., 1975). The freezing rate was then increased to 8°C per hour. Samples were removed from the freezing unit upon reaching their designated test temperatures and thawed overnight at 1°C. Thawed crowns were planted into flats containing a soil-peat-vermiculite (1:1:1 by volume) potting mixture which was always kept moist. The flats were placed in a growth room maintained at 18°C with a 16 h day/8 h night. After three weeks, plant recovery (alive vs. dead) was rated and LT_{50}s were calculated for each sample.

iii) *Tissue-water determination* : Individual plant crowns and leaves were prepared for water determination by removing all but 0.1 cm of the roots and separating the shoot into a 2.0 cm crown portion and the remaining leaves. Any frozen surface water was blotted off with dry paper towelling. Water content was then determined either by oven-drying or Karl Fischer titration.

Oven-drying

Fresh leaves or crowns were weighed, and dried in aluminium or glass dishes for 48 h at 50°C in a forced-air convection oven. After drying, samples were cooled in a glass desiccator, then weighed. Per cent water content was determined as follows:

$$\text{Per cent water} = \frac{\text{Fresh weight} - \text{dry weight}}{\text{Fresh weight}} \times 100$$

Karl Fischer Titration

Fresh samples were weighed, their individual crowns placed in glass vials containing 5 ml methanol while leaves of the same plant were cut and placed in glass vials containing 10 ml of methanol. The vials were sealed with stoppers and parafilm, then continuously agitated on a rotating table at 100 rpm for 18 h. A 1-ml aliquot from each sample was then titrated with an automatic Karl Fischer titrator using a 1-min conductivity end point. The per cent water content of the sample was calculated as follows:

$$\text{Per cent water} = \frac{(\text{KFR}_s - \text{KFR}_b) \times \text{METH}_s \times f \times 100}{\text{Weight of sample (mg)}}$$

where, KFR_s = ml of Karl Fischer reagent required to neutralize 1 ml of sample;

KFR_b = ml of Karl Fischer reagent required to neutralize 1 ml of methanol (blank);

METH_s = total ml of methanol in each sample (5 for crowns; 10 for leaves)

$$f = \frac{\text{ml of } H_2O \text{ in standard}}{\substack{\text{ml of Karl Fischer reagent} \\ \text{required to neutralize the standard}}}$$

3. Other Tests

Some other indirect tests are described below.

Soluble-protein content: Soluble-protein content in tops and crowns of unfrozen wheat plants, for example, have been found to correlate significantly with percentage survival after freezing. This is determined by a lixiviation test. High electrical conductivity of the leachant indicates more extensive cellular disruption and freezing damage.

Plasmolysis test: Healthy cells plasmolyse in a hypertonic solution of salts, such as calcium chloride. When cells are injured by freezing, selective permeability of cell membrane is lost and the cells are unable to plasmolyse. Microscopic examination of this tissue reveals the extent of damage and type of cells that have been damaged.

Tetrazolium test (TZ): TZ test is a useful test for rapidly obtaining an indication of living cells. The living cells are made visible by reduction of an indicator dye. The indicator used in the TZ test is a colourless solution of a tetrazolium salt (2,3,5 triphenyl tetrazolium chloride) imbibed by the living cells. Within the living cells it interferes with the reduction processes of living cells and accepts hydrogen from the 2,3,5-triphenyl tetrazolium chloride, a red, stable and non-diffusable substance, triphenyl formazan, is produced in living cells.

For quantitative assay, it is extracted from the cells with 95% ethanol and quantified photometrically.

However, the initial reducing capacity of cells may vary with plant part, age and genotype which makes relative comparisons difficult.

Practical experience has demonstrated however, that all laboratory tests are of limited value in breeding programmes because of inefficiency and high experimental error due to various reasons. Therefore, from a plant breeding viewpoint field screening remains the only means for evaluation of germplasm and breeding materials for direct use in improvement for winter hardiness/cold tolerance.

Method of Field Scoring

Comparative ratings should be recorded in the following manner after the susceptible check is completely killed (Table 26.4).

Table 26.4. Method of field scoring for cold tolerance

Score	Percentage leaf damage	Percentage damage to branches, tillers, etc	Percentage of plants killed
1. Immune	No visible symptoms of damage		
2. Highly tolerant	up to 10%	–	Nil
3. Tolerant	11-20	up to 20%	Nil
4. Moderately tolerant	21-40	up to 20%	Nil
5. Intermediate	41-60	21-40	5
6. Moderately susceptible	61-80	41-60	6-25
7. Susceptible	81-99	61-80	26-50
8. Highly susceptible	100	81-99	51-99
9. One hundred percent plants killed.			

Breeding Approaches

The ability of plants to withstand low temperatures is a latent trait, which exhibits an annual periodicity. It is only through the interaction of appropriate environmental cues and the genetic potential of a species that an increase in cold hardiness is manifest. A knowledge of what constitutes freezing damage is essential. After analysis of the target environment the breeder is required to define the combination of mechanisms of resistance.

The genetic variation of most plant responses is rather small and most of the cold-tolerance traits are genetically unrelated. Simultaneous selection for all the traits is practically impossible and ineffective. The best approach, therefore, is to select source material for the various traits separately and to recombine the selected sources. Advance populations might be selected for a whole complex of features in order to recombine and to accumulate the desired genes.

27

Breeding for Specific Traits

LODGING RESISTANCE

Lodging

The term lodging is applied to a crop which has partly or completely lain over the ground before harvest and which may have become more or less entangled. In some cases, the stems may simply lie over with no perceptible bend at the base of the plant, while in other cases there may be an abrupt bend in the stem at lower internodes, with even crumpling of internodes. Lodging usually results from rank growth, storms, inherently weak stems and waterlogged conditions. When lodging is due to a weak root system, it is called root-lodging and, when it is due to weakness of stem and weight of ripening grain and a buckle or break occurs in the lower internodes, it is called stem-lodging.

Depending upon the stage of crop growth at which lodging occurs plants may become nearly upright with the return of favourable conditions. So long as the straw is soft and green, and the cells of the nodes and thick basal portion of the leaf sheath retain their vitality, the ears of a laid crop may to a greater extent regain their upright position by forming a crinkle joint or elbow joint through geotropic stimulus. When the nodal tissues are dead or dying, the plant remains lodged.

Mechanism of Lodging Resistance

The degree/magnitude of reduction in yield due to lodging is determined by the type of lodging. Culm breakage causes the most severe loss compared to bending and root-lodging. Lodging is attributed to many factors, namely culm length, thickness of culm wall, culm diameter, height of the centre of gravity, culm weight, quality of culms and total weight of plants. Lodging is initially induced by the weight of leaves and culms themselves, and the heavy load of grains at ripening. Serious lodging results when strong winds and a heavy load of raindrops cause further loading on the leaves and culms.

To reduce the stress on culms, the centre of gravity of the plant needs to be lowered. Dwarfing genes are widely used to achieve this objective, which provides stiff straw and tolerance to heavy fertilization. Lodging-resistant genotypes are usually less affected by cultural practices, such as heavy nitrogen, deepwater and dense planting than susceptible ones.

Assessment of Characters Associated with Lodging

Lodging is a complex character influenced by genetic as well as environmental factors, such as soil fertility, rate of seeding, disease infection, abundant rainfall, wind storms, etc. For breeding purposes we shall, however, limit our discussion to genetic factors alone. The genetics of various associated characters is rather well studied. Among morphological characters, short stiff straw and semi-dwarf stature have been the most successfully used for incorporating lodging resistance in cereals such as wheat, barley and rice. Features such as breaking strength of straw and culm density have also been used by many workers.

The conventional breeding methods described earlier are used for selection of lodging-resistant varieties.

PHOTOSYNTHETIC AND RESPIRATORY EFFICIENCY

Crop yields can be simply defined as a function of net amount of carbon fixed by the photosynthetic system, and allocation of more carbon into appropriate sinks (economic yield versus total biomass production). Photosynthetic efficiency includes the proportion of sunlight that falls on the organism, the efficiency with which light is harvested and the efficiency of the enzyme reactions involved in electron transfer and carbon and nitrogen fixation. Crop species differ in their ability to fix carbon dioxide (CO_2). On theoretical grounds, therefore, there may be possibilities for yield improvement. Let us examine these possibilities.

CO_2 Fixation in C_4 Plants

Plants such as corn, sugar-cane are known as C_4 plants; they are more efficient in CO_2 fixation compared to plants such as wheat, known as C_3 plants. It has been proposed that this is due in part to the lack of photorespiration in plants and in part to differences in the metabolic pathway in C_4 plants.

Tissue anatomy: In C_4 plants there are two concentric layers of cells around vascular bundles, the 'bundle sheath', and a sheet forming the mesophyll part. Both types of cells have chloroplasts, but the chloroplasts of the bundle sheath have very few grana and therefore have a low level of partition membranes. For this reason, the level of photosystem II[1] is also low.

Metabolic pathway: CO_2 fixation in C_3 plants, referred ..⌐ as the Calvin cycle, is regulated in the stroma of chloroplasts by several enzymes. But the main site of control is the enzyme ribulose 1,5 bisphosphate carboxylas⌐ (RBPC), the first reaction of the cycle. The carboxylase activity of RBPC is controlled by CO_2 concentration and inhibited by oxygen. Briefly, 6 carbons in the form of CO_2 react with 6 molecules of RBPC, and the two give rise to glucose ($C_6H_{12}O_6$) which is taken out of the cycle.

---

[1] Photosystem I is situated mainly in the stroma, and photosystem II is associated with partition membranes.

CO_2 fixation in C_4 plants. on the other hand. follows a different cycle (Hatch and Slack pathway). The CO_2 is fixed in the cytosol of the mesophyll cells by the enzyme phosphoenol pyruvate (PEP) carboxylase and the oxaloacetate so formed is reduced to malate by $NADP^+$—in linked malate dehydrogenase. Malate[2] is subsequently transported into the chloroplasts of the bundle sheath cells, and via the action of malic enzyme CO_2 is regenerated in the chloroplast of the bundle sheath cells where it is used by the Calvin cycle. Pyruvate is also generated and transported back into the mesophyll cells where the action of pyruvate phosphodikinase regenerates PEP. As PEP carboxylase in the mesophyll cells has a higher affinity for CO_2 than RBPC in the bundle sheath cells, this mechanism has the effect of concentrating CO_2 in the bundle sheath cells for the Calvin cycle. This concentrating effect probably accounts for the 2-3 times higher rate of photosynthesis in C_4 plants compared to C_3 plants, because such a mechanism enables RBPC to function more efficiently. This system of CO_2 fixation is very efficient at producing hexose at high light intensities and at high temperatures. The plants grow rapidly due to their fast rate of hexose biosynthesis.

The other important differences between C_3 and C_4 systems are:

(1) CO_2 fixation in C_4 plants requires extra energy. For 6 CO_2 molecules to be fixed into glucose an extra 6×2 ATP molecules are required.

(2) In C_4 plants, there is an effective transfer of reducing equivalents (NADPH) from mesophyll cells to the bundle sheath cells by the action of malate dehydrogenase, then malic enzymes. And since the level of photosystem II is low, little NADPH is produced in the bundle sheath cells. So. this transfer tops up the amount required for CO_2 fixation.

Photorespiration

We have already stated that CO_2 fixation by RBPC is inhibited by oxygen. When oxygen is present and the RBPC activity is inhibited on this account, the oxygen is metabolized by RBPC and the enzyme then referred to as RBPC oxygenase to remove inhibition. This is known as photorespiration.

C_4 plants do not exhibit photorespiration. This may be because the Calvin cycle, and therefore RBPC, is in the bundle sheath cells (the inner layer of cells) in these plants. CO_2 is concentrated there and the oxygen concentration is low, so photorespiration is unlikely to operate. C_4 plants, however, do contain peroxisomes and it is thought that when the stomata are closed in these plants under dry conditions. the CO_2 concentration becomes too low for the Calvin cycle to operate and phosphoglycolate is formed by the action of RBPC oxygenase. The CO_2 then formed by photorespiration, using the peroxisome and mitochondrial enzymes, is subsequently fixed by the Calvin cycle.

---

[2] In some C_4 crops aspartate may form in place of malate.

Current Status of Research

Research is currently being carried out on the effects of inhibition of photorespiration in C_3 plants, as this could increase crop yields in these plants. Photorespiration is inhibited by lowering the $O_2:CO_2$ ratio. Crop yields can be nearly doubled by a fivefold increase in CO_2 concentration.

Attempts to use inter- and intraspecific genetic variability in the rate of net leaf photosynthesis in breeding programmes have not generally been satisfactory. Production of hybrids between C_3 and C_4 species with the objective of obtaining higher rates of photosynthesis than the parental C_3 plants have not yielded expected results.

SYMBIOTIC NITROGEN FIXATION

Soil bacteria of the genus *Rhizobium* can invade the roots of particular host plants and stimulate the development of root nodules. The bacteria inhabit the nodules as endosymbionts called *bacteroids*. In this state they fix nitrogen (N) into ammonia, a form that the plant can assimilate. The process of nodulation is complex and requires the active participation of both *Rhizobium* and plant. An important feature of this symbiotic relationship is the specificity of *Rhizobium* for a particular group of plants. Almost without exception, the hosts for *Rhizobium nodulation* are members of the family *Leguminosae*. Within that family, groups of plants are nodulated by particular *Rhizobium* species. Not only this, different strains of a *Rhizobium* species may express differential host reaction (in this case varieties, genotypes of a species).

Host Range and Cross-inoculation Groups

The bacterial strains which infect a certain plant are found also to typically infect, a certain set of other host plants. This set is called the cross-inoculation group (Table 27.1).

Table 27.1. *Rhizobium* species and their host legumes*

Rhizobium species	Host plant
Rhizobium leguminosarum	
bv. *viceae*	Pea, vetch, broad bean
bv. *trifolii*	Clover, bird's-foot trefoil
bv. *phaseoli*	French bean, *Phaseolus*
bv. *lupini*	*Lupinus*
R. meliloti	Alfalfa, *Melilotus*, *Trigonella*
R. fredii	Soybean
R. ciceri	Chick-pea
Bradyrhizobium japonicum, *B. elkanii*	Soybean, some cowpea
B. parasponiae	*Parasponia*, cowpea, others
Azorhizobium canlinodans	*Sesbania* spp.

* Source: Rhijn and Venderleyden (1995).

Many bacteria infect hosts outside their official group, or fail to nodulate all members of that group. Also, some plants, appear to nodulate promiscuously.

Rhizobium strains may be categorized as 'slow growers' and 'fast growers'. Besides growth rate, they also differ in other features, such as carbohydrate preference, and acid or alkali production.

Measurement of Nitrogen Fixation

Usually, there is a good correlation in legumes between the nitrogen fixation and dry matter production when they are grown under low fertility (low nitrogen status) soils. Measurements of simple parameters, such as dry weight, grain yield and total nitrogen are of direct relevance. However, the role of a plant in fixing nitrogen is confounded with its capacity to use it; this necessitates use of other screening techniques described below.

1. *Acetylene reduction*: This technique involves incubating fixing systems in a closed vessel of known volume containing 10 per cent acetylene and measuring ethylene accumulation by gas chromatography. This technique has a high potential for error when used for comparisons between genotypes, or even the same genotypes if subjected to differential disturbance.

2. *Nitrogen difference*: This method relies on comparing total nitrogen accumulation in the fixing plant with that in a non-fixing reference plant. The accuracy of measurement depends on whether the soil N pool under both plants is equal and their quantitative recovery of soil N also comparable.

3. *Use of radio isotopes (^{15}N fertilizer addition)*: This approach is similar to the nitrogen difference method, but in addition fertilizer enriched with ^{15}N is used to label the nitrogen in the rooting medium. This method has the advantage that its accuracy is less affected by differences in quantitative recovery so long as both treatments recover ^{15}N and ^{14}N in the proportions available in the soil.

However, the assumptions on which this method is based are not easily validated in the field.

Conventional Breeding Approach

It is generally understood that the expression of *Rhizobium* genes concerned with fixation is quantitative rather than qualitative. There are two possible approaches for improving the symbiosis, namely general symbiotic competence and specific symbiotic competence.

1. Improvement of General Symbiotic Competence

This involves screening a number of plant genotypes to a range of rhizobia over a number of locations to identify general symbiotic competence. When such testing is done and the data analyzed for mean performance and stability one can readily observe that no single variety is outstanding with all the strains of rhizobia, or no single strain of rhizobia is excellent with all varieties. Nevertheless, some varieties may do reasonably well with all rhizobia populations or vice versa, indicating a potential in the species for development of general symbiotic competence. Once the varieties and rhizobia strains that exhibit good general symbiotic competence are identified, further improvement in symbiotic competence can be made by use of established selection and hybridization procedures.

2. Improvement of Specific Symbiotic Competence

Phenotypic differences in nitrogen fixation are not wholly vested in either the plant or *Rhizobium* genotype but also result from an interaction between the variety and a specific *Rhizobium* strain. This is due to the presence of a genetically determined mutual recognition mechanism operating between symbionts. Finding and enhancing this trait within a species should help to ensure that 'right' rhizobia renodulate the 'right' host. The various steps involved are:

Step 1. (a) The plants of a species are grown in soil which is relatively low in soluble nitrogen but well supplied with other nutrients so that nitrogen fixation thus correlates with productivity. The natural rhizobial population in the soil may be further supplemented by introducing rhizobial strains from other locations. Phenotypic selection is done on the basis of dry matter production although a check is kept on protein content. The best nodules are also selected from the best plants. Rhizobia are isolated, multiplied and reintroduced into the soil as each screening cycle proceeds.

(b) An identical complementary programme should be carried out by screening in soil supplemented with non-limiting applications of soluble nitrogen.

Step 2. Progeny from both programmes are tested with and without soluble nitrogen.

Step 3. Further selections are made in F_2 and F_3 populations.

The base population contains two types of potentially high-yielding plants, those which are significantly limited by fixation, and those which have little capacity to exploit additional nitrogen without further improvements in carbon fixation capacity. Carbon-limited plants impede the rate of genetic advance under selection and must be eliminated by screening under high nitrate. The breeding programme is carried out by growing plants in soil with a natural population of rhizobia. Both plant and *Rhizobium* genotypes are selected and carried forward for further cycles of hybridization and reselection. Rigorous field assessment of novel genotypes over a large number of environments needs to be done to ensure that environmental factors have not altered the response.

Molecular Approach to Breeding

Plant-Rhizobium Interaction

MECHANISM OF NODULATION

Nodulation includes all the events starting from initial contact of the two symbionts (*Rhizobium* and the plant) and ends with the stimulation of cell division in the host. Nodular development is a complex process. In association, bacterium and host bind, and the epidermal root hairs of the host often show characteristic deformations. Bacteria infect some of these root hairs; their invasion path is delimited by a host-produced infection thread which appears to be cell wall material. The developing root hair cells are most susceptible to infection.

As the bacteria invade, host cells in the root cortex begin dividing. Cell divisions occur in the nodule during early organogenesis and then cease, or localize in a meristem which continues activity throughout the life of the nodule depending on the host species. Plant cell division and enlargement produce the body of a nodule. Infection threads penetrate into some of the host cells, and release bacteria. The bacteria differentiate morphologically and biochemically within the host cell and are referred to as 'bacteroids'. The plant also differentiates in terms of ultrastructure and gene expression. In this differentiated state the two partner organisms co-operate to fix nitrogen. The bacteria produce the enzymes of the nitrogenase complex, which reduces nitrogen into ammonia. The plant provides energy from the photosynthate and regulates local supply of oxygen (which the bacteria require but which also inactivates nitrogenase enzyme) by producing leghaemoglobin, an oxygen-binding haemoprotein. Uninfected plant cells in the nodule may also differentiate, producing some of the enzymes which assimilate ammonia into transport compounds such as ureides.

GENETICS OF PLANT NODULATION

Genes of both *Rhizobium* and plant host are required in order for symbiotic nodules to form. New techniques of genetic engineering, particularly the recombinant DNA technology, have been extremely useful in studying the *Rhizobium* role in symbiosis.

1) *Nitrogen-fixation genes* : The nitrogen-fixation bacterial genes, *Sym* genes, consist of *nif, fix* and *nod* genes.

Nif genes : *Nif* genes have a homologous counterpart from the *nif* gene cluster of *Klebsiella pneumoniae*, for example, *nif, H, D, K* and *A*. The first three encode the three subunits of the nitrogenase enzyme complex and the last one encodes a key transcriptional regulator responsible for the activation of *nif, H, D, K* and other *nif* genes. Expression of *nif A* is in turn a response to changes in the intracellular O_2 concentration via a complex regulatory circuitry.

Fix genes : These genes, controlling the ability of rhizobia to fix nitrogen within a nodule, have no homologue in *Klebsiella pneumoniae*, for example *A, B, C*, the gene products of which may be involved in diverting electrons to nitrogenase. It is not yet known whether most of the *fix* genes are specifically involved during symbiosis or whether some of these are also expressed in free-living bacteria.

2) *Nodulation (nod) genes* : Bacterial genes involved in nodulation are collectively known as *nod* genes. Example are genes such as *ro C* (root colonization), *ro a* (root adhesion), *ha b* (hair branching), *ha d* (hair deformation), *ha c* (hair curling), *hs n* (host specificity of nodulation), *inf*

(infection), *noi* (nodule initiation), *in b* (infection thread branching), *ba r* (bacterial release) and *ba d* (bacterial development). The regulatory *nod D* encodes a polypeptide which when combined with certain flavonoids excreted by the legume root system, activates the other *nod* gene operons (mentioned above) responsible for the synthesis of an oligosaccharide signal molecule which affects these characters, these being the visible signs of nodular formation. Host specificity is built into this signal molecule through modifications to the oligosaccharide structure. Based on their phenotypic properties, *nod* genes are of two types:

i) *nod* genes whose defects can be complemented by *heterologous Rhizobium* strain, for example *nod A, B* and *C* genes;

ii) *nod* genes whose defects cannot be complemented by genes from a heterologous species of *Rhizobium*, for example, *hs n* (host specificity gene). The transfer of these genes is associated with the transfer of host range specificity. Two of the *hs n* genes, namely, *hs n A* and *hs n B* (*nod F* and *E*) are homologous among *R. leguminosarum*, *R. meliloti* and *R. trifolii*. The *hs n C* and *D* (*nod G* & *H*) are also implicated in host-range specificity in *R. meliloti*. In addition to these, there are five other *nod* genes, namely, *nod I, J, L, M* and *X* identified by DNA sequence analysis.

PLANTS GENES

Plant genes are involved in almost all aspects of symbiosis which includes number and size of nodules, morphogenesis and morphology of nodules and the rate of nitrogen-fixation activity. The host range and host-bacterial interactions are rather intricate. There are bacterial-host pairs which are unable to form nodules with each other, although each is capable of nodulating with other partners in the same cross-inoculation group.

Induced plant genes (nodulins): Products of inducible genes or nodule-specific host proteins are called nodulins. Nodulins common to all legumes are called *C nodulins*, and those associated with specific species are called *S nodulins*. In legumes, 30-40 proteins have been identified as being synthesized only in developing and/or functioning root nodules. Nodulins synthesized in the early stages of nodular formation are called early nodulins, for example *E nod 2* found in pea, alfalfa and soybean root nodules. *E nod 2* appear to be involved in either the infection process or in nodular morphogenesis. Nodulins synthesized at the onset of nitrogen fixation activity are called late nodulins. The majority of nodulins are late nodulins, for example leghaemoglobin. The function of leghaemoglobin is to facilitate the diffusion of O_2 in the infected region of the nodule from the intercellular spaces into the cells.

Production of Transgenic Plants

Transgenic legumes are obtained by using *Agrobacterium tumefaciens* or *A. rhizogenes*. In the process of tumour or hairy root formation induced by these bacteria, part of their *Ti* or *Ri* plasmid (T-DNA) is transferred to the host-plant genome. A limitation to this approach is that a few legume species are amenable to *Agrobacterium*-mediated transformation. Progress has been made most significantly with pea. However, the only legume routinely used for this purpose is bird's-foot trefoil regenerated from hairy root cultures induced by *A. rhizogenes.*

Breeding for New Nitrogen-fixing Plants

There are two ways to achieve this objective:

(i) introduction and manipulation of host-plant genes involved in nodulation and fixation to make a new host plant susceptible to *Rhizobium* nodulation;

(ii) introduction of *nif* genes to a host plant.

These methods are based on the assumption that some of the legume host nodulin genes have homologous counterparts in non-leguminous plants. Possibly, many non-legumes possess most nodulin genes so that only regulation of their expression needs to be modified. However, although the prospects of achieving these objectives have brightened, there are still important areas of information which remain obscure at present, precluding achievement of the goal of engineering new nitrogen-fixing plants.

CROP IDEOTYPES

An ideotype literally means a character symbolizing the idea of a thing without indicating the sequence. In our context it means a concept about what our 'plant type' should be for facilitating selection in the breeding programme of a crop to achieve the targeted objectives. The term 'Crop ideotype' is attributed to Donald (1968).

Defining an Ideotype

All plant breeders have an ideotype in mind when they evaluate their material and select plants. The ideotype, however, is likely to be of more value if it is based on good understanding of the complex growth and development related characters, their interrelations with each other, heritability and genotype-environment interactions. According to Donald (1968) crop ideotypes are plants with model characteristics known to influence photosynthesis, growth and production, such that they are expected to perform or behave well and in a predictable manner under defined (known) environments.

Several publications have explained the association of yielding ability with certain plant types. Tsunoda (1959, 1960, 1962) attempted to relate yielding ability and fertilizer response to plant type and has reported that in some crops, namely sweet potato, soybean and rice, varieties having erect, short, narrow, thick, dark green leaves and short sturdy stems were associated with high nitrogen response. Since then a number of the studies have dealt with the concept of plant ideotypes. Numerous traits have been suggested to be of value in breeding programmes. Some can be easily and rapidly assessed, and if significant for yield would be worthwhile for routine selection. Others, though purportedly significant for yield, have proven to be of little value.

Principles of Designing an Ideotype

All the attributes of the ideotype are morphological characters, but all are based on physiological considerations. Taking wheat as an example, the model plant characters (Donald, 1968) that may be helpful in making better yielding selections are described below:

1) a relatively short and strong stem,
2) erect leaves: near-vertical leaves,
3) few small leaves,
4) a large ear,
5) an erect ear,
6) simple awns,
7) a single culm[3].

The extent to which an ideotype is defined has to be a matter of judgment. Other characters which are left unspecified should also be considered during the selection.

Limitations of Ideotype

1. There seems to be such a wide array of compensating mechanisms or routes towards high yields, and multiplicity of suggestions offered by plant physiologists that it becomes difficult to directly apply them in practical plant breeding. Also, there can be no immediate certainty of success. The designed model needs to be tested for performance.

2. Assessment of the proposed characters should be simple and rapid. Besides, there needs to be heritable variation in the suggested character(s).

Notwithstanding these limitations, the value of physiological concepts lies in providing concepts which permit appropriate decisions within a breeding programme.

Application of Plant Ideotype Concept

Swaminathan (1981) has suggested breeding for following specific situations based on the plant ideotype concept.

1. Breeding for Dense Populations

The underlying plant type concept here is that increase in leaf area index (ratio of leaf area: area of soil) in earlier stages of crop growth would increase yield. The important consideration in designing a plant type is to increase penetration of light (erect leaves) and improved leaf arrangement in the canopy (small, thick leaves and closed leaf canopy).

---

[1] No concrete evidence is available to support this single culm ideotype.

2. Breeding for High Soil Fertility Conditions

The suggested desirable 'plant type' in breeding for high-fertility conditions should possess good early growth vigour, or high tillering ability, resistance to lodging, diseases and pests, (the intensity of which may increase due to increased nitrogen availability) and improved harvest index. Increased response to nitrogen is reportedly associated with thick leaves and early maturity.

3. Breeding for Low Soil Fertility Conditions

The suggested 'plant type' for low-fertility conditions should show good stand establishment, thin pale green leaves, long growth duration, extensive root system, a low harvest index, and a short grain-filling period.

4. Breeding for Multiple Cropping, Mixed Cropping, Relay Planting, Intercropping

Development of varieties for a multiple cropping system involves development of short-duration varieties through breeding, which are relatively insensitive to photoperiod and temperature so as to fit in promoting multiple cropping system involving two-three and even four crop sequences. The per day yield has to be used as a selection criterion in segregating generations, besides factors such as seed dormancy (if a crop being grown ripens before the monsoon rains have ceased).

Development of multilevel or three-dimensional cropping for garden lands where a wide variety of plantation crops are grown, involves the effective use of both horizontal and vertical spaces that could be achieved by breeding varieties for use in three-dimensional crop canopies. For example, coconut, cocoa and pineapple form a good combination. Breeding efforts should be directed towards varieties that generate complementarity among crops, particularly in the root system of companion crops.

Development of varieties for mixed cropping and intercropping involves development of varieties of various crops in mixed cropping and intercropping on the principles of complementarity between companion crops. The components of sunlight, ability to tap nutrients and moisture from different soil depths, non-overlapping susceptibility to pests and diseases and inclusion of legume crop to promote biological N fixation. The suggested 'plant type' for multiple cropping should show early maturity, uniform (synchronous) ripening, photoperiod insensitivity[4], high yield per unit per day and relatively small crop residue. In addition, suitability for mechanized harvesting and processing may be desirable for a relay-intercropping. Other features, namely provision of shade at the seeding stage, tolerance to shade at a later stage (for reducing evapotranspiration of shorter stature crop), elastic straw adapted to trampling, or tolerance to drought or waterlogging are also important. Plant characteristics, namely resistance to common diseases and insects, different rooting depth, and varying nutrient requirements will benefit both crops.

---

[4] In some instances a photoperiod sensitive genotype may suit specific growth periods.

5. Breeding for Minimum Cultivation

The suggested 'plant type' should show rapid germination, quick and vigorous stand establishment and plant growth, strong root growth, good competitive ability against weeds (rapidly expanding and luxuriant leaf canopy), low sensitivity to phytotoxicity of herbicides. and a balance between vegetative growth and grain production.

6. Breeding for Rainfed Farming

This involves development of varieties which can be grown in flood-free seasons in chronically flood-prone areas and drought-escaping varieties in drought-prone areas. The various aspects of drought and flood resistance have already been discussed in Chapter 26.

It may be mentioned here that conventional breeding procedures are employed to breed/develop varieties suited to specific adaptations as described above. Also, it may not be possible at the present state of our knowledge to rapidly screen the material and study heritable variations in respect of several to many of the plant characters as conceptualized above.

28

Plant Tissue Culture

The term plant tissue culture has been rather generally applied to all forms of *in vitro* grown plant cultures. There are five specific applications of plant cell and tissue culture in plant breeding:

1) mass multiplication of plants,
2) elimination of pathogens,
3) wide hybridization,
4) production of haploids,
5) genetic engineering of crop plants.

These objectives can be achieved through plant cell and tissue culture technologies, namely micropropagation, callus culture, embryo culture and anther/ovule culture, which are discussed in this chapter. The basic paradigm of genetic engineering is discussed in Chapter 29.

MICROPROPAGATION

Mass cloning (propagation) of selected genotypes using tissue culture techniques (*in vitro* techniques) is known as micropropagation. Micropropagation gives true-to-type plants, that is, genetically pure stocks. The techniques (protocols) of mass *in vitro* propagation (commercial-scale production) have been standardized for some of the important economic crops, while for other important crops efforts are underway. The success so far achieved is largely in respect of ornamentals, cut flowers, some fruit tree species and a number of plantation crops (Table 28.1).

General Method of Micropropagation

Micropropagation is usually done in the following stages (Dixon, 1994)

1. Preparative Stage

For micropropagation, the selection of hygienic starting material in good physiological condition is very important. Not only this, the starting material should be disease-free as well. To obtain desirable starting material the selected 'stock plants' should be maintained in good environmental conditions in a greenhouse at 25°C temperature and low relative humidity (70%). The minimum duration needs to be determined for each species, though for

Table 28.1. List of important crop species for which micropropagation techniques have been standardized*

Plantation crops	Fruit crops	Medicinal crops	Ornamentals	Forest trees
Sugar-cane	Citrus	Dioscorea	Rose	Teak
Turmeric	Pineapple	Glycorrhiza	Bougainvillea	Eucalyptus
Ginger	Pomegranate		Bamboo	
Rubber	Almond		Sandal	
Mustard	Banana		Rosewood	
Cardamom	Apple		Pine	

*Source: Chopra, V.L. (1995). Proc. Inter-Center Seminar on IARCs & Biotechnology, IRRI, pp. 395-400.

most of the species a period of 3 months seems to be sufficient. To improve the physiological fitness of stock plants, plant growth regulators may be sprayed, injected or applied. The photoperiod may be kept constant to control flowering and rejuvenation of stock plants may be done by appropriate techniques such as bottom heating, pruning, etc. Similarly, techniques such as thermotherapy may be employed to eradicate viruses, if needed. The important thing is that the starting material be disease and/or pest-free.

2. Culture Initiation

Choice of explant: The regeneration capability of a plant is usually inversely proportional to the age and size of the explant and to the age of the explant source. Choice of young plant parts with active meristematic tissue is therefore imperative. The risk of contamination is much less when the interior plant parts/aerial plant parts are given priority over exterior or underground parts. Also, the smaller the explant size the lesser is the risk of contamination.

For most micropropagation work, the explant of choice is an apical or axillary bud. In some species leaf pieces are used on which adventitious buds are induced. The use of adventitious buds, however, involves the risk of increasing chances of somaclonal variation. When the objective is to produce virus-free plants from infected individuals, it becomes imperative to start with submillimetre shoot tips.

Sterilization of selected material: The harvested shoots (with three to six buds) with pedicle stumps (to protect the buds against sterilization damage) are washed with running tap-water and put in a vessel. Working in a laminar flow hood, the material in the vessel is rinsed in 95% ethanol for a few seconds to a minute and the alcohol poured away. Then 0.1-1.0% mercuric chloride ($HgCl_2$) solution plus detergent (2 drops/100 ml) is added and poured away after 3-5 minutes. The material is rinsed with autoclaved water. Then 7-15% NaOCl plus detergent (2 drops/100 ml) is added and poured away after 10-30 minutes. The material is again rinsed three times with autoclaved water. Afterwards the damaged base of the shoot is cut off; nodal explants are cut or meristems are isolated for inoculation in a separate tube.

CULTURE INITIATION MEDIA

Tissue culture base media consist of a mixture of inorganic salts, organic salts, amino acids, sugar and vitamins. The most commonly used defined culture medium is that of Murashige and Skoog (1962). The components of this medium are given below.

Constituent	Concentration in culture medium (mg/litre)
KNO_3	1900
NH_4NO_3	1650
$MgSO_4 \cdot 7H_2O$	370
$CaCl_2 \cdot 2H_2O$	440
KH_2PO_4	170
$MnSO_4 \cdot 4H_2O$	22.3
KI	0.83
H_3BO_3	6.2
$ZnSO_4 \cdot 7H_2O$	8.6
$CuSO_4 \cdot 5H_2O$	0.025
$Na_2MoO_4 \cdot 2H_2O$	0.25
$CoCl_2 \cdot 6H_2O$	0.025
$FeSO_4 \cdot 7H_2O$	27.8
Na_2EDTA	37.3
Nicotinic acid	0.5
Pyridoxine-HCl	0.5
Thiamine-HCl	0.1
Myo-Inositol	100
Glycine	2.0
Sucrose	30,000

Agar 0.6% (W/V) and gelrite 0.12% (W/V) are used as gelling agents. The pH of the medium is 5.8.

Culture media contain growth regulators which help maintain dedifferentiated cell growth and promote cell division, respectively. Some variation may be needed in their concentration and choice depending the material being used and the crop species concerned for which standardization of specific protocols is desirable.

ENVIRONMENTAL CONDITIONS

Culture initiation is done under controlled environmental conditions. The usual conditions are: temperature, around 25°C (night temperature could be 1 or 2°C lower); relative humidity (usually not controlled), and photoperiod (0 h the first days after initiation of the meristem, and 16 h. in most other cases (Cool white fluorescent tube set point for light quantity, 30 μE. m^{-2}. s^{-1})

3. Shoot Multiplication

After four to eight weeks the original explant is transformed into a mass of ramified shoots or a cluster of basal shoots. These miniature shoots or clusters are excised and planted on a fresh medium in which the cytokinin level can be raised. The shoot multiplication cycle can be repeated. It may be noted, however, that a higher cytokinin-to-auxin ratio promotes shoot formation and a higher auxin-to-cytokinin ratio favours root differentiation. The exogenous requirements for hormones depend on the endogenous levels in the plant. In a number of cases cytokinin alone is enough for optimal shoot multiplication. For some species only axillary bud development should be aimed for. Axillary buds are usually present in the axil of each leaf and every bud has the potential to develop into a shoot. Apical dominance is overcome by growing shoots in the presence of a suitable cytokinin. An excessive dosage or inappropriate choice of phytohormones needs to be avoided, to prevent formation of epigenetic off-types. Some plants are propagated by promoting elongation of shoots which are subcultured by taking nodal cuttings. For some species adventitious buds, arising from any place other than the leaf axil or the shoot apex may be chosen. This enables substantially faster multiplication.

Subcultures: On the same medium, stage 2 cultures, originally yielding axillary shoots, can produce abundant adventitious shoots after a number of subcultures. Usually 10-12 subcultures are considered maximum.

4. Shoot Elongation and Root Induction or Development
Stem elongation: For most plants, good elongation can be obtained by transfer to a medium devoid of cytokinin or a weaker cytokinin. A cluster of shoots may be transplanted. The medium should allow good elongation of all shoots.

Root induction: Adventitious and axillary shoots which develop in the presence of a cytokinin often lack roots. Root development can be performed *in vitro* or *ex vitro*.

In the *in vitro* system, instead of transplanting the shoots to a fresh medium, liquid media can be added to the established cultures. For most species, auxins such as NAA or IBA (0.1-1 mg/litre) are needed to induce rooting. Most often a lower salt concentration is required. Higher sugar concentrations (3-4%) improve the rooting and quality of plants. Activated charcoal, added to the liquid medium, eliminates the residual effects of cytokinins by absorption. Also, it is possible to achieve both shoot elongation and root induction with one (liquid) medium. At this stage bottom cooling may be required to prevent hyperhydricity.

5. Transfer to Greenhouse Conditions
Micropropagated plantlets need to be gradually acclimatized to the environment of the greenhouse or field. This is necessitated by the physiological and anatomical characteristics of the micropropagated plantlets. The process of acclimatization can start *in vitro*. Bottom cooling reduces the relative humidity in the head space of the container and this can initiate the weaning process. The uncapped culture vessels are put in the greenhouse several days (not more than a week) prior to removal of the plants from the

culture medium. Subsequently, the plants are transplanted carefully to reduce the risk of wounding on a substrate such as unfertilized peat. Maintaining a high relative humidity for the first few days is critical.

PRODUCTION OF VIRUS-FREE PLANTS

Virus-free plants of many species and/or varieties may be produced by culture of meristematic tissue. The specific protocols to eliminate viruses from different plants vary widely, however.

General Method of Production of Virus-free Plants

The various steps involved are as follows.

1. Detection of Virus

Several immunodiagnostic techniques such as immuno-stained electron microscopy (ISEM), enzyme-linked immunosorbent assay (ELISA) and radio immunosorbent assay (RISA or RIA) etc., are now available for the detection of plant viruses. After detection, the type and extent of therapy required to eliminate the virus and maintain tissue viability are assessed.

2. Heat Therapy

Heat therapy has long been used to rid plant material of infectious agents. The temperatures and duration and length of illumination for heat therapy vary with crop species and need to be standardized for each. The underlying principle is to allow conditions that permit plant growth but maximize virus elimination. New growth may be free of virus and can be cultured. Heat therapy may not be successful but nonetheless is a useful treatment of material prior to tissue culture.

3. Isolation and Culture of Meristems for Virus Elimination

Use of material from plants grown as cleanly as possible to reduce the chances of contamination of cultured tissue is important. Meristems on shoots covered by developing leaves and leaf primordia are considered aseptic. The selected tissue is dipped in 75-95% ethanol or 0.1-0.5% sodium hypochlorite for a few seconds to minutes. This is followed by several rinses in sterile water. In some cases disinfection by both treatments in succession may be needed. In the laminar flow hood, using frequently sterilized instruments and aided by a binocular dissecting microscope, carefully remove the outer leaves and leaf primordia to expose the meristematic area. Once the meristematic area is exposed, it is removed and placed on the surface of the medium in a culture tube.

Culture : Culturing is done in the medium under standardized conditions and suitable environmental conditions for the species and/or variety. The conditions generally applicable have already been discussed under micropropagation.

Elimination of virus : Heat treatment and/or meristem culture may not lead to culture of virus-free plants. Many chemicals, such as Ribavirin, Virazole at 0-100 mg/litre have been successfully added after sterilization by filteration to cooling culture media before solidification to try to enhance production of virus-free plants. The reaction of species to Ribavirin and subsequent success in producing virus-free plants vary. Following establishment of plants from the culture in soil, they are tested for virus using appropriate technique.

Detection by infectivity: An indicator host that produces characteristic symptoms after infection by mechanical inoculation can be very useful, especially if the virus has not been identified.

A virus-free plant is not virus resistant. Care must be taken so that plants produced do not become infected again. Virus-free plants can serve as a source of propagation of other virus-free plants.

EMBRYO CULTURE

In interspecific and intergeneric hybridization wherein fertilization has taken place but the embryo fails to develop, there is always a chance that the young developing embryo excised from the parent tissue before abortion sets in may be successfully *in vitro* cultured into a mature hybrid plant. This is termed embryo rescue, or zygotic embryo culture, or embryo culture.

Uses and Application of Embryo Culture

Embryo culture, that is, the isolation and culture of embryos from defective seeds before embryo abortion sets in, is the primary means of obtaining viable seedlings from such crosses. It has been successfully used in a large number of wide crosses to obtain viable seeds as a means of overcoming post-fertilization barriers to crossability.

Method of Embryo Culture

STAGES OF EMBRYOGENESIS

For zygotic embryos, embryogenesis starts at zygote formation. The first division of the zygote is transverse. The cells near the micropyle develop into a structure known as the 'suspensor'. The suspensor serves as a pathway of nutrient transfer during the heterotrophic phase in which the growing embryo is dependent on the endosperm for nutrients. It further serves as an important source of plant-growth regulators. The other cells resulting from initial division of the zygote develop into an embryo. The embryo progresses through a succession of stages, namely undifferentiated (globular stage), differentiation stage (heart stage) at which the embryo switches from heterotrophic phase to an autotrophic phase in which it becomes capable of independent growth, torpedo stage and cotyledonary stage of embryo development. Embryogenesis ends at seed maturation.

EMBRYO RESCUE

The tissue type, physiological state of the tissue and genetic make-up of the donor plant have considerable bearing on tissue-culture response. The stage of embryogenesis at which the embryos are excised is therefore important. Embryo(s) that has reached the differentiation stage is more difficult to culture, possibly due to the fatal damage caused to it during excision. Selection of an appropriate time and technique for embryo rescue requires some knowledge of when and why the embryo aborts *in vivo*. In many instances the embryo needs to be excised from the cultured ovule when it has reached a sufficiently advanced stage to be capable of surviving in culture on its own. Excision and transfer of growing embryos should be performed under sterile conditions to avoid contamination. When the embryo is of a rather large size, it can often be dissected out with a needle while the seed is held between the fingers. With smaller embryos a dissecting microscope is necessary.

MEDIA AND CULTURE CONDITIONS

The range of media used to culture excised embryos is extensive and involves numerous modifications to otherwise standard media depending on the crop species of concern. Usually, media are liquid or partly solidified with low agar concentration and contain mineral salts, a carbohydrate source, vitamins, hormones and sometimes an organic nitrogen source. The nutritional requirements of embryos and the osmotic potential of the medium may depend on the stage of embryonal development. Younger embryos generally require higher osmotic values to continue development, with lower osmotic values promoting germination. To illustrate the general principles the embryo culture of wheat is described below.

Embryo Culture in Wheat

1. Collect the mid-sections of spikes, containing 15-20 developing seeds, at 16-18 days after pollination and disinfect by dipping in 95% ethanol for 2 minutes followed by three washes with sterile distilled water.
2. Dissect out the embryo from a developing seed under a dissection microscope on an ethanol-swabbed surface. For dissection, peel back the surrounding glume, lemma and palea from the developing grain and remove the smooth green grain.
3. Holding the grain (smooth side up) with asceptic forceps, and with the coleorhizal end towards the viewer, make a transverse incision with an asceptic scalpel across the grain at mid-point. Make a second incision along the side of the grain toward the coleorhizal end. Peel back the epidermal flap with the forceps and carefully lift out the white oval-shaped embryo with the tip of the scalpel.
4. Immediately transfer the embryo to a 60×20 mm plastic petri dish containing agar solidified (0.8% Bacto agar) MS medium lacking growth regulators and modified to contain 5×10^{-5} M ABA. Culture embryos, five per dish, at 25°C under a 16 h photoperiod.
5. Place cultured embryos on sterile filter paper in a 60×20 mm plastic petri dish and transfer to a plastic desiccator[1] with controlled humidity (75%, saturating solution, NaCl).

Desiccation of somatic embryos serves two related purposes: (1) it breaks the dormancy associated with the somatic embryos in some species and (2) it promotes the accumulation of nutrients, such as storage protein, which will benefit the embryo when germination is allowed to proceed.

6. Every 24-48 h, transfer the dishes to desiccators with progressively lower humidity levels (62.5% RH, saturating solution, NH_4NO_3; 50.5% RH, saturating solution $Ca(NO_3)_2 4H_2O$; 43% RH, saturating solution $K_2CO_3 2H_2O$.

7. To rehydrate, place dried embryos directly on filter paper on MS medium lacking growth regulators and modified to contain GA (5 x 10^{-5} M GA_3).

8. Assess germinating embryos for survival one week after their placement on the medium.

9. After germination, hybrid plants with one or more roots and two to three leaves can be transplanted to a sterile soil consisting of 1/2 sand : 1/2 sphagnum to which liquid organic fertilizer is applied. Plants should be protected from strong sunlight for the first few days.

Limitations of Embryo Culture

Until reasons for embryo failure in wide crosses are understood cytologically, physiologically and genetically, success in attempts to develop specific crosses will continue to be limited.

ANTHER CULTURE

Immature anthers or pollen grains, mostly at the stage of uninucleate microspores, may be cultured on a defined artificial agar nutrient medium to induce androgenesis[2]. A very small proportion may grow into entire plants. The usual practice is to grow anthers in culture tubes on defined medium in growth chambers under suitable light and temperature conditions. Under these conditions the young microspores instead of growing into mature pollen grains, either grow directly into embryoids and eventually into plants, or may produce undifferentiated calluses which under right conditions may give rise to shoots and roots and eventually entire plants. Since the pollen grains are haploids, the resultant plants are also haploids. Through induced or spontaneous doubling the chromosome number of haploids is doubled to obtain homozygous diploid plants. The major advantages of anther culture are the rapid advance to homozygosity, easy expression of recessive characters and the creation of novel types that are usually difficult to obtain through conventional methods.

Application and Uses of Anther Culture

1. Development of Improved Varieties

Although haploids have been produced through anther culture in nearly 60 species, most of them are from Solanaceae and Gramineae; only a few

---

[2] Refers to plants derived from pollen grains. The female cytoplasm is not involved. In some relatively rare instances in nature, the male nucleus enters the egg cell and the egg cell is eliminated; the resultant plant contains the chromosome segment of the male gamete only together with the cytoplasm of the female parent. This is known as ovule androgenesis.

examples are available from Chinese work, Wherein it has been possible to obtain improved varieties (Rice: Huayu-1, Huayu-2, and Taufong-1; wheat: Huapei-1; tobacco: Tanyu No. 1, Tanyu No. 2, and Tanyu No. 3).

2. Selection of Resistant Lines

Single-cell cultures raised from androgenic haploids of tobacco have been used to select lines resistant to methionine analogue, methionine sulfoxide (MSO), which is similar in structure and effect to the toxin produced by the pathogen *Pseudomonas tabaci*. High alkaloid-containing mutants of *Hyoscyamus niger* were similarly obtained by subjecting the anther-derived haploids to UV and X-ray treatments.

3. Development of Supermales in Asparagus

Haploids in *Asparagus officinalis* have been developed by anther culture, which could be diploidized to form homozygous males and females. Crosses between the homozygous males (supermales) (MM) and females (mm) yielded homogeneous F_1 hybrids with all male seeds.

4. Development of the components of a cytoplasmic male sterility hybrid system in rye through anther culture (Bicar and Darvey, 1997).

Australian rye varieties were crossed reciprocally to the variety 'Luchs' which carries the Pampa male sterile cytoplasm (cms - P). Anthers of the F_1s in the cms-P cytoplasm (primary cross) and their reciprocals in the normal cytoplasm (reciprocal cross) were cultured in a modified C17 medium. Male sterile and male fertile doubled haploids were obtained from the anther culture of F_1s in the cms-P cytoplasm. Test crosses indicated that the male sterile doubled halpoids were A lines (male sterile) and the male fertile doubled haploids were R lines (restorer lines). The anther culture of genotypes in the normal cytoplasm (reciprocal cross) gave all male fertile doubled haploids. Test crosses indicated that the male fertile doubled haploids were R lines in the normal cytoplasm.

The results obtained are significant and especially important in the production of hybrids in several species.

Limitations of Anther Culture

Practical application of androgenic haploids is still limited due to the following reasons:

1. Only a very small proportion of the androgenic grains develop into full sporophytes; this is a serious limiting factor in obtaining the full range of genetic segregation of interest to plant breeders.

2. Emergence of plantlets of higher ploidy levels, wherein the precise mode of origin is not known, necessitates extensive screening and selection of such anther-derived plants for further research. In cereals in which androgenesis involves pollen callusing followed by plant formation, the occurrence of chromosomal aberrations is very common.

3. A high degree of karyological instability of haploid cells in callus and suspension cultures seriously affects its potential use in mutant selection and other genetic studies.

4. Consecutive cycles of androgenesis increase growth depression and abnormalities. Doubled haploid plants usually contain increased amounts of total DNA and increased proportion of highly repeated DNA sequences.

5. Albino plantlets are of common occurrence in cereal crops. The frequency is very high (barley 99%, rice 85%, durum wheat 82%). Since these plants cannot survive in nature, they are of no agronomic value. This necessitates measures to improve the percentage of green plants.

To illustrate the general principles the anther culture of rice is described be'ow.

A.ither Culture in Rice

Donor plants: The donor plants should be in good physiological condition. Plants exposed to high temperatures at meiosis easily form albinos. Plants with high nitrogen content have enhanced chances of callus formation and green plant regeneration.

Identification of correct development phase of microspore (late uninucleate stage): The ideal time for anther collection' for anther culture is when the pollen cells are at the late uninucleate stage. A rice floret at this stage can be distinguished by examining lemma width and colour (width almost what it should be at maturity but still yellowish), and stamen length (33-50% longer than the glume flower).

Pretreatments: Pretreatments such as placing the rice raceme at 10°C for 2 days, or cold pretreatment (floating the anthers on liquid N_6 medium at 8°C for 4 days) have been suggested for increasing the frequency of callus formation. After pretreatment, the anthers are transferred to agar medium and cultured at 30°C.

Culture media: N_6 basic medium (Table 28.2) supplemented with 2 mg/l 2,4-D, 1-2 mg/l NAA and 1-2 mg/l kinetin (KT) has reportedly been found suitable for calli induction and high differentiation rate in China (Zhang, 1989)[3]. Addition of plant extracts, for example 15% V/V coconut milk to the medium, has been reported to facilitate green plantlet formation and subsequent strong seedlings.

Induction of callus: Anthers are grown in culture media in tubes in growth chambers at 25°C. Two to three weeks after planting some pollen grains start enlarging two to three times their initial volume. In another week division takes place within the confines of the exine. Stimulation to undifferentiated growth is associated with varying degrees of differentiation at the first pollen mitosis. It appears that the differentiated growth directly from the microspore to the primary sporophyte (embryoid) is possible only if the daughter nuclei are fully differentiated at the first mitotic division. If this fails only undifferentiated (callus) growth occurs. After four to five weeks the exine ruptures and a callus mass is liberated, which resembles a globular embryo. The callus may give rise to secondary sporophytic development through the

---

[3] The total concentration of auxins should not exceed 5 mg/l.

Table 28.2. Constitution of N_6 basic medium

Component	mg/l
$(NH_4)_2SO_4$	463.00
KNO_3	2,830.00
KH_2PO_4	400.00
$MgSO_4 \cdot 7H_2O$	185.00
$CaCl_2 \cdot 2H_2O$	166.00
$MnSO_4 \cdot 4H_2O$	4.40
$ZnSO_4 \cdot 7H_2O$	1.50
H_3BO_3	1.60
KI	0.80
Iron*	5.57
Glycine	2.00
Thiamine HCl	1.00
Pyridoxine HCl	0.50
Nicotinic acid	0.50
Sucrose	50,000.00
Agar	8,000.00
pH	5.80

* 5 ml of solution obtained by dissolving 5.57 g $FeSO_4 \cdot 7H_2O$ and 7.45 g Na_2EDTA in 1 litre distilled water.

production of plantlets or sequentially of roots and shoots. Thus plantlets may be produced without having passed through the typical embryoid phase. The age of the callus also influences the differentiation rate. A 51-60-day-old culture appears to be the optimum stage.

Different genotypic combinations differ in their callus induction and differentiation rates. The anther culture ability of a genotype is determined by dividing the callus induction rate (%) by the callus differentiation rate (%). At present the anther culture ability of *japonica* rice is around 8% and that for *indica*, 1%.

The individual plantlets or shoots obtained from the callus are separated and transferred to a medium which will support good development of the root-shoot system. The plantlets are then transferred to a sterilized potting mix. Plantlets arising from the different pollen grains in an anther are genetically heterogeneous. Rapid clonal multiplication of the desired genotypes can be achieved by micropropagation through shoot multiplication.

Viability of pollen-plant offspring and stability of their characters : The stability and uniformity of pollen-plant offspring is an important criterion in anther culture rice breeding. Research in China has revealed that there is no progressive decrease in vigour of plants in advanced generations and that the characters are quite stable. Selection can therefore be practised in the early generations.

Diploidization of pollen plants : Plants regenerated from pollen callus differ in ploidy but most are haploid and diploid. The rate of diploids in pollen plants always exceeds 50% (Chen and Li, 1978).

Genetic analysis has indicated that 90% of the progeny of diploids derived from hybrids are homozygous. Therefore, it can be concluded that diploid pollen plants are caused by spontaneous chromosome doubling during anther culture and do not originate from somatic anther cells.

For breeding purposes, haploid pollen plants need to be diploidized through colchicine (0.2% colchicine); they may also be diploidized by somatic callus culture. Use of anther culture in rice breeding consists of three steps, namely (i) culture of pollen plants, (ii) evaluation and selection of pollen-plant offspring and (iii) comparison of varieties.

Rehybridization and Anther Culture Application in Multiple Crosses

This involves selection of pollen plants derived from various hybrid combinations, crossing them and repeatedly culturing F_1 anthers to obtain pollen plants. This way characters of several parents can be accumulated to produce pollen plants which are homozygous diploids uniform in genotype and phenotype.

OVULE CULTURE

Like anthers or pollen grains, ovaries or ovules (gynogenesis[4]) may also be successfully cultured on artificial medium in many crops of plant breeding interest, namely barley, wheat, rice, maize, sunflower, tobacco, sugar-beet, onion, etc.

Application and Uses of Ovary/Ovule Culture

Until now the practical application of ovary/ovule culture has been limited to plant species in which anther culture has not been successful, for example sugar-beet and onion. The reason lies in the following limitations of ovary/ovule culture.

Limitations
1. The frequency of haploid induction with the exception of potato is very low.
2. Calli or embryos may be produced simultaneously from the gametophytic and sporophytic cell. This poses a major problem of distinguishing them and precluding those having a somatic origin. When only haploids are produced, a clear discrimination is possible through chromosome counts in the regenerated plants. But when both haploids and diploids are produced, the origin of the non-haploids, that is, whether they have doubled spontaneously or are of somatic origin, can only be distinguished in the offspring.

Methods of Ovule Culture

Many of the regular plant tissue-culture methods are applicable to ovary/ovule culture. The developmental stage of the gynogenetic cells inside the

---

[4] Refers to the production of an embryo (plant) from an unfertilized egg cell.

ovule is also important for successful culture. In most plant species culturing can be successfully done over a broad range of the ovary/ovule growth stages. Usually only one embryogenetic structure develops at the micropylar end of the embryo sacs.

29

Genetic Engineering

What is genetic engineering?

Genetic engineering, recombinant DNA technology, or gene cloning are terms which refer to genetic alteration of cells or the genotype of an organism involving *in vitro* modification of the DNA (deoxyribonucleic acid). In its simplest form it involves: (i) purification and splicing of vector DNA, (ii) joining foreign DNA fragments to the vector, (iii) insertion of recombinant molecules (transformed DNA) into the bacterial host cell, (iv) selection of clones that carry the desired sequences, (v) infection of plants/plant cells with the recombinant bacteria, (vi) culture of transformed cells of the plant and (vii) regeneration of plant. The new technology differs from conventional plant breeding in its ability to introduce a single gene into a plant and to manipulate its expression without changing any other constituent of the plant's genome. In addition, genetic engineering makes gene transfer possible between plant species which cannot interbreed.

Crop improvement by genetic engineering is based on the molecular manipulation of relevant genes and availability of vectors for transformation of the plant cells. The resultant plants are known as transgenic plants.

For the sake of easy and clear understanding we shall discuss the various aspects of genetic engineering under the following broad headings:

1) Gene cloning,
2) Gene transfers in plants,
3) Application and uses of genetic engineering in plant breeding.

GENE CLONING

A. General Features

Cloning, analysis and modification of DNA fragments are accomplished with a small set of simple but powerful techniques.

1. Isolation of High Molecular Weight Nuclear DNA

Nuclear DNA is isolated by disrupting fresh plant material in the presence of nuclear membrane stabilizing agents (e.g. Triton x-100), filtration to remove non-disrupted tissue and isolation of nuclei by differential centrifugation, lysis, deproteinization and DNA purification on Cesium chloride/Ethidium bromide (CsCl/EtBr) gradient ultracentrifugation.

2. Production of Defined DNA Fragments

DNA fragments are obtained by splicing DNA samples with restriction enzymes. The target DNA is digested with an appropriate restriction enzyme under conditions that achieve partial digestion. Generally, this involves varying enzyme concentrations in test reactions and then scaling the reaction up in order to generate large quantities of DNA fragments for cloning purposes. The DNA is then fractionated by sucrose density gradient centrifugation and size-selected DNA fragments are isolated and dialyzed into a low-salt buffer in preparation for ligation to vector DNA.

ROLE OF RESTRICTION ENZYMES

Restriction enzymes used in genetic engineering are nucleases that cleave DNA wherever it contains a particular short sequence of nucleotides matching the restriction site of the enzyme. Most restriction sites consist of four or six nucleotides, within which the restriction enzyme makes two single-strand breaks, one in each strand, generating 3'-OH and 5'-P groups at each position. The sequences recognized by restriction enzymes are identical in both strands of the DNA molecule (palindromes). The breaks need not be directly opposite one another in the two DNA strands. The cuts in the DNA strand could be staggered, producing single-stranded ends called sticky ends or cohesive ends that can adhere to each other because they contain complementary nucleotide sequences. The fragments produced by many restriction enzymes can spontaneously form circles. These circles can be made linear again by heating. However, if after circulization, they are treated with E. coli DNA-ligase, which joins 3'-OH and 5'-P groups, the ends became covalently joined. A number of restriction enzymes cleave both DNA strands at the centre of symmetry, forming blunt ends. Blunt ends can also be ligated by DNA ligase. However, whereas ligation of sticky ends recreates the original restriction site, any blunt end can join with any other blunt end and not necessarily create a restriction site.

Several hundred restriction enzymes with different restriction sites have been isolated from micro-organisms. Most of these recognize their unique restriction sequence without regard to the source of the DNA. Thus, DNA fragments obtained from one organism will have the same sticky ends as the fragments from another organism if they have been produced by the same restriction enzyme. The number of cuts made in the DNA from an organism by a particular enzyme is limited compared to random cuts. Smaller DNA molecules, such as viral or plasmid DNA may have only from one to several sites of cutting (or even none) for particular enzymes. Plasmids containing a single site for a particular enzyme are especially valuable.

3. Cloning Vectors

Vectors are genetic elements such as plasmids or bacteriophages that allow the production of many copies of the isolated DNA fragment. Their maintenance in the cell does not necessarily require integration into the host genome and their DNA can be isolated independently from the host genome.

A vector should possess the following properties.

  (i) The vector DNA can be introduced into a host cell.

  (ii) The vector contains a replication origin and so can replicate inside the host cell.

  (iii) Cells containing the vector can usually be selected by a straightforward assay, most conveniently by allowing growth of the host cell on a solid selective medium.

At present, the most commonly used vectors are *E. coli* plasmids and derivatives of the bacteriophages I and M13. Plasmids are the most convenient for cloning relatively small DNA fragments (5-10 kb[1]), while somewhat larger fragments can be cloned with bacteriophage I (15-20 kb). Still larger fragments can be inserted into cosmid vectors (30-40 kb). Recombinant DNA can be detected in host cells by means of genetic features or particular markers made evident in the formation of colonies or plaques. Plasmid or phage DNA can be introduced into host cells by a 'transformation' procedure in which cells gain the ability to take up free DNA by exposure to a $CaCl_2$ solution. Recombinant DNA can also be introduced into cells by application of electric pulses (electroporation). After introduction of the DNA the cells containing recombinant DNA are plated on a solid medium.

    If the added DNA is a plasmid, colonies consisting of bacterial cells containing the recombinant plasmid are formed, and the transformants can usually be detected by the phenotype that the plasmid confer on. If the vector is phage DNA, the infected cells are plated in the usual way to yield plaques.

4. Joining DNA Fragments

    We have earlier stated that a particular restriction enzyme produces fragments with identical sticky ends, without regard for the source of the DNA. Therefore, DNA fragments from DNA molecules isolated from two different organisms can be joined. When the donor DNA fragments and the linearized plasmids are mixed the recombinant molecules are formed by base pairing between the complementary single-stranded ends. At this point, DNA is treated with DNA ligase to seal the joints, and the fragments become permanently joined in a combination that may never have existed before. The ability to join a donor fragment of interest to a vector is the basis of recombinant DNA technology. Joining sticky ends however, does not always produce a DNA sequence that has functional gene. This is due to changes in the arrangement of different fragments. The problem of scrambling of vector fragments is minimized by the use of a vector having only one cleavage site for a particular restriction enzyme. Plasmids of this type are available. Many vectors contain unique sites for several different restriction enzymes, but often only one enzyme is used at a time. DNA molecules lacking sticky ends can also be joined. A direct method uses the DNA ligase made by *E. coli* phage T4. This enzyme differs from other DNA ligase in that it not

---

Kb = kilo base pairs, 1000 base pairs

only seals single-stranded breaks in double-stranded DNA, but can also join molecules with blunt ends.

5. Insertion of a Particular DNA Molecule into a Vector

A collection of fragments obtained by digestion with a restriction enzyme can be made to anneal with a cleaved vector molecule, yielding a large number of recombinant molecules containing different fragments of donor DNA. However, if a particular gene or segment of DNA is to be cloned, the recombinant molecule possessing that particular segment needs to be isolated from among all the recombinant molecules containing donor DNA.

GENE LIBRARIES

A collection of recombinant molecules that contain DNA fragments that represent the entire genome of a given organism is called a genomic library. The number of clones that are required to have a complete gene library depends on the size of the genome and the type of cloning vector. Specific DNA fragments can be isolated from genomic libraries or a cDNA libraries.

Construction of cDNA Library

Direct isolation of a plant gene from a mixture of DNA fragments separated by electrophoresis is not feasible due to the very large number of cleavage sites for a typical restriction enzyme. On the other hand, direct cloning of any DNA coding sequence is feasible provided corresponding mRNA can be isolated in almost pure form. This is possible for many plant genes.

The cytoplasm of specialized cells contains specific mRNA molecules which together constitute a large fraction of the total mRNA synthesized in the cell. Samples of mRNA can usually be obtained that consist predominantly of a single mRNA species. The purified mRNA serves as a starting point for creating a collection of recombinant plasmids, many of which contain coding sequences of the genes of interest. The technique depends on unusual polymerase, reverse transcriptase, which can use a single-stranded RNA molecule (mRNA) as a template and synthesize a double-stranded DNA copy, called complementary DNA or cDNA. The corresponding full-length cDNA contains an uninterrupted coding sequence. It is used for synthesizing the gene product in bacterial cells. Joining cDNA to a vector is accomplished by procedures for joining blunt ends.

The various steps involved are given below (Xoconstle-Cozares et al., 1993).

(i) Isolation of plant RNA

The procedure developed by Logemann et al. (1987) is convenient and widely used to isolate total plant RNA. It consists of homogenizing the tissue in the presence of the strong protein denaturing agent guanidine hydrochloride and 2-b mercaptoethanol. RNA is separated from the proteins, DNA and polysaccharides by several ethanol precipitation steps, or by sedimentation through a discontinuous caesium chloride gradient.

(ii) Isolation of poly (A) + mRNA

The mRNA is separated from other RNA species in a total RNA preparation by affinity chromatography based on the presence of a polyadenosine tail (poly (A)) present at the 3'end. When a total RNA preparation is passed through an Oligo (dT) cellulose or poly(U) sepharose column, poly (A) + RNA will be retained, whereas tRNA, rRNA and other RNAs that lack a poly (A) tail will not bind to the column and may be displaced by washes. Subsequently, the mRNA fraction can be recovered by washing with low salt buffers.

(iii) cDNA synthesis

The cDNA first strand is synthesized by an RNA-dependent DNA polymerase, called reverse transcriptase (Fig. 29.1). This enzyme is used to make a DNA copy of poly (A) + mRNA using oligo (dT) as a primer. The product of the synthesis is a hybrid, cDNA-mRNA molecule from which the RNA-component is eliminated by chemical or enzymatic hydrolysis. Remanents of mRNA, random oligonucleotides or specifically designed oligonucleotides are usually used as primers for synthesis of second strands, which is carried out by the Klenow fragment of *E. coli* DNA polymerase I. Finally, the double-stranded products of the reaction are ligated to a cloning vector by bacteriophage T_4 DNA ligase.

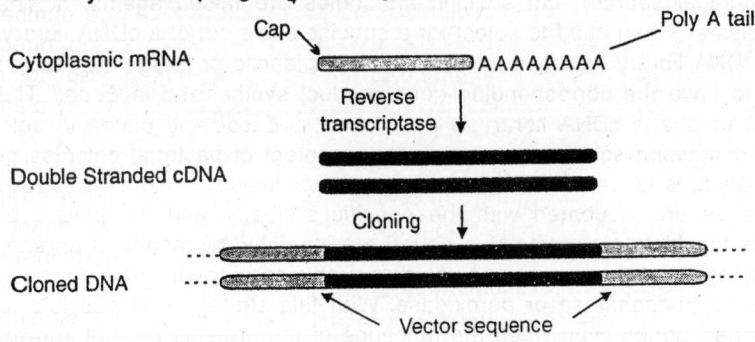

Fig. 29.1. Cloning of cDNA molecules derived from mRNA by enzymes reverse transcriptase

POLYMERASE CHAIN REACTION (PCR)

PCR is a powerful and sensitive *in vitro* method to produce large quantities of a particular DNA sequence without cloning. It uses DNA polymerase and short, synthetic oligonucleotides, usually about 20 nucleotides in length, that are complementary in sequence to the ends of any DNA fragment of interest. The oligonucleotides are called primer sequences as they anneal to the ends of the sequences to be amplified and serve as primers for chain elongation by DNA polymerase. In the first cycle of PCR amplification, the DNA is denatured to separate the strands and then renatured in the presence of a vast excess of the primer oligonucleotides. Then DNA polymerase is added and elongation of the primers produces double-stranded molecules; the first cycle in PCR produces two copies of each molecule containing sequences complementary to the primers. The steps of denaturation, renaturation and replication are repeated from twenty to thirty times, and the number of molecules of the amplified sequence is doubled in each cycle. Thus repeated rounds of DNA replication increase the number of molecules exponentially. The amplified DNA can often be used without further purification for DNA sequencing, as probe DNA in hybridizations, or even for cloning.

Limitation of PCR

1. The DNA sequences at the ends of the region to be amplified must be known so that primer oligonucleotides can be synthesized.
2. The sequence to be amplified must be smaller than 3000 nucleotides to allow the replication steps to occur efficiently and remain error free.

B. Gene Identification

Molecular identification of a gene of interest amongst all the recombinants in a cDNA library can be accomplished in the following ways.

1. Identification by the Gene Product

a) *Use of antibodies*: A very successful approach is the use of antibodies against the protein encoded by the gene of interest. The protein is purified (from original source) and specific antibodies are raised against it. The antibodies are then used to select for a specific clone out of a cDNA library.

A cDNA library is constructed in either a plasmid or phage expression vector to have the corresponding gene product synthesized in *E. coli*. The plasmid or phage cDNA library is introduced in *E. coli* and plated in petri dishes containing solid media. The protein content of bacterial colonies or phage plaques is transferred to nylon or nitrocellulose membranes. These membranes are incubated with the specific antibody, and the plaque or colonies to which this antibody is bound are identified by means of an anti-antibody that is linked to an enzyme that produces a colour reaction, such as alkaline phosphatase or peroxidase. With this strategy it is possible to identify new genes even when the functions of the proteins are not known.

b) *Use of amino acid sequence of the gene product to produce synthetic oligonucleotide probes*: The amino acid sequence of a protein, obtained after its purification and subsequent microsequencing directly from two-dimensional gel electrophoresis can be used to design the nucleotides version of this sequence in the form of synthetic oligonucleotides. These oligonucleotides can be radioactively labelled and used as specific probes to find the corresponding gene with a cDNA library.

Oligonucleotides can also be designed to use the PCR technique.

2. Identification of Genes Based on Sequence Characteristics

(a) *Use of heterologous probes*: Proteins with the same functions in different organisms often have a similar primary structure. In this approach, genes isolated and characterized in other organisms are used for probes (heterologous probes) to screen for similar genes in a cDNA library of the organism of interest.

b) *Differential screening*: The objective of differential screening is to select for a population of mRNAs (cDNAs) which are specific to a particular growing condition, development stage or tissue. Selection is based on the substraction of RNA populations that are common to the two RNA samples by first preparing cDNA copies of one of the populations and then allowing the hybridization of these cDNA molecules with the other RNA population. The cDNAs for which homologous sequences exist in this hybridization mixture will form double-stranded molecules that can be separated from the RNA molecules that are present in only one of the two populations by chromatography on hydroxyapatite columns. The isolated specific mRNAs are used as probes to screen a cDNA library or to construct a directed cDNA library.

3. Identification of Genes Based on the Function of the Gene Product

In this method the cDNA of interest is expressed in bacterial or yeast systems where the enzymatic or structural activity of the gene product is measured or observed as a distinguishable phenotype. Identification of the target gene is carried out by complement of the mutant bacteria or yeast.

4. Molecular Tagging of Genes

Mobile genetic elements cause mutations when they transpose within or near the coding sequence of a gene. For the case in which the element has been characterized, they become a molecular tag that can be used to isolate the mutated gene by hybridization/PCR experiments. Two types of mobile genetic elements have been used for molecular tagging of genes.

a) *Transposon tagging*: Insertion of a copy of transposable elements (e.g. Ds, Ac and others) may inactivate a gene producing a mutation. Phenotypic screening of mutants allows the function of the interrupted gene to be identified. The mutated gene can be cloned using the Tn element as a hybridization probe. The technique is called 'transposon tagging'.

b) *Agrobacterium T-DNA*: *Agrobacterium tumefaciens* has the capacity of transferring a defined segment (T-DNA) of its tumour-inducing (Ti) plasmid to cells of plant species that are susceptible to infection by this bacterium. Several vectors for the transfer of foreign genes based on the Ti plasmid of *A. tumefaciens* have been developed. Upon insertion in the genome of plant cells the T-DNA can cause mutations. Genes mutated by the T-DNA can be identified and isolated using the T-DNA as a hybridization probe. This strategy is known as the T-DNA tagging procedure.

To isolate a gene of interest a population of transgenic plants obtained by transformation with *A. tumefaciens* is screened for the plants showing the desired phenotype. After screening, the affected gene is isolated from a genomic library of the mutant plant using the T-DNA as a probe. The isolated mutated gene can later be used to isolate the wild-type gene in a gene bank from a normal plant. In this case the probe will be the plant DNA found on the left or right side of the T-DNA. Confirmation of the identity of the isolated gene can be carried out by complementation of mutant plants after transformation with the wild-type gene.

MAP-BASED CLONING

A general strategy (map-based cloning) for the isolation of genes for which no biochemical or molecular information is available about the gene or the gene product was recently developed. Map-based cloning relies on the knowledge of the phenotype conferred by the gene of interest and its genetic map position. The various steps involved are described below.

(1) Identification of tightly linked DNA sequences located at both sides of the target gene. The linked sequences are used as starting points for cloning of the chromosomal region where the target gene is located. Restriction Fragment Length Polymorphism (RFLP) or Random Amplified Polymorphic DNA (RAPD) markers are the most useful flanking sequences, since saturated maps of the target region using these markers can be obtained with relative ease.

(2) Isolation of DNA clones covering the entire region between two markers by a process known as chromosome walking.[2] The DNA sequences progressively farther away from the RFLP or RAPD markers are identified by a series of overlapping cloning steps, based on DNA hybridization experiments. Then walking from one of the cloning sequences to the one on the other side will result in cloning all the DNA between all these sequences including the target gene. Chromosome walking is based on vectors, such as lambda or cosmid, which are capable of harbouring genomic inserts of up to 50 kb. Overlapping clones up to 200 kb, can be isolated in this manner.

(3) Pinpointing the clone harbouring the target gene among all the overlapping clones identified during chromosome walking. This is accomplished for dominant genes by transformation and complementation of recessive individuals for the phenotype conferred by the target gene with the clones derived from the chromosome walk. The clone containing the target gene is identified by production of transgenic individuals expressing the appropriate phenotype. This identification is feasible, provided the introduced gene is functionally expressed. If transformation is not feasible or transforming sequences are not expressed, other more complex identification strategies would need to be used.

ANTISENSE RNA INACTIVATION

Alteration of gene expression can also be achieved using antisense mRNA technology. The various steps involved are:

1) isolate a cDNA encoding an unknown function and then construct a

---

[2] A chromosome walk (walking) results when the alternate libraries are used for identifying overlapping clones through hybridization

clone having gene in opposite orientation to direct the production of the antisense strand of mRNA;

2) introduce the above into plant cells by gene transfer procedures.

Whatever is the mechanism of inactivation involved, the result is the lack of corresponding gene product.

5. Detection of Recombinant Molecules

When a vector is cleaved by a restriction enzyme and renatured in the presence of many different restriction fragments from a particular organism, many types of molecules result. The desired recombinant molecules are screened in the following manner.

1) Screening for plasmid-containing cells: Usually plasmids with marker genes, such as antibiotic-resistance gene(s) are used as vectors. The transformed bacteria are grown on the medium containing the antibiotic. Only cells in which a plasmid is present can form a colony. A good example is the plasmid p^{BR322}. It is a small plasmid and contains two genes for resistance to different antibiotics-tetracycline (tet^r) and ampicillin (amp^r). Therefore, transformed bacterial cells that contain p^{BR322} are easily selected by growth of such cells on a medium containing either tetracycline or ampicillin.

2) Screening for cells with donor DNA: The presence of two antibiotic-resistance markers in p^{BR322} enables screening for detecting insertion called insertional inactivation. In p^{BR322}, the tet-gene contains sites for cleavage by restriction enzymes Bam HI and Sal1. Thus, insertion of donor DNA at either of these sites yields a plasmid, that is, tet^s, because the insertion interrupts and hence inactivates the tet-gene. However, the recombinant plasmid remains amp^r. If the bacterial cells are transformed with a DNA sample in which the cleaved p^{BR322} and restriction fragments have been joined, and the cells are plated on a medium containing ampicillin, all surviving colonies must be amp^r and hence must possess the plasmid. Some of these colonies will be tet^r and others tet^s and these can be identified by replica plating onto a medium containing tetracycline. Because the unaltered p^{BR322} contains the tet^r allele, the amp^r colonies will also be tet^r unless the tet^r allele was inactivated by insertion of donor DNA. Thus, the bacterial cells that are amp^r tet^s must contain recombinant p^{BR322} plasmids that have donor DNA inserted into the tet^r gene.

3) Screening for particular recombinants: The procedure of colony hybridization allows detection of the presence of any gene for which DNA or RNA labelled with radioactivity or some other means is available. Colonies to be tested are transferred from a solid medium onto a nitrocellulose or nylon filter by gently pressing the filter onto the surface of the solid medium. A part of each colony remains on the agar medium, which constitutes the reference plate. The filter is treated with NaOH, which simultaneously breaks open the cells and denatures the DNA. The filter is then saturated with the labelled 'DNA or RNA, complementary to the gene being sought, and the cellular DNA is renatured. The labelled nucleic acid used in the hybridization

is called the probe. After washing to remove the unbound probe, the positions of the bound probe identify the desired colonies. For example, with a radioactively labelled probe, the desired colonies are located by means of autoradiography. A similar assay is done with phage vectors but in this case plaques are lifted onto the filters. If transformed cells can synthesize the protein product of a cloned gene of cDNA, immunological techniques may allow the protein-producing colony to be identified. In one method the colonies are transferred as in colony hybridization, and the transferred copies are exposed to a labelled antibody directed against the particular protein. Colonies to which the antibody adheres are those containing the gene of interest.

GENE TRANSFERS IN PLANTS

Numerous different methods have been evolved for the production of transgenic plants. The most commonly used methods are. described below. Interested reader is referred to Potrykus (1993), Shaw (1968) and Kosuge and Nester (1984).

1. *Agrobacterium*-mediated Gene Transfer

The soil bacterium *Agrobacterium tumefaciens* has the ability to induce tumours on some plant species of agricultural interest, for example, tobacco, petunia, tomato, potato and alfalfa. A large plasmid called the Ti, or tumour-inducing plasmid, has been shown to be responsible for tumour formation, due to a remarkable natural capacity to transfer, insert and express a particular segment of DNA in the plant cell genome. The segment of the Ti plasmid DNA which is transferred from the bacterium and becomes stably integrated into the plant genome has been called the T-DNA. Upon integration , the T-DNA encoded genes are expressed in the plant cell nucleus (Fig. 29.2). The mode of transfer of T-DNA to plant cells is not as yet completely understood. However, directly repeated 25 bp sequences, known as border sequences, have been shown to be essential for T-DNA transfer and integration. The 25 bp border repeats, in the correct orientation, are sufficient to promote DNA transfer when complemented by a functional *virulence* region. This property forms the basis for producing vectors for plant cell transformation.

a. Introduction of Gene Constructs into Agrobacterium

The general process for manipulating genes to be transferred into the genome of plant cells is carried out in two phases:

   i) all the cloning and DNA modification steps are done in *Escherichia coli;* and
   ii) the plasmid containing the gene construct of interest is transferred into *Agrobacterium.*

The resultant *Agrobacterium* strain is finally used to transform plant cells.

b. Plant Cell Transformation

   i) Growth of sterile plant tissues: To transform plant cells and obtain mature plants from them it is necessary to start with axenic plant tissues.

1. The kanamycin-resistance gene is inserted into an *E. coli* plasmid between two regions which are homologous to the T-DNA of the Ti plasmid (shaded regions).

2. This plasmid is then mobilized into *Agrobacterium* where it can integrate under homologous recombination with the resident Ti plasmid. Selection for kanamycin-resistant *Agrobacteria* identifies those which have recombined and now have a kanamycin-resistance gene in the T-DNA of the Ti plasmid.

3. These recombinant bacteria are used to infect plants or plant cells in culture.

4. The tumorous cells can be cultured and a plant regenerated from them. Analysis of the DNA in the transformed plant cells shows that they have integrated the T-DNA including the bacterial kanamycin resistance into the nuclear genome (depicted on the right).

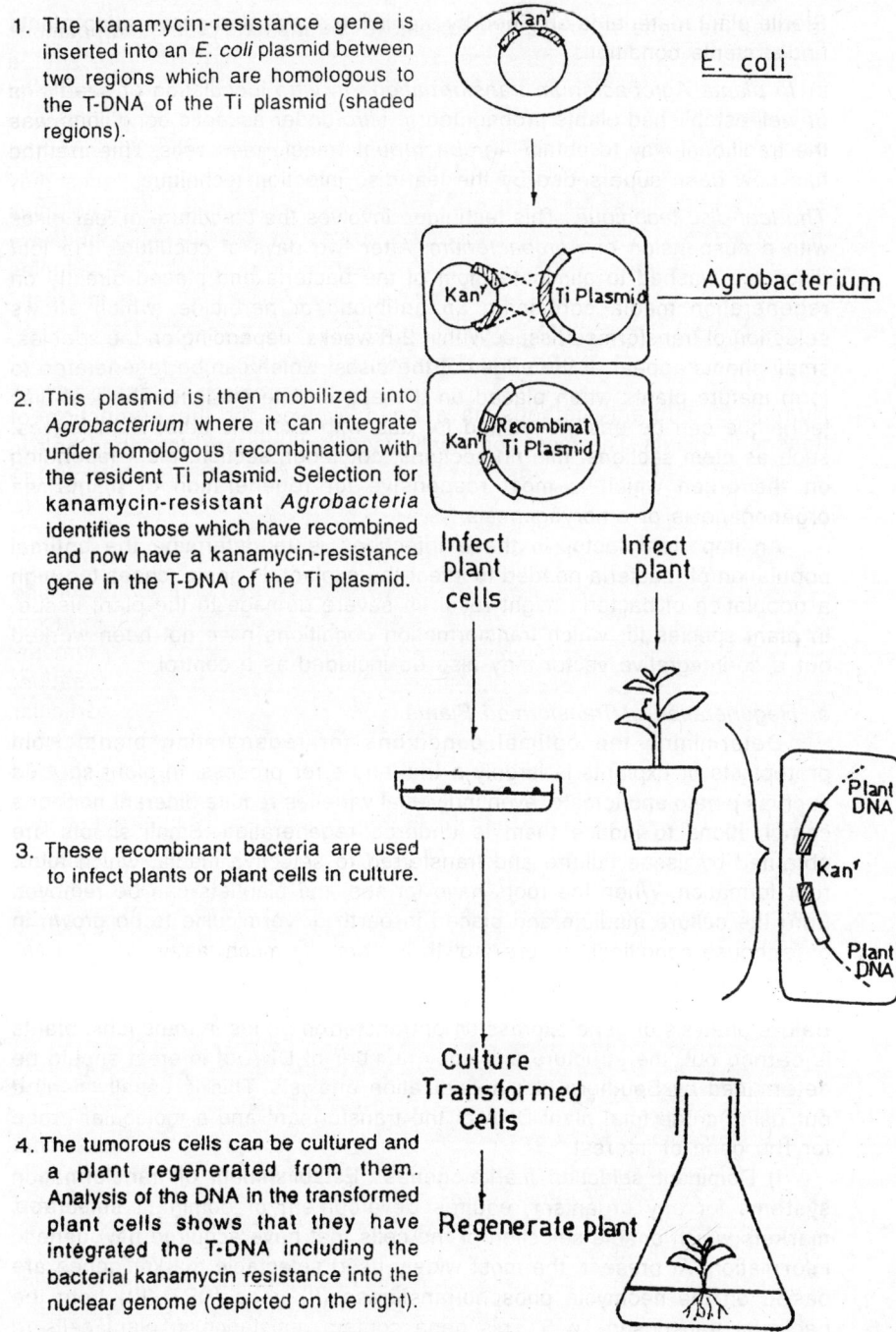

Fig. 29.2. Steps in genetic engineering using an *Agrobacterium* vector.

Sterile plant material is obtained by sterilizing seeds and germinating plants under sterile conditions.

ii) *In planta Agrobacterium transformation* : *In vitro* inoculation of seedlings or well-established plants propagated *in vitro* under asceptic conditions was the traditional way to obtain *Agrobacterium-* transformed cells. This method has now been superseded by the leaf-disc infection technique.

The leaf-disc technique : This technique involves the co-culture of leaf discs with a suspension of *Agrobacterium.* After two days of coculture, the leaf discs are washed to eliminate most of the bacteria and placed directly on regeneration media containing an antibiotic or herbicide, which allows selection of transformed tissue. Within 2-6 weeks, depending on the species, small shoots appear at the edges of the discs, which can be regenerated to form mature plants when placed on the appropriate medium. The leaf-disc technique can be easily modified for use with explants other than leaves, such as stem sections, mid-rib sections, cotyledon section, etc., depending on the organ which is most responsive for regeneration of plants via organogenesis or embryogenesis.

An important factor in these infections is to determine the optimal population of bacteria needed to infect the explant. In some cases too high a population of bacteria might result in severe damage to the plant tissue. In plant species for which transformation conditions have not been worked out a co-integrative vector may also be included as a control.

c. Regeneration of Transformed Plants

Determining the optimal conditions for regenerating plants from protoplasts or explants is largely a trial-and-error process. In plant species such as potato and tomato, even individual varieties require different hormone combinations to induce them to undergo regeneration. Small shoots are obtained by tissue culture and transferred to selective media, which allow root formation. When the roots have formed, the plantlets can be removed from the culture medium and placed in earth or vermiculite to be grown in greenhouse conditions, where growth is normally much faster.

GENE EXPRESSION ANALYSIS IN TRANSGENIC PLANTS

Before analysis of gene expression of transferred genes in transgenic plants is carried out, the structure and copy number of DNA of interest should be determined by Southern blot hybridization analysis. This is usually carried out using crude total plant DNA of the transformant and a molecular probe for the gene of interest.

i) Dominant selection marker genes : Establishment of transformation systems for any organism requires development of dominant selectable markers which enable selection of the cells that have acquired new genetic information. At present, the most widely used selectable marker genes are based on the neomycin phosphotransferase (II) gene (NPT (II)) from the bacterial transposon Tn 5. This gene confers resistance in plant cells to aminoglycoside antibiotics, such as kanamycin and G 418. Other selectable markers are the hygromycin phosphotransferase (HP) gene from *E. coli*, the bleomycin-resistance gene. It is important to note that selection depends on determination of susceptibility of the plant to the different selectable agents, and the minimal concentration of this agent which effectively kills most, if not all of the non-transformed cells. Resistant strains are selected by growth.

ii) *mRNA analysis*: When intact genes are transferred from one plant species to another, gene expression can be analyzed at the RNA level by hybridization techniques, or at the protein level by the use of specific antibodies. One of the easiest and more sensitive techniques which requires small amounts of RNA is the dot-blot hybridization technique. Techniques which are more complicated but produce more precise data about the size or structure of the mRNA are the Northern blot and nuclease S_1 mapping techniques.

iii) *Reporter genes*: For detailed analyses of the transcriptional regulatory signals of a given gene, it is more straightforward to link this sequence to a reporter gene. Reporter genes are those which are well characterized both genetically and biochemically, and have a long coding region which can be easily fused to the regulatory sequences of other genes. The enzymatic activity of most reporter genes is not normally present in the host plant, or is easily distinguishable from other endogenous gene activities. This enables study of the regulation of a gene under different environmental conditions or in different organs of a plant. Examples are octopine synthase, chloramphenicol acetyl transferase gene, neomycin phosphotransferase (II), (b)-galactosidase and luciferase genes.

FACTORS INFLUENCING GENE EXPRESSION IN TRANSFORMED PLANTS

Many factors may influence gene expression in transformed plants. The important ones are the position and number of integrations of the foreign sequence within the plant genome. Several authors have reported significant differences in the levels of expression between different transformation events utilizing the same foreign sequences. It has also been shown that not only quantitative, but also qualitative changes may occur, resulting in altered organ-specific patterns of expression, albeit at a lower frequency. These effects are presumably due to regulatory sequences around the site of insertion of the foreign sequence which can influence genes over long stretches of DNA. The copy number of the transferred DNA may also be a factor affecting the level of expression, although no correlation between copy number and level of expression has been found so far.

Other factors that have to be taken into consideration when studying gene expression are the developmental stage of the plant, light intensities, and additives such as sucrose and hormones included in the culture media. Final analysis of the expression of genes should be performed, if feasible, in the progeny of transgenic plants, to eliminate the effects of passing through tissue culture on the original transformed plant.

2. Protoplast-based Direct Gene Transfer

Protoplasts are plant cells from which the wall has been digested enzymatically. In this state they are used as recipients for insertion of new DNA. Methods using naked DNA have become known by the general term of 'Direct gene transfer'. Protoplasts of different species may also be fused to create hybrids even between species that cannot be hybridized sexually. The protoplasts will synthesize new walls and can regenerate into entire plants, thereby perpetuating the changes induced (Fig. 29.3).

There are a number of ways in which protoplasts can be induced to take up DNA, including the use of polyethylene glycol (PEG) and electroporation. These methods have the limitation that DNA can only be transferred into protoplasts and these protoplasts must be able to divide and form colonies. For this to be of use, one must then be able to regenerate a normal plant from the tissue produced by the process. While this is the case with *Solanaceous* plants, such as tobacco, potato and petunia, this is not generally true for most of the crops of plant-breeding interest, although a few researchers have been successful in obtaining transgenic plants from *japonica* and *indica* type rice, maize and wheat.

Protoplast Isolation

i) Source Tissue

The plant material which is often chosen as a source of protoplasts in dicots is leaf mesophyll. This is because this tissue is available in large amounts and yields fairly uniform protoplasts with a good yield. In some cases the leaf material is taken from shoot cultures maintained axenically in a sterile culture system. Protoplasts can also be isolated from cell cultures (normally suspension) or from other portions of the whole plant. The use of roots and seedlings as a source of protoplasts has been successful in forage legumes. In the case of monocots, the protoplast normally comes from non-morphogenic suspension cultures. In the case of rice, only protoplasts from embryogenic rice suspension cultures have been reported to be regenerable to plants in a repeatable way.

ii) Protoplast Isolation

Tissue is taken in before digestion by osmotically active solution with an osmotic potential of 400-700 m Os/Kg H_2O. Either a salt or sugar (or sugar alcohol) solution or a mixture of the two is normally used for the purpose. For the enzyme digestion step, this can also be a salt solution. Typical solutions consist of mannitol (0.4 M) with a low concentration of calcium chloride to stabilize the membranes. The enzymes used are some cellulose and pectinases free of protease, DNase and RNase activities. In order for the enzymes to penetrate the tissue, it is usually necessary to slice the tissue in the enzyme solution and/or to vacuum infiltrate the enzyme into the tissue, neither of which is required for suspension culture cells. After digestion of the tissue, the protoplasts are separated from undigested material, the enzymes themselves, and any toxic material produced during the digestion. This is achieved by a 2-step process consisting of filtration through a 50-100(μ)m stainless steel sieve, followed by a series of washing steps. A

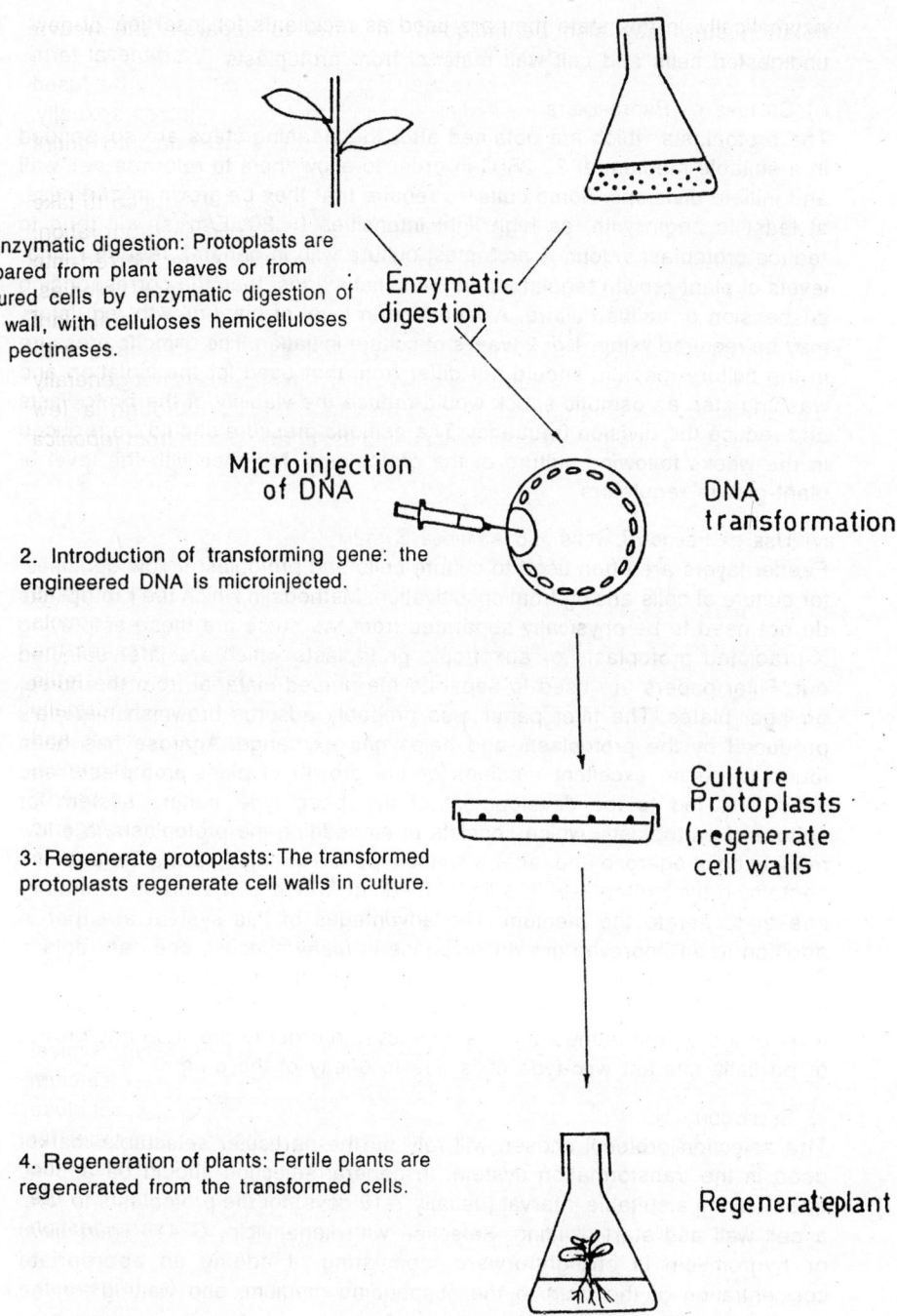

Fig. 29.3. Steps in genetic engineering by DNA-mediated gene transfer.

flotation step may also be included, in order to separate the heavier undigested cells and cell wall material from protoplasts.

iii) Culture of Protoplasts

The protoplasts which are obtained after the cleaning steps are suspended in a suitable medium at 22-28°C in order to allow them to reform a cell wall and initiate divisions. Some cultures require that they be grown in darkness, at least to begin with, as high light intensities (> 20 $\mu E/m^2/s$) will tend to reduce protoplast vigour. A protoplast culture will, in general, require higher levels of plant-growth regulators, in the initial stages, than the corresponding suspension or callus culture. A reduction in level of plant-growth regulators may be required within 1 or 2 weeks of culture initiation. The osmotic pressure in the culture medium should not differ from that used for the isolation and washing step as osmotic shock would reduce the viability of the protoplasts and reduce the division frequency. The osmotic pressure should be reduced in the weeks following culture of the protoplasts, together with the level of plant-growth regulators.

iv) Use of Feeder Layers and Agarose System

Feeder layers are often used to culture cells and protoplast in low densities, for culture of cells arising from cocultivation. Methods in which the protoplasts do not need to be physically separated from the nurse are those employing X-irradiated protoplasts or auxotropic protoplasts which are later selected out. Filter papers are used to separate the nursed material from the nurse, on agar plates. The filter paper also probably adsorbs brownish materials produced by the protoplasts and helps gas exchange. Agarose has been found to be an excellent medium for the growth of plant protoplasts and cells. This led to the development of the 'bead type' culture system for culture of protoplasts, which consists of embedding the protoplasts in a low melting point agarose and, after a suitable period of time, placing the agarose containing the protoplasts in a liquid medium, which is shaken on a rotatory shaker to aerate the medium. The advantages of this system are that in addition to an improved growth response in many species, one can replace the medium surrounding the growing cells without disturbance. This is particularly valuable when carrying out selection in that one can maintain the level of a drug in the medium at a high level in order to preclude any growth of partially tolerant wild-type cells due to decay of the drug.

v) Selection

The selection protocol chosen will rely on the particular selectable marker used in the transformation system. In general selection should be applied after leaving a suitable interval (usually 7-10 days) for the protoplasts to form a cell wall and start dividing. Selection with kanamycin, G 418 (*geneticin*) or hygromycin is straightforward, consisting of adding an appropriate concentration of the drug to the suspending medium, and waiting for the resistant colonies to appear.

vi) Regeneration of Plants

Plant regeneration in model plants, such as petunia and tobacco, is achieved by manipulation of cytokinin: auxin ratio. For most other species the process is rather more complicated.

Cybrids and the production of male sterile lines

Cybrids, that is, cytoplasmic hybrids, are the hybrids that possess nuclear genes from one parent only and cytoplasm from both parents. The organelle DNA in the cytoplasm, namely, the chloroplast DNA (cp DNA) and mitochondrial DNA (mt DNA) are known to control expression of male sterility. The cytoplasmic male sterility (CMS) trait is caused by sequence rearrangements and mutations in the mt DNA. The economic importance of the CMS trait in hybrid seed production led to attempts to transfer male sterility from one species to another, especially in incompatible combinations through protoplast fusion. The usual procedures consist of enucleation of protoplast of one parent by X-ray irradiation or by centrifugation through the percoll gradient and subsequent fusion with normal protoplast; or elimination of chromosomes of one parent during proliferation of hybrid cells; and fusion of normal and subprotoplasts or microprotoplasts.

Cybrids have been used in the transfer of male sterility in many species. A few examples are cited below.

Brassica : Sterilizing cytoplasm from radish (Ogura CMS System) was transferred to *Brassica napus* and other *Brassica* spp. However, chlorophyll deficiency, low nectar production and lack of complete fertility restoration prevented its utilization. The use of protoplast fusion involving exchange of radish chloroplast with that from *B. napus* solved the problem of chlorophyll deficiency and low nectar production (Pelletier et al., 1983).

Rice : Cybrid plants involving the nucleus of a fertile variety and mt DNA derived by recombination from both parents have been generated. The regenerated plants retained the CMS trait from the male sterile partner (Yang et al., 1988, 1989).

Carrot : Cybrids involving transfer of the CMS trait (brown anther type) to normal varieties have been reported. (Tanno-Suenaga et al., 1988).

Tobacco : Cybrid plants involving transfer of the CMS trait from *Nicotiana tabacum* to *N. sylvestris* through fusion of X-ray irradiated protoplast of the former with the normal protoplast of the latter have been reported (Zelcer, 1978).

3. Direct Gene Transfer

Direct gene transfer has proved to be a simple and effective technique for the introduction of foreign DNA into plant genomes.

General Features

i) Vector : The plasmid used for direct gene transfer should produce large quantities of plasmid without undue effort, for example pABD1 is constructed using PVC8 as the basic replicon.

ii) Marker : The selectable marker to be used varies with the plant species and the markers available. The most effective markers for stable transformation are NPT (neomycin phosphotransferase, of APH (e') II from transposon 5 (Tn 5) and hygromycin resistance. Kanamycin has often been replaced by G 418 for selection for NPT II activity in Graminaceous monocots.

iii) Promoter : The choice of promoter to which the gene is to be attached depends on the desired effect. The 35 sRNA promoter from cauliflower Mosaic virus (CaMV) is being increasingly used. The nopaline synthase promoter in its improved from is also being used.

iv) Carrier DNA : In plants plasmid DNA as carrier has been successfully used. The role of carrier DNA in promoting DNA uptake and/or integration is not known.

Transformation of Protoplasts

i) PEG-mediated transformation: The use of PEG has now been developed as a method of transfer of DNA into protoplasts. PEG-based methods can also be used to carry out transient expression.

ii) Electroporation: Electroporation has proved to be an easy and efficient method for introducing foreign genes into plant protoplasts in a way which allows transient expression and integrative transformation to be achieved. There are two basic methods of electroporation used to stably transform plant cells. These use either a short or long pulse of an electric field to induce pore formation in the membrane of protoplasts. The long-pulse method uses a low voltage electric field and the short-pulse method a very high voltage field. It is not clear which system is the most efficient but observations are that the short-pulse system is more efficient for stable transformation and the long-pulse for transient expression. The precise parameters must be determined for each individual type of protoplasts.

TRANSIENT EXPRESSION

Following treatment of the protoplasts with the DNA in suitable uptake conditions, certain genes are expressed only for short periods usually in the region of 1-7 days, either the activity of the gene product is measured or the presence of the desired protein in the cells is determined by use of antibody techniques to ensure the transformations.

Biolistics or Particle Gun

Acceleration of heavy microparticles coated with DNA has been developed into a technique to carry genes into virtually all types of cell and tissues. One shot can lead to multiple hits (transfer of genes into many cells) and genes may be transported to many cells at nearly any desired position in an experimental system without much ado. The cells survive the intrusion of particles and the genes may be coated on particles resume biological activity. They can be located at the surface or in deeper layers of organs. The method depends on physical parameters only.

Despite success in obtaining transgenic plants, for example in corn and soybean, use of this technique is likely to be limited considering the low conversion rate of transient events (hits) to stable integrative events, due to many biological problems.

Microinjection

A sophisticated microinjection apparatus has been developed for mammalian systems which allows DNA to be introduced directly into the nucleus of an individual cell. Although only a small number of cells can be microinjected, the frequency of successful gene transfer is sufficiently high up to 20%) and no selection for transformation is necessary. These introduced genes are sometimes expressed in the adult animal, and they are inherited in a Mendelian fashion.

The interest generated by these experiments has prompted a number of groups to investigate microinjection of plant cells (both with and without cell walls). If successful transformation could be achieved in fertilized embryos, or even in the pollen or eggs cells, new genes could be introduced directly into crop plants without the need for regeneration of plants from cultured tissues of vegetative cells. This would bypass one of the biggest obstacles in the application of gene transfer to crop plants. Microinjection has yielded rare transgenic events but so far has not been developed into a routine procedure. Compared to the particle gun, it has the disadvantage that only one cell receives DNA per injection and handling requires more skill and instrumentation. The advantages are that the quantity of DNA delivered can be optimized, delivery is precise and predictable even into the cell nucleus, and is under visual control.

Microtargeting

Microtargeting is the ideal procedure for DNA delivery into shoot meristems. A novel device (microtargeter) has been developed which enables targeting of populations of microprojectiles to individual shoot meristems of cereal seedlings. This device combines advantages of the microinjection technique (the capacity to predict the position of DNA-delivery) with the advantages of the biolistic technique (achieving multiple hits with one event) for biolistic targeting of shoot meristems. It has the potential to develop into a generally applicable method, if the cells of shoot meristems are competent for integrative transformation.

The identical size particles are accelerated to target areas of variable size (100-200 nm) which arrive at the target as individual particles with predictable but variable penetration. The particles have to carry DNA which is freely accessible within the cell. The particles are carried by a method developed for preparation of gold particles from solutions of gold salts by the addition of photographic developer. This procedure allows the production of homogeneous particles of any desired size. Clumping of particles is prevented by precluding any kind of macroprojectile. Instead an aerosol is created from suspension of gold particles in DNA containing water. The aerosol is then accelerated in a pilot tube by the same gas pressure pulse that created the aerosol. Acceleration occurs in a capillary, the length of which determines the impulse of the particles and the width of the target area. Accurate aiming is achieved with the help of crosslines in the microscope.

As each physical parameter of the system is variable within a wide range, the microtargeting system can easily be adjusted and optimized to suit the specific requirements of shoot meristems from different plant species. Many particles can be targeted to a single meristem, thus allowing for the delivery of DNA to statistically every cell. Treated shoot meristems regenerate with high frequency to fertile plants. Integrative transformation in competent cell cultures has been reported to be very high (Sautter et al., 1991).

Engineered Commensals

Micro-organisms can be transformed with no apparent effect to contain genetic information that will be beneficial to the plants they infest. DNA determining the production of a substance toxic to insects has been incorporated in such an organism and is now being tested as a corn-borer protection strategy for corn.

APPLICATION AND USES OF GENETIC ENGINEERING

The long-term goals of genetic engineering in crop improvement are the same as those of conventional breeding methods. It should be clearly understood that advances through genetic engineering must ultimately be integrated into conventional breeding programmes. Genetic engineering, however, offers many new opportunities to effectively solve many plant-breeding problems. These are discussed below. For a more detailed account the interested reader is referred to Salamini and Motto (1993).

Use of Single Genes

1. Virus Resistance

· Different viral DNA sequences appear to function in transgenic plants to confer resistance to viral infection.

a) VIRAL COAT PROTEIN (CP) GENES

Transgenic plants with TMV coat protein gene (driven by CaMV 35S promoter in transformed tobacco plants), when artificially inoculated with varying concentrations of TMV, showed delay in symptom development; further up to 50% of the plants showed no symptoms of TMV during the duration of the experiment (Abel et al., 1986). Transgenic tomatoes carrying the TMV coat protein gene have been shown to be highly resistant to TMV infection (Nelson et al., 1988).

This approach produced similar results in transgenic tomato, tobacco and potato against a broad spectrum of plant viruses, including alfalfa mosaic virus (Tumer et al., 1987). The level of coat protein in the engineered plants varied from .01-0.5% of the total proteins which is well below the level reported in plants infected with this endemic virus. This should eventually facilitate commercialization of virus-resistant cultivars.

Homozygous plants of cantaloupe line CZW-30 containing coat protein gene constructs of cucumber mosaic cucumovirus (CMV), Zucchini yellow mosaic poty virus (ZYMV), and watermelon mosaic virus 2 potvirus (WMV-2) were found to be highly resistant in field trials in that they never developed systemic symptoms as did the non transformed plants but showed few symptomatic leaves confined close to the vine tips. The control plants were severely stunted compared to transgenic plants. (Fuchs et al., 1997).

Transgenic *Carica papaya* plants carrying coat protein gene of papaya ringspot virus (strain HA 5-1) remained symptomless and ELISA negative for 24 months after inoculation with Hawaiian strains of papaya ring spot virus under field conditions. These results indicate that the pathogen-derived resistance can provide effective protection against a viral disease over a significant portion of the crop cycle of a perennial species. (Lius et al. 1997).

b) ANTISENSE RNA

It has been reported that the expression of parts of a viral genome in the form of antisense RNA confers disease resistance. This is so because genes that express antisense RNA are regulatory genes. These genes direct the synthesis of RNA that by itself exhibits negative control on the expression of other genes at the level of transcription or translation. The antisense RNA contains base sequences complementary to the target (sense) RNA transcript, and the formation of RNA-RNA hybrids blocks the function of sense RNA. Antisense RNA may also interfere with mRNA activity, presumably by hybridizing to sequences at or close to the ribosome binding site. The term mic RNA (mRNA interfering complementary RNA) has been proposed for this specific regulatory RNA species.

For the present, experimental data to define the effectiveness and optimal antisense RNA structure to be used in transformation experiments of crops is lacking.

c) VIRAL SATELLITE RNA

Satellite RNA are species of RNA associated with some strains of certain plant RNA viruses. They are not necessary for virus replication and exhibit no base sequence homology to the genomes of either virus or plant host, and are replicated by the helper virus-coded RNA-dependent RNA polymerase. One aspect of their parasitic association with plant viruses is the fact that they can affect disease and symptom development in virus-infected plants. Transgenic plants expressing satellite RNA present a unique approach to genetically engineered virus-resistant crops in that the protective mechanism seems to be triggered to high efficiency only when the viral pathogen is invading the host plant. Therefore, in all such instances wherein the presence of satellite RNA attenuates the disease caused by the virus, satellite RNAs have been used as a biological control in crop protection. Examples are the satellite RNA of cucumber mosaic virus (CMV) and satellite RNA of tobacco ring spot virus (Tob RV) that confer resistance to infection by satellite-free helper viruses CMV and Tob RV respectively.

2. Insect Resistance

Progress in engineering insect resistance in transgenic plants has been achieved through the use of insect control protein genes of *Bacillus thuringiensis* (*Bt*). *Bt* toxins are active against Lepidoptera, but some are specific for Diptera and Coleoptera. The insect toxicity of *Bt* resides in a large protein with no toxicity to beneficial insects. The mode of action of *Bt* toxins is exerted by disruption of ion transport across brush border membranes of susceptible insects. Several genes encoding lepidopteran type toxins have been isolated.

a) BT SSP. KURSTAKI HD-1 (WATRUD ET AL., 1985)
It contains an open reading frame of 3468 bp encoding of protein of 1156 amino acids. Truncated forms of the gene still encode active toxins. Chimeric *Bt Kurstaki* genes with a CaMV 35 S promoter and a sequence coding for an active truncated variant as well as the full-length gene have been constructed and expressed in tomato plants. The insecticidal protein was expressed at a level sufficient to kill larvae of *Manduca sexta, Heliothis virescens* and *H. zea*. Progenies of transgenic plants showed that *Bt Kurstaki* gene segregates as a single dominant Mendelian marker.

b) BT STRAIN BERLINER 1715 (HOFTE ET AL., 1986)
It produces *Bt* 2 protein, 1155 amino acids long, which generates a smaller polypeptide with full toxic activity. Vaeck et al. (1987) have used chimeric genes containing the constitutive 2' promoter of the *Agrobacterium* T_R-DNA, the entire coding sequence of *Bt* 2 as well as the truncated toxin gene, to transform tobacco plants. Other constructs contained a chimeric toxin-*NPT II* gene. The *Bt-NPT II* fusions were found to be particularly useful because they allowed the selection of transformants with levels of toxin sufficiently high to be insecticidal. In transformed plants their insect toxicity correlated directly with the level of kanamycin resistance, the selectable marker used in the transformation experiments. A correlation was also found between the quantity of the toxin and the insecticidal activity of transgenic plants. Transgenic tobacco plants were protected from feeding damage by larvae of *Manduca sexta*.

c) COWPEA TRYPSIN INHIBITOR GENE
This inhibitor is an antimetabolic agent against the bruchid beetle and other insects of the genera *Heliothis, Spodoptera, Diabrotica* and *Tribolium*. Constructs containing the CaMV 35 S promoter and a full length of cDNA clone 550 bp long were mobilized into *Agrobacterium* and used to transform leaf discs of tobacco. Transgenic plants contained 3-7 copies of the constructs and a variable level of inhibitors. Western blotting of plant extracts showed that a polypeptide was produced. And processed bioassay for insecticidal activity was done with young tobacco plants infected with emerged larvae of the lepidopteran *Heliothis virescens*. Insect survival and plant damage were decreased on transgenic tobacco compared to the controls, although considerable variability among plants was observed.

d) Other plant-derived genes
Transgenic potato plants expressing snowdrop lectin (GNA) gene showed an enhanced level of resistance to larvae of the tomato moth (*Lacanobia*

oleracea). Leaf-damage was reduced by more than 50% compared to controls, total insect biomass per plant was reduced by 45-65%, but larval survival was only slightly reduced (20%). These results indicated that GNA has a significant anti feedant effect against tomato moth (Gatehouse et al., 1997).

3. Herbicide Resistance

The use of broad-spectrum herbicides is restricted because they often lack selectivity. It is possible now to expand their possible applications by introducing herbicide-resistance genes in susceptible crops. There are three mechanisms now available for incorporation of herbicide resistance in susceptible crops.

a) OVERPRODUCTION OF A HERBICIDE-SENSITIVE BIOCHEMICAL TARGET

Glyphosate, a broad-spectrum herbicide, interferes with aromatic amino acid biosynthesis by inhibiting the enzyme EPSP (5-enol-pyruvylshikimate-3-phosphate synthase) normally expressed in the chloroplasts of plants. This herbicide is rapidly degraded by soil micro-organisms. Tolerance to this compound is achieved by the overproduction of the target enzyme (EPSP) or by engineering plants with an altered enzyme. Both types of mutations have been selected in *Salmonella typhimurium* and the gene *aro* A encoding EPSP has been cloned and sequenced. A high level of herbicide tolerance has been achieved by accumulating EPSP in the chloroplast.

Herbicides, namely Bialaphos and Phosphinothricin (PPT) are non-selective herbicides and competitive inhibitors of glutamine synthase (GS). Inhibition of GS causes a rapid accumulation of ammonia which leads to death of the plants. Increased synthesis of the enzyme GS has been found sufficient to overcome the toxic effects of the inhibitor.

Metz et al. (1997) reported occasional loss of expression of phosphinothricin (PPT) tolerance in sexual offspring of transgenic oilseed rape (*Brassica napus* L.). They found that in within variety and between variety cross no transgenic inactivation was observed. However, after selfing and backcrosses with non-transgenic oilseed rape infrequent loss of the expression of the PPT tolerance transgene was observed, independent from its homozygous or hemizygous nature. Molecular analysis of PPT-susceptible plants showed that the loss of expression was due to gene inactivation and not to the absence of transgene. Methylation and cosupperession are the mechanisms that might have caused reduced or even loss of expression of the transgene in later generations.

The commercial and economic value of genetically modified crops is determined by a predictable, consistent and stable transmission and expression of transgenes in successive generations. No gene inactivation is expected after selfing or crosses with non-transformed plants of homozygous transgenic oilseed rape plants if the expression of the transgene in homozygous or hemizygous nature in such plants is stable.

b) Detoxifying Enzymes

Detoxifying enzymes inactivate the herbicides before they are able to inhibit the target enzyme. *Bar* gene: The bar gene conferring resistance to PPT was isolated from *Streptomyces hygroscopicus*. It encodes phosphinothricin acetyl transferase (PAT), an enzyme that inactivates the herbicidal compound PPT by acetylation of the free NH_2 group of the molecule. The transgenic plants are protected against the applications of glucophosinate and *bialaphos*.

GSTs: GSTs (Glutathione-S-transferase) are multifunctional enzymes. They are responsible for the modification of the triazine herbicides by glutathione. cDNA clones corresponding to genes *GST I* and *GST II* of corn have been isolated. Treatment of maize seeds with safeners (herbicide protectant chemicals) results in three- to fourfold activation of *GST m* RNA in etiolated tissue. Similar results have been obtained in metribuzin-tolerant tomatoes, in which detoxification involves enhanced activity of the enzyme N-glucosyltransferase, and tolerance is governed by a single locus.

Mixed function oxidases: Mixed function oxidases are involved in the detoxification of 2,4-D in pea.

Soil micro-organisms: Several soil micro-organisms involved in herbicide degradation have also been characterized as potential sources of herbicide-resistant genes. Examples are *bxn gene*: Bromoxynil is a potent photosynthetic inhibitor in plants. The gene *bxn*, encoding a specific nitrilase was cloned from natural soil bacterium *Klebsiella ozaenae*. The *bxn* gene under control of a light-regulated promoter was transferred to tobacco plants to which it conferred resistance to high levels of commercial formulation of bromoxynil.

tfd A gene: Transgenic tobacco plants resistant to 2,4-D were recently produced through *tfd A* gene of the soil bacterium *Alcaligenes eutrophus*, encoding the first 2,4-D monoxigenase (DPAM) enzyme involved in a 2,4-D degradative pathway. Tobacco plants expressing this enzyme exhibited resistance when sprayed with levels of herbicide up to 8 times the field application rate. Introduction of the gene for DPAM into broad-leaved crop plants may eventually allow 2,4-D to be used as an inexpensive post-emergence herbicide in economically important dicot crops.

c) Structural alterations

Herbicides, namely sulphonyl ureas and imidazolinones target enzyme acetolactate synthase (ALS). They block branched chain amino acid biosynthesis inhibiting the metabolic routes leading to valine, leucine and isoleucine. Mutations conferring resistance have been isolated.

Transgenic tobacco plants expressing a mutant ALS gene from tobacco or *Arabidopsis* tolerate herbicide concentrations four times that of typical field application rates. Maize plants homozygous for the resistance gene are tolerant to application of imidazolinone herbicides. Herbicides, atrazine and simazine inhibit electron transport through photosystem II (PS II) by binding to chloroplast thylakoid membranes. The herbicides bind to a 32 KDa thylakoid membrane polypeptide encoded by the chloroplast *psb A* gene. Resistance is maternally inherited in all atrazine-resistant weeds which also carry the highly concerned *psb A* genes. A single amino acid substitution (serine to glycine) at position 264 in the 32 kDa protein results in decreased herbicide binding. Transfer of resistant chloroplasts or development of a chloroplast transformation system are possible ways to obtain atrazine resistance in valuable crops.

Resistant chloroplasts of *Brassica campestris* have been introduced by standard genetic backcrosses into crops such as oil-seed rape. More recently, chloroplasts of a terburtryn-resistant mutant of *Nicotiana plumbaginifolia* were transferred into the *N. tabacum* nuclear background by protoplast fusion. The regenerated plants were resistant to high levels of atrazine (10 kg/ha).

4. Protein Quality

Manipulation of protein quality (for example increased lysine in cereals; methionine and cysteine in legumes) through genetic engineering is being vigorously researched. One major obstacle is that these storage proteins are the products of multigene families. The addition of one gene to plant genomes modified by transformation may therefore not be very effective in improving the plant phenotype due to the expression of the rest of the multiple gene family. Use of cloned regulatory sequences should such become feasible in the near future, may give a new tool for breeding cereals with a better protein quality.

5. Hybrid Breeding

In recent years several possibilities for the control of plant reproduction either through manipulation of incompatibility or by inducing male sterility have emerged. Two systems for inducing male sterility have been developed. The first is used as a molecular construct based on the gene *rol c* of *rhizogenes* under the control of CaMV 35 S promoter, flanked by a selectable marker. Tobacco plants transgenic for *rol c* are sterile and phenotypically different from the wild type. The possibility of selecting in favour of a flanking marker would enable isolation of homozygous male sterile populations for use in hybrid-seed production.

A novel genetic method for inducing pollen sterility is a gene construct that couples a promoter with a highly tissue-specific gene expression to a bacterial or fungal RNase gene. The promoter TA 29 ensures that the coding sequence to which it is attached is transcribed only in the tapetum at a particular stage of anther development. Transgenic tobacco plants retain the specificity of TA 29. The same specificity of expression and the same level of sterility were observed when gene constructs were introduced in oil-seed rape plants. The approach has been applied successfully to crops such as cabbage, cotton and corn. The ability of the TA 29 RNase gene to introduce male sterility provides a new strategy for the production of hybrid crop plants. By coupling the chimeric TA 29 RNase gene to a dominant herbicide gene (for example, *bar* gene), breeding systems could be devised to select for uniform populations of male sterile plants. In crop plants such as lettuce, carrot, cabbage, etc. in which fruits are not harvested, male sterile plants can be crossed with any pollinator line to produce hybrid seeds.

Antisense RNA technology and the existence of *bar star*, a specific protein inhibitor of *barnase*, should facilitate the development of strategies for male fertility restoration, a requirement for hybrid varieties in crops such as wheat, rice and corn. Oil-seed rape pollinator lines capable of producing fully fertile hybrid seeds on the male sterile plants have been obtained by expressing *bar star* in their anthers. Hybrid progeny express both *barnase* and *bar star* proteins in the tapetum which form a molecular complex resulting in the full restoration of fertility.

6. *Genetic Engineering of Simple Biochemical Pathways (Metabolic Engineering)*

In the days to come, it is likely that the products produced in plants by metabolic engineering will become increasingly important in agriculture. The identification of genetic basis of developmental or metabolic pathways and the control systems of complex cellular functions will have a profound impact on our basic understanding of plant science, and also lead to the creation of next generation of improved crops. The current state-of-the-science was recently reviewed at 'The Metabolic Engineering in Transgenic Plants' conference held at Copper Mountain, Co, USA, 12-16 April, 1997. (Dixon and Arntzen, 1997). The sessions were organized to highlight following specific biosynthetic pathways or cellular processes that influence the genetic modifications of these pathways.

1) Plant-gene suppression and disruption.
2) Signal-transduction pathways for plant defense responses.
3) Genetic manipulation of natural-product pathways.
4) Carbohydrate biosynthesis and regulation.
5) Genes and controls for lipid synthesis.
6) Cell-wall polymers.
7) Protein engineering.
8) Protein targeting and engineering.

7. Disease Resistance

Genetically engineered resistance to fungal pathogens and to bacteria is in the early research stages.

8. Stress Tolerance

Alteration of gene expression during environmental stress has permitted the isolation of stress-related genes. However, considerable research effort is still needed to define the role of stress proteins in the acquisition of stress tolerance, including clarification of the structural and enzymatic functions that these proteins may possess. Several model systems are being developed.

9. *Diagnostics*

The molecular probe may be used to specifically monitor the level of stress or disease reactions as well as the presence of pathogenic organisms in the breeding material, such as the potato spindle tuber viroid (PSTV) in potato.

Genetic Bit Analysis (GBA)

Genetic Bit Analysis is a relatively new technique developed to score single. nucleotide polymorphisms among alleles. The technique is based on hybridization capture of a single stranded PCR product to a sequence-specific microplate-bound primer, followed by enzyme-mediated single base extension of the capture-primer across the polymorphic site, enabling direct determination of the base composition of DNA sequence polymorphisms (SNPs) through simple colorimetry (Reynolds et al., 1995). It is currently being used as a paternity test as well as pedigree analysis of farm animals.

The identification of a single nucleotide polymorphism which distinguished the plastome of cytoplasmic male sterile onion varieties from the plastome of fertile lines provided a model system for testing the utility of GBA in plants (Alcala et al., 1997). They demonstrated that GBA permits rapid and accurate allele determination in onion breeding lines resulting in accurate prediction of sterility at the seedling stage.

The results obtained by Alcala et al. (1997) suggest that GBA can be incorporated into marker assisted breeding programs for rapid and efficient diagnostic analysis when large number of samples must be rapidly and accurately scored. By facilitating the screening of important traits using new DNA-based diagnostic techniques such as, GBA^{TM} breeders can increase the probabilities of finding desirable genotypes with inherent savings in time, labor and economic resources.

10. Improved Fertilizer Efficiency

Introduction of the *nif* genes from *Klebsiella pneumonia* into a suitable non-legume host plant may be achieved through plant transformation. These are discussed in Chapter 27.

Recovery of Transgenic Plants

Transgenic plants in which the foreign gene is inherited according to Mendelian rules have been recovered in many agriculturally important crops, namely rice, corn, wheat, soybean, cotton, potato, tomato, rape-seed, pea, cabbage, and a number of fruit and forest trees. The genes transferred include not only model genes, such as those governing resistance to antibiotics or colour markers, but also agronomically interesting traits, such as virus resistance, insect resistance, herbicide resistance, male sterility, fruit ripening, etc.

LIMITATIONS OF GENE TRANSFER

Gene transfer and regeneration of transgenic plants is hampered by a variety of biological problems.

a) Plant cell wall is a perfect barrier and trap for DNA molecules.
b) Egg cells, sperm cells and zygotes are virtually inaccessible.
c) Proembryos are extremely small and enclosed within solid tissues.
d) The tiny cells of hidden meristems which contribute to the germline may not be competent for the integration of functional cells.
e) There are no known retroviruses which could help to spread and integrate foreign DNA systematically.

f) The one functional biological vector system that is available does not work with the agronomically most important groups or varieties of crop plants.

g) Regeneration of transgenic plants mostly depends on expression of totipotency of somatic cells, and, although some plant cells are totipotent, the majority probably are not.

The production of transgenic plants is restricted to some 'model plant' species and varieties. Efficient application of gene technology to plant breeding is still limited by the fact that routine and efficient procedures for the recovery of sufficient numbers of independent transgenic plants which retain their varietal identity are still missing.

The success achieved so far, though impressive, is rather limited.

PART III

SEED PRODUCTION

30

Release and Maintenance of Crop Variety

The development of new varieties has value only if their seeds in genetically (varietally) pure form become available to farmers. This involves release of varieties and scientific multiplication of seeds.

RELEASE AND REGISTRATION OF VARIETIES

The farmer must get a variety that is adapted to the agro-climatic conditions of his area. One way to ensure this objective is the system of release of varieties. The release of a crop variety implies that the variety has been adequately tested in the areas for which it is eventually released, and the results obtained through testing have been evaluated by a team of scientists.

A variety is ready for release and distribution when it has been proved to be distinctly superior to and different from existing commercial varieties in at least one or more characteristics, and satisfactory in all other important respects. The superiority should have been proved in comparative trials. The multilocation testing provides reliable information on the range of variety adaptation. Generally, the decision to release a cultivar (variety) is made on the basis of a recommendation made by a Cultivar Review and Release Committee. The committee is usually an advisory group representing various interests and technical personnel. The committee makes its recommendation after reviewing the history and performance record of proposed new cultivar.

In USA, new crop varieties may be registered by the Crop Science Society of America. A description of each variety registered by the society is published in *Crop Science*. This becomes the official description of the variety.

MAINTENANCE OF CROP VARIETIES

The responsibility of a plant breeder does not come to an end with the release of his variety and supply of a small quantity of initial seed, called the parental seed material, to a seed-multiplication agency. Usually the breeder is required to maintain a continued supply of small quantities of genetically pure seed of the variety for further multiplications so long as the variety remains under cultivation. This needs to be done in an exact manner so that genetic qualities of the seeds of the variety do not unduly change.

Genetic Principles of Maintenance of Varieties

Causes of Genetic Deterioration

The genetic purity of a variety may deteriorate during the seed-production cycle due to several factors.

(1) *Mechanical admixtures*: Mechanical admixtures with seeds of other varieties often take place at the time of sowing, harvesting, threshing and storage, as well as through admixture of seeds from volunteer plants of other varieties that may be present in the seed fields. Mechanical admixtures lower the varietal purity of seeds and are considered an important source of variety deterioration.

(2) *Natural outcrossing*: Natural outcrossing with unintended parents, namely plants of other varieties, diseased plants, undesirable plants that may be present in the seed field or nearby fields, is another important source of genetic contamination and deterioration of crop varieties.

(3) *Developmental variations*: When the seed crops are grown in different climatic conditions or under stress environmental conditions or at different elevations for several consecutive generations, developmental variations may arise sometimes as differential growth response and result in genetic shifts in the varietal populations.

(4) *Other factors*: Factors such as mutations, minor genetic variations, cytogenetic irregularities, breakdown of male sterility (in male sterile lines) and selective influence of diseases are other factors which adversely affect and lower the genetic purity of a variety in varying proportions.

Principles of Maintenance of Genetic Purity

The genetic purity of varieties is maintained through adoption of the following principles.

1. Generation System

Restricting multiplication to a limited number of generations from the breeder's seed is an important principle for maintaining varietal purity. In a seed certification system, designed to maintain genetic purity and other qualities of seeds, seed production (multiplication) is usually limited to four generations described as classes of seeds.

Breeder's seed: Breeder's seed is seed or vegetative propagating material which is directly controlled by the originating or, in certain cases, sponsoring breeder or institution and which provides the initial and recurring increase in foundation seed.

Foundation seed: Foundation seed, including *Select* in Canada, is seed stock so handled as to most nearly maintain specific genetic identity and purity and that may be designated or distributed by an agricultural experiment station. Production must be carefully supervised or approved by representatives of the station. Foundation seed is the source of all other certified seed classes, either directly or through registered seed.

Registered seed: Registered seed is the progeny of foundation or registered seed that is so handled as to maintain satisfactory genetic identity and purity, and that has been approved and certified by a certification agency. This class of seed should be of a quality suitable for production of certified seed.

Certified seed: Certified seed is the progeny of foundation, registered or certified seed that is so handled as to maintain satisfactory genetic identity and purity and that has been approved and certified by the certifying agency.

2. Growing Seed Crops of Respective Crop Varieties in Areas of Their Adaptation Only

Seed multiplication of crop varieties should usually be undertaken in the areas of their adaptation only to avoid undue risk of variety deterioration due to developmental variations resulting in genetic shifts when grown in different environments for several consecutive generations. The tendency for a shift in genetic make-up is much greater in varieties bred for extreme conditions, such as winter hardiness, drought resistance, maturity, etc.

However, in many instances it may be economical and desirable to undertake seed production (for example, for disease and/or weed-free seed multiplication) in different agroclimatic regions. It can be done provided necessary care is taken to ensure that the basic seed is invariably produced in the areas of adaptation only, and further seed multiplication in a different agroclimatic region is restricted to one or possibly two more generations.

3. Preventive Measures

The following preventive measures greatly help to minimize genetic contamination.

i) *Preceding crop requirements*: Raising seed crops in fields which are free from volunteer plants that may grow is one good way to prevent genetic contamination arising through admixture of seeds from volunteer plants. This is ensured by fixing preceding crop requirements in a seed-multiplication system, for example, through seed certification.

ii) *Isolation of seed fields*: Isolation of seed fields from potential sources of contamination is a necessary measure to prevent genetic contamination that may otherwise take place due to natural outcrossing with unintended parents.

iii) *Roguing*: Roguing, that is, removal of off-types from the seed fields, is also a necessary measure to prevent genetic contamination that may otherwise take place due to natural outcrossing with off-type plants and the mechanical admixture that they would cause if not removed.

4. Quality Control Measures

The following quality control measures greatly help to achieve the desired objectives.

i) *Seed certification*: Genetic purity in commercial seed production is often maintained through a system of seed certification. The principal objective of seed certification is to maintain and make available crop seeds, tubers, or bulbs and sometimes turf grasses which are of good seeding value and true to variety. To accomplish these purposes, qualified and well-experienced personnel of the seed-certification agency carry out field inspections at appropriate stages of crop growth. They also make seed inspections to verify that the seed crop/seed lot is of the requisite genetic purity and quality after harvesting, and at the processing plants draw samples for seed testing and sometimes for grow-out tests also. In addition to inspections, the seed-certification agency also ensures that seed crop and seed lot respectively must conform to prescribed standards to obtain approval as certified seed. Field standards include land requirements, isolation requirements, maximum permissible off-type, tassel shedding (in the case of hybrid maize production), etc.

The genetic purity of seed is thus assured if the certification agency has approved the seed. Seed certification implies that both the seed crop and seed lot have been duly inspected and that they meet requirements of good-quality pedigree seeds.

ii) *Grow-out tests*: Varieties grown for seed production should periodically be tested for genetic purity by grow-out tests, to make sure that they are being maintained in their true form.

METHODS OF MAINTENANCE OF NUCLEUS AND BREEDER'S SEED IN SELF-FERTILIZED CROPS

The initial handful of seeds obtained from selected individual plants of a variety, for the purposes of maintaining, and or purifying that variety by the plant breeder under his own supervision is called the nucleus seed. Breeder seed is raised from Breeder's stock seed from the nucleus.

The varieties of self-fertilized species should theoretically be completely homogeneous. However, in practice the stage of complete homogeneity is seldom reached and a considerable amount of variation may still occur during the seed-production cycle, especially in newly released varieties. Purification of such varieties during maintenance of nucleus/breeder's seed may therefore be necessary.

Methods of maintaining nucleus seed/breeder's seed can be conveniently divided into the following two groups:

1) maintenance of newly released varieties,
2) maintenance of established varieties.

Maintenance of Nucleus Seed of Prereleased or Newly Released Varieties

The procedure outlined by Harrington (1952) for the maintenance of nucleus seed of prereleased or newly released varieties is described below:

a) *Sampling of the variety to obtain nucleus seed*. New numbers, lines or selections, which are highly promising on the basis of performance in breeding nurseries and yield trials, should be sampled for seed purification. These samples provide a beginning for purifying new varieties and for possible increase and distribution to farmers. Not more than fifteen new varieties in any one crop at a station should be sampled in one year.

The sampling of these promising varieties should be done as follows:

Obtain approximately two hundred plants from the central three metres of the border rows of replicates of the new variety in one of the yield tests.

Discard poor plants and those with few tillers. Retain the remaining plants.

These plants should be pulled four to five days before the grain is fully mature to avoid shattering. Tie all these plants in a bundle after wrapping in cloth or paper. This is necessary to prevent breakage and loss.

Check each bundle and store properly until the final yield results are available.

After data availability discard that bundle of any of the new varieties which analysis indicates to be less worthy than others.

b) *Table examination of samples* : The two hundred plants of each sample should be threshed separately and the seed should be examined in piles on a table. Discard any pile appearing obviously off-type, diseased or otherwise unacceptable. The seeds of each two hundred plant samples or less are now ready to be sown in a variety purification nursery called a nucleus.

c) *Locating and seeding of nucleus* : Each nucleus seed should be grown on clean fertile land at an experimental station in the region or area in which this new variety would be grown in the event of its release. The land must not have had a crop of the same kind in the previous year.

The two hundred progenies constituting a nucleus should be sown in a block of two hundred double row plots in four series of fifty double-row plots each. The rows should be seeded with a single-row machine seeder or by hand. The seeds should be sown sufficiently apart in rows. The plot-to-plot distance should be at least 45 cm to facilitate examination of rows during growth.

The nucleus seed plot must be properly isolated to prevent contamination by natural outcrossing and the spread of diseases from neighbouring plots.

d) *Inspection of nucleus two-row plots and removal of off-types* : Throughout the season of growth, from the seedling stage until maturity, the nucleus plot should be critically examined. Differences in habit of early plant growth, leaf colour, rate of growth, time of heading, height, head characteristics and disease reactions should be looked for. If a plot differs distinctly from the average in preheading stages of growth, it should be removed before heading.

As the plots mature the awn and chaff colours develop fully and off-types not observable at earlier stages of growth are clearly observable and should be removed by cutting. In addition, because of the probability of contamination by natural outcrossing with nearby plots, the two plots on each side of the removed off-type should also be removed. If the crop of variety being purified is known to have a rather high incidence of natural outcrossing, say four per cent, it is safer to remove all material growing within three metres of an off-type plot, unless, of course, that plot was removed before its pollen ripened.

If an individual plant progeny is removed after flowering has occurred, all plants within a metre of it should be pulled and discarded.

e) *Harvesting and threshing of nucleus* : Each remaining plot, of which there should be at least 180 out of the original 200, should be harvested individually with a sickle and tied in a bundle. The total bundles of each nucleus should be labelled and stored until the current year's yield result for trials are obtained. The nucleus bundles of any new variety should be discarded if it is found unworthy of being continued.

The nucleus bundles of the remaining new variety, or varieties, should be threshed with an individual plant thresher. Great care should be taken to complete work on one nucleus before commencing work on another, so as to avoid mixing the seed of the two. The seed of each plot should be kept in a separate paper bag.

Later, the seed should be cleaned in a fanning mill or by hand methods, the grain from each nucleus plot being placed in a pile on the seed table. The 180 or more piles of seed of one nucleus must be examined for approximate uniformity of seed appearance, and any pile which appears to be off-type discarded. All the remaining piles of seed should be massed together in one lot. This should be treated with fungicide and insecticide, bagged, labelled and stored as 'Breeder's Stock Seed' for use in the next year. Breeder's stock seed is the original purified seed stock of a new variety in the hands of the plant breeder.

MAINTENANCE OF BREEDER'S SEED OF PRERELEASED OR NEWLY RELEASED VARIETIES
The following steps are involved in the maintenance of breeder's seed.

a) Breeder's stock seed from the nucleus should be sown on clean, fertile land which did not grow a crop of the same kind in the previous year. The space required for seeding the breeder's stock is about 1.2 ha in the case of wheat and as much as 3 ha in the case of transplanted rice.

b) The field should be properly isolated.

c) The best farm procedures should be used in the sowing, raising and harvesting of breeder's stock.

d) The stock should be produced at the experimental station in the area in which the new variety has been bred.

e) Seeding should be done in such a way as to make the best use of the limited amount of seed available and to facilitate roguing. Row spacing should be sufficient to permit examination of plants in rows for possible mixtures or off-types.

f) Roguing. All plants not typical of the variety should be pulled and removed. There should be very few plants to rogue out if the previous year's nucleus breeder's stock seed was well protected from natural outcrossing, careful roguing was done and there were no impurities during cleaning, etc. Roguing should be done before flowering, as was done for the nucleus/breeder's stock seed. Where plants have been removed after flowering and the pollen has escaped, all surrounding plants within one metre should also be pulled and discarded.

g) Harvesting the breeder's stock. When the breeder's stock is harvested and threshed, the equipment used must be scrupulously clean and free from seeds of any other varieties. This cleanliness should be extended to carts and bags as well as the threshing machine itself.

The seed should now be about 99.9% pure as to variety. This breeder's seed is thus ready for increase of foundation seed. A portion of this breeder's seed should be retained by the breeder to sow a continuation breeder's seed of the variety.

SECOND INCREASE OF BREEDER'S SEED (CONTINUATION BREEDER'S STOCK)

A continuation breeder's stock may be maintained by the plant breeder each year to furnish a fresh stock of seeds to foundation-stock growers. This may be continued until the new variety is within two or three years of being replaced by a still newer variety. Each year this continuation breeder's stock is sown from the bulk seed of the preceding year's stock and handled in the same manner described above for breeder's seed.

REPETITION OF PURIFICATION OF BREEDER'S STOCK

Some varieties can be kept reasonably pure in the continuation breeder's stock without difficulty, whereas others, through natural outcrossing and mutation, soon develop noticeable off-types. Repurification is necessary in the latter case. Repurification is commenced by taking at random a lot of 100 to 200 plants from a continuation breeder's stock and sowing a nucleus with individual plant seeds. The nucleus is handled as described earlier. A smaller number of nucleus progenies than two hundred is satisfactory if there is no great urgency in replacing the existing seed stocks of the variety. The situation differs when the variety is new and the size of the nucleus definitely limits the important first distribution of the variety.

On the other hand, when the results of a breeding programme may immediately save farmers from the ravages of a certain insect, disease or weather situation, the original samples for the nucleus need not be restricted to two hundred plants, but can be much larger.

Maintenance of Breeder's Seed of Established Varieties

The breeder's seed of established varieties can be satisfactorily maintained by any one of the following methods.

a) *Raising the crop in isolation*: The breeder's seed of local varieties can be maintained by growing them in isolated plots and by rigorous roguing during various stages of crop growth, when the various plant characters are observable. The method of handling the breeder's seed crop is the same as described earlier for breeder's seed of newly released varieties.

b) *Mass selection*: The genetic purity of established varieties can be satisfactorily improved by mass selection. In this method 2000 to 2500 plants typical of the variety are selected, harvested, and threshed separately. The seeds from each plant are examined and any pile which shows any obvious off-types, or otherwise appears dissimilar, is discarded. The remaining piles of seed are bulked to constitute the breeder's seed. The other practices of handling remain the same.

How long a particular method should be used depends on the rate of deterioration in a variety. Careful production ought to insure genetic purity for several generations. However, it is important to note that during purification and increase there should be no change in important plant characteristics of economic importance. It is, therefore, necessary to include breeder's seed in yield tests.

Carry-over Seed

The breeder must carry over at least enough seed to safeguard against loss of the variety should there be a complete failure during the foundation-seed multiplication phase. In addition, the breeder should further safeguard his variety by arranging to have a portion of the seed originally released stored under ideal conditions.

Those responsible should plan to produce sufficient breeder's seed at one time to meet the requirements of two to three productions of foundation seed. Production of breeder's seed is a very expensive process, with the associated risk of contamination by repeated multiplication and of loss due to adverse growing conditions. These risks can be reduced and the continuity of the seed programme better assured by the carry-over breeder's seed. Such carry-over seed must be stored under optimal conditions in order to maintain its vigour and viability.

METHODS OF MAINTENANCE OF NUCLEUS AND BREEDER'S SEED IN CROSS-FERTILIZED CROPS

Maintenance of varieties of cross-fertilized crops is generally much more complicated. The specific methods adopted for maintenance of parental material depend on the method of breeding the variety.

Maintenance of Nucleus Seed of Inbred Lines

After a hybrid has been thoroughly tested and its desirability ascertained, the seed of parent inbred lines must be increased in the following manner.

(a) *Hand-pollination*: Methods of maintaining nucleus seed of inbred lines involve self-pollination, sib-pollination or a combination of the two procedures. Generally, maintenance by sibbing is preferred by some breeders because it does not reduce the vigour excessively. However, if a change in breeding behaviour is noted, then selfing should be used as a means of stabilizing the inbred lines. It is preferable to maintain some parental material by alternate selfing and sibbing from one generation to the next. There is little or no experimental evidence on the ideal population required in a maintenance programme. Obviously such a programme should not be limited to only a few ears. Individual selfed or sibbed off-types, or inferior in any regard, or differing in any characters, e.g., texture and colour, seed size, chaff colour, size and shape of ear, should be discarded. Individual selfed of sibbed ears may then be threshed separately and planted ear to row, or all the ears from an individual inbred line may be composited for increase in the next season. The obvious advantage of ear-to-row planting is that the off-types from individual ears may be more easily detected and discarded than in bulk plantings.

(b) *Seeding hand-pollinated seeds*: Hand-pollinated seeds should be sown on clean fertile land and on soil in which the same kind or variety has not been sown the previous year (this is not necessary for corn). The increase should be in the area in which the hybrid seed produced from these shall eventually be used.

(c) *Isolation*: It is very important, at this early stage, that the crop be very well isolated. Isolation requirements vary from crop to crop. The isolation distance depends on various factors, such as the nature of material to be protected by isolation, nature of contaminant from which isolation is sought and direction of the prevailing wind. Isolation can be effected by distance, or also by time in the case of corn. What is considered to be adequate isolation is mostly based on practical experience in particular situations, rather than on experimental evidence. In principle, however, it is generally agreed that much more stringent isolation is required at this stage than at later stages.

The methods used for growing inbred lines do not differ much from those used for raising a good crop, except in the isolation requirements. However, it seems to be most important that the inbred lines be given good growing conditions to enable them to show their genetic potential.

(d) *Roguing*: Despite all the efforts made to maintain purity in inbred lines by hand-pollination and by adequate isolation, it is still not possible to achieve perfection by these means alone. The isolated fields of inbreds must also be carefully rogued and checked for off-types, prior to shedding pollen. It is very easy to recognize the outcrossed rogues because normally they are much more vigorous and stand out quite clearly in an inbred field. There are, however, other off-type plants which are not easily detected, particularly in bulk planting, for which careful inspection is necessary.

(e) *Harvesting, drying and shelling*: The nucleus seed crop can be harvested soon after it attains physiological maturity if artificial drying facilities exist. It is better to harvest the ear-to-row lines separately and piles made in front of each progeny. These piles should be critically examined for ear characteristics and all off-coloured, off-textured or diseased or otherwise undesirable ears sorted out. If the overall percentage of off-colour and off-textured ears is more than 0.1% hand-pollination should again be done for production of the second year's breeder's seed.

After examining the ears critically and discarding the undesirable ones, the remaining ears may be bulked and dried in a clean dry bin at temperatures not exceeding 43°C. After drying, shelling should be done. Before use, the shelling machine must be thoroughly cleaned to avoid any mechanical mixtures at this stage. After shelling, the seed may be cleaned, treated and stored under ideal storage conditions. This will constitute the breeder's stock seed.

Maintenance of Breeder's Seed of Inbred Lines

To increase the breeder's seed, the breeder's stock obtained from nucleus seed is planted in an isolated field. During increase of breeder's seed, adequate attention must be paid to land requirement, isolation, roguing, harvesting and drying, sorting of ears, shelling of ears, etc. so as to maintain the maximum possible genetic purity. The cultural practices, isolation, roguing and harvesting, etc. are similar to those used for raising a nucleus crop.

Maintenance of Nucleus Seed of Non-inbred Lines

Hand-pollination. The number of plants to be pollinated for such increase should be large enough so as not to alter the genetic make-up of the varieties by narrowing the genetic base by sibbing only a few plants. No definite number, however, can be suggested. This depends on the genetic make-up of the line. Preferably, it should be large enough and may even go up to 5000 or so, if practically feasible. The sibbed ears must be examined very carefully and all those having off-colour, off-texture, or diseased, or otherwise undesirable must be sorted out. The remaining ears should be bulked, dried, shelled, cleaned, treated and stored. The other practices for seeding sibbed nucleus seed are similar to those discussed earlier for inbred material. However, roguing in these lines must be done very carefully and by those knowledgeable about the material. Excessive strictness, or attempts to bring uniformity, may lead to changes in the genetic make-up and hence are best avoided.

The seed obtained from these plots constitutes the breeder's stock seed.

Maintenance of Breeder's Seed of Non-inbred Lines

The breeder's stock seed produced from nucleus seed should be used for increasing breeder's seed. During increase, attention must be paid to land requirements, isolation, roguing, harvesting and further handling so as to maintain maximum genetic purity. These practices are similar to those used for raising a nucleus crop.

Maintenance of Seed of Established Varieties

The breeder's seed of established cross-fertilized crop varieties can be maintained by any one of the following methods :

Raising breeder's seed crop in isolation : The breeder's seed of established varieties can be maintained by raising the seed crop in isolation. Isolation requirements vary from crop to crop. The seed crop is thoroughly rogued at various stages, such as vegetative, flowering and at maturity and all off-types removed as and when noticed. The other practices remain the same as described earlier for breeder's seed production.

Mass selection: The breeder's seed of cross-fertilized crops is often purified by mass selection. The crop is raised in isolation and rogued carefully as described earlier. At maturity, approximately 2000 to 2500 true-to-type plants are selected. The selected plants are harvested separately and after careful examination bulked to constitute the breeder's seed. The other practices remain the same. Modified mass selection, in which the whole field is divided into several sectors and an equal number of true-to-type plants selected from each sector, can also be used for maintaining breeder's seed.

Carry-over Seed

The handling of breeder's seed must be efficient because of the great importance of genetic stability and its likely influence on hybrid seed production. Nevertheless, seed production of inbred lines can be prone to accidents and unpredictable disasters. Hence, a system of carry-over seed is the best insurance against failure. Essentially, carry-over seed is extra seed which is retained for a year, or longer if needed, as a safeguard against unforeseen shortages.

METHODS OF MAINTENANCE OF BREEDER'S SEED OF APOMICTIC SPECIES

All plants in varieties of apomictic species should be genetically alike for reason of asexual reproduction. In many of these species however, there is a certain degree of sexual seed formation. As in these varieties the plants are normally highly heterozygous, each plant derived from a seed formed in the sexual way will differ from typical plants of the variety. Mutations may also produce aberrant types. The natural admixtures from other varieties will also cause deterioration of the variety. In order to maintain varietal purity in such species, a large number of plants should be planted or sown thinly enough for each plant to be observed separately. From this material all inferior plants should be removed before flowering and the seed from the remainder should be bulked and used as new breeder's seed.

METHODS OF MAINTENANCE OF BREEDER'S SEED OF ARTIFICIAL POLYPLOIDS

In the maintenance of varieties of some artificial polyploids, admixture with diploid plants has given rise to specific problems. In red and alsike clover, in which the seed-setting capacity of the tetraploids is much lower than that of the diploids, an admixture of diploids tends to increase from one generation to another very rapidly; after a few generations the tetraploid material could possibly disappear completely (Julen, 1956). An admixture of a few per cent of diploids can easily occur as a result of volunteer plants. It is important, therefore, that every seed lot intended for further multiplication be thoroughly checked for its content of diploids. If there is a small proportion of diploid seeds, it can be removed mechanically by sieving, because of the differences in seed size between diploids and tetraploids. If the proportion of diploids is high, such a mechanical separation would hardly be possible. In any case, only a very small part of tetraploid seed could, in this case, be retained. Precaution must therefore be taken to avoid admixture of diploids in the treatment of the seed and in the field.

PRODUCTION OF VIRUS-FREE NUCLEUS STOCK IN POTATO

The technique of producing virus-free stock consists of tuber indexing and selection of single pile units for freedom from viruses and mycoplasms (Upreti, 1977).

Detection of Viruses and Mycoplasms

Latent viruses such as X and S are detected by serological methods. For detection, a drop a antiserum on a glass slide is mixed with a drop of sap from the leaf of the plant to be tested. The reaction shows agglutination of the chloroplast if the virus is present.

The viruses Y and A are detected by the A6 test using detached leaves of *Solanum demissum* x *Solanum tuberosum* var. *aquila*. Detached leaves of A6 are placed in a box containing a moist mass and are dusted with 500 mesh carborundum powder and a few drops of phosphatic buffer solution. Sap from test plants is inoculated and the residual carborundum powder is rinsed off with water. In combined tests for viruses Y and A, the leaves are kept for seven to eight days at about 20°C to watch for development of symptoms.

The leaf roll virus (LR) is detected by the phloroglucinol test. In a modified technique, a piece is taken from near the base of the stem and then transverse sections from the nodal region are put into phloroglucinol solution in concentrated 10 to 11 N HCL. After one minute, the sections are removed and mounted in water. The primary phloem of leaf-roll infected plants will be stained orange-red.

The mycoplasms, viz., Witch's Broom (WB), Marginal Flavescence (MF) and Purple Top Roll (PTR) are detected visually on the basis of symptom expression. The PTR is characterised by rolling and purple/pink pigmentation of the basal parts of the leaflets of the top leaves. MF induces flavescent margins of younger leaves and stunting. WB causes stunting and produces filamentous stems with small leaves and small-sized tubers.

Methods of Nucleus Stock Production

Tuber indexing: A large number of single-hill units of different varieties raised at the Nucleus Seed Station, are inspected for visual symptoms of diseases, varietal purity and high yields. Healthy plants are harvested individually. Four tubers of each selected plant are numbered individually for glasshouse testing. One eye from each tuber is scooped out and planted in small pots and the counterpart is stored. The problem, however, becomes complicated given the frequent presence of healthy and infected parts in the same tuber. In such situations, the main objective is to detect and eliminate from the base material as much viral/mycoplasmal infection as possible at the time of tuber indexing, so that infection may be diluted in subsequent stages of nucleus stock multiplication.

Each plantlet from a single leg is serologically tested in the glasshouse simultaneously for PVX and PVS by the use of bivalent antiserum. Plants free from S and X are further tested biologically for PVY and A, by using detached leaflets of A6 and for LR by histochemical test, and visually for WB, MF and PTR.

Maintenance of Virus-tested Stocks

Stage I: The counterparts of the single eyes showing freedom from the above-mentioned viruses are subsequently planted in the field during the low aphid period, in plains or hills (2400 m above sea level). Virus-free counterparts are separately planted in rows, with a distance of 1 x 1 m so that the foliage may not touch each other. Each plant is tested serologically for X and S viruses by taking a composite sample of one leaf from each stem. Whenever a positive agglutination reaction is observed, the entire clone is destroyed. The plants are visually examined for other viral and mycoplasmal diseases.

Stage II: In the second stage, all produce of a hill, unit or clone, is separately planted in a plot. For serological testing against viruses X and S, two leaves from each individual plant are taken, and six leaves constitute a sample. Visual inspection is carried out for other diseases. Any positive serological reaction or visual symptom in any of the samples makes it obligatory to rogue out the entire clone.

Stage III: The seed material of each clone selected under stage II is planted in bulk. Viral infection in the third stage is determined by sample testing (serologically for X and S) or visually (for viral and mycoplasmal diseases).

Virus-free nucleus stocks are subsequently utilized for production of breeder's seed.

31

Plant Variety Protection

Plant variety protection, also termed *plant breeder's* right is an independent *sui generis* (of its own kind) form of legal protection of new plant varieties[1]. It confers an exclusive right to a plant breeder[2] to market a variety that he has developed. Any other person who wishes to market the protected variety must be licensed by the developer or owner of that variety and is required to pay a royalty. It is a form of intellectual property right. It has certain features in common with patents but concomitantly certain fundamental differences also (Table 31.1).

PURPOSE OF PLANT VARIETY PROTECTION

The purpose of plant variety protection is to encourage the development of novel varieties by granting right to profit from investment in terms of time, labour and money to all those who breed, develop or discover them. However, the right granted in *sui generis* does not cover use of the material for research purposes nor personal sowings by a farmer. And yet the right to market, or grant of a license to market on royalty payment is good enough incentive to private seed companies and individuals to make long-term investment in infrastructure and to specialize in breeding specific crops, and more particularly hybrid varieties. A system of plant variety protection thus is of interest of any country which believes that a system of incentives will increase the effectiveness of plant-breeding research and greater private-sector participation.

There is yet another aspect of plant variety protection. In most advanced countries plant breeders are eligible for certain rights similar to patent rights. Viewed in this context, one can readily visualize that the flow of propagating material of new plant varieties from any of these advanced countries to any

---

[1] A plant variety means a plant grouping within a single botanical taxon of the lowest known rank that can be defined by the expression of the characteristics resulting from a given genotype or combination of genotypes, distinguishable from any other plant grouping by the expression of at least one of the said characteristics, and considered as a unit with regard to its suitability being propagated unchanged.

[2] Breeder means the person (employer of the aforementioned person, or he who has commissioned the latter's work, or the successor in title, as the case may be) who bred, or discovered and developed a variety.

Table 31.1. Fundamental differences in protection through patent and plant variety protection*

	Patent protection	Plant variety protection
1. Object	Industrial invention	Plant variety
2. Scope of protection	Scope is determined by the claims of patent. Use of propagating material for purposes such as research and personal sowings may require authorization of the patentee	Scope is fixed by national legislation. Use of propagating material for research and personal sowings is exempt and hence no authorization from the right holder is needed.
3. Conditions of protection	(a) novelty, (b) industrial applicability (c) Unobviousness (inventive step) and (d) an enabling disclosure	(a) commercial novelty (b) distinctness, (c) uniformity, (d) stability, (e) an appropriate denomination (f) testing, field examination and documentary examination are required to confirm (a to e above)
4. Term of protection	20 years from the date of application	25 years for trees and vines and 20 years for other species from the date of grant of right.

*Source: From a Lecture delivered by Barry Greengrass, Vice Secretary General, UPOV, Geneva, Switzerland. National Seminar on the nature and rationale for the protection of plant varieties under the UPOV convention. Organised by UPOV in cooperation with Ministry of Agriculture of India. New Delhi, September 12, 1996.

other country would depend on whether the breeder's rights are protected in the importing country or not. If one looks at the progress made in advanced countries, more particularly in the areas such as development of hybrid varieties of agricultural and vegetable crops, new plant varieties of fruit and ornamental crops, and varieties already bred or being bred through sophisticated techniques of genetic engineering and tissue culture, both of which have had and continue to have profound implications for seed, nursery plants and plant-breeding industries worldwide, the importance of importing and introducing such propagating materials cannot be overemphasized. Enactment of the Plant Variety Protection Act is therefore an impending necessity for any developing country, if it wants to derive the benefits of outstanding plant-breeding work done elsewhere.

INTERNATIONAL UNION FOR PROTECTION OF NEW PLANT VARIETIES (UPOV), GENEVA, SWITZERLAND

UPOV was established in 1961 by the international convention for the protection of new varieties of plants. It came into force in 1968. At present 31 countries, namely, Argentina, Australia, Austria, Belgium, Canada, Colombia, Czech Republic, Chile, Denmark, Finland, France, Germany,

Hungary, Ireland, Israel, Italy, Japan, Netherlands, New Zealand, Norway, Poland, Portugal, Slovakia, South Africa, Spain, Sweden, Switzerland, Ukraine, U.K., USA and Uruguay, are its members. Many more countries are expected to join UPOV in the years to come.

Principal Functions of UPOV

The two principal functions of UPOV are (UPOV, 1996):
1. Promotion of international co-operation between member countries in matters related to plant variety protection and assistance to countries willing to introduce legislation on plant-variety protection, and
2. Promotion of co-operation between member countries in such matters, as variety testing to enable member countries to restrict both the time and cost involved in checking whether a variety qualifies for protection.

Organization of Work

UPOV accomplishes its tasks through three committees: Consultative Committee, Administrative and Legal Committee, and Technical Committee. The Technical Committee is further assisted by five working parties, namely, for agricultural crops, vegetables, fruit crops, ornamentals and forest trees, and for automation and computer programmes.

Function of UPOV Convention

The convention has two main functions :
1. To prescribe minimum rights that must be granted to plant breeders by the member countries; in other words, it specifies a minimum scope of protection;
2. To establish standard criteria for grant of protection.

The Convention held in 1961 revised its objectives in 1972, 1978 and 1991.

Responsibilities of Member Countries and Membership

The member countries are required to protect the varieties of breeders of member countries in a manner similar to that exercised for breeders of their own country. Any country with appropriate plant variety legislation may become a member of UPOV and benefit from the combined experience of the member countries and also contribute to the worldwide promotion of plant breeding.

MAIN FEATURES OF PLANT-VARIETY PROTECTION IN UPOV, 1991 ACT

Plant genera and species covered in the Act : Plant-variety protection systems are designed to encourage plant-breeding innovations. For any member country, it would therefore be appropriate to extend plant-variety protection to all plant genera and species to which the UPOV 1991 Act is applicable, albeit in phases, if necessary.

Breeder Rights (UPOV, 1991 Act)

Subject to exceptions and exhaustion of the breeder's right, the following acts in respect of the propagating material of the protected variety[3] shall need the breeder's authorization. The breeder may make his authorization subject to conditions and limitations.

1(a) PROPAGATING MATERIAL

(i) multiplication of propagating material, (ii) conditioning for the purposes of propagation, (iii) offering for sale, (iv) selling or other marketing, (v) exporting, (vi) importing and (vii) stocking for any of the purposes mentioned in (i) to (vi) above.

(b) HARVESTED MATERIAL/CERTAIN PRODUCTS

The Act referred to in items (i) to (vi) above in respect of harvested material and or the products made from harvested material of the protected variety, including entire plants and parts of plants, obtained through the unauthorized use of propagating material of the protected variety shall require the authorization of the breeder. Unless the breeder has had reasonable opportunity to exercise his right in relation to the said propagating material, the same cannot be used for propagation purposes.

(c) Acts other than those mentioned in a(i) and to (vi) above, if granted by the Competent Authority under the provisions of the National Law on Plant Variety protection, shall also need the breeder's authorization.

2. ESSENTIALLY DERIVED[4] AND CERTAIN OTHER VARIETIES

The provisions of paragraphs 1(a) to (c) also apply in relation to (i) varieties which are essentially derived from the protected variety, whenever the protected variety is not itself an essentially derived variety, (ii) varieties which are not clearly distinct (distinguishable), and (iii) the varieties whose production requires repeated use of the protected variety.

Exceptions to Breeder's Right

1. Compulsory exception : The breeder's right in respect of protected varieties described earlier does not include acts such as (i) non-commercial or privately done acts, (ii) experimental purposes and (iii) plant-breeding purposes, barring those leading to development of essentially derived varieties.

---

[3] Breeder rights are independent of other regulatory laws in a country for the production, certification and marketing of seeds.

[4] A variety is deemed to be essentially derived from the initial variety or from another variety that is itself predominantly derived from the initial variety when except for the differences which result from the act of derivation, it conforms to the initial variety in the expression of essential characteristics that result from the genotype or combination of genotypes of the initial variety, and is clearly distinguishable from the initial variety. Essentially derived varieties may be obtained by selection of natural of induced mutant, or of a somaclonal variant, backcrossing, or transformation by genetic engineering.

2. Optional exception : A country may within reasonable limits choose to restrict the breeder's right in relation to any protected variety in order to permit farmers to use for propagating purposes on their own holdings, the product of the harvest which they have obtained by planting on their own holdings.

Exhaustion of Breeder's Right

The breeder's right does not extend to acts concerning any material of the protected variety or of the essentially derived variety, which has been sold or otherwise marketed by the breeder or with his consent in the territory of the concerned country, or any material derived from the said material, unless such acts—

(1) involve further propagation of the variety;

(2) involve an export of material (propagating material, harvested material and any product made directly from the harvested material) of the variety, which enables propagation of the variety, into a country which does not protect varieties of the plant genus or species to which the variety belongs, except where the exported material is for final consumption purposes.

Nullity and Cancellation of Breeder's Rights

Breeder's rights granted in relation to a protected variety are declared null and void by the Competent Authority under the following conditions:

1. If it is subsequently found that the variety lacks novelty and distinctness, and/or is not found to be uniform and stable, since these are essential requirements for grant of protection.

2. That the breeder's right has been granted to a person who is not entitled to it, unless it is transferred to a person who is so entitled.

No other reason except the aforesaid is applicable in declaring a breeder's right null and void.

Cancellation of Breeder's Right

A breeder's right may be cancelled by the Competent Authority on the following grounds:

1. The variety is subsequently found to be not uniform or stable.

2. A breeder's right may also be cancelled on request within a time limit under the following conditions:

   i) the breeder does not provide the Competent Authority with the information, documents or material deemed necessary for verifying maintenance of the variety;

   ii) the breeder fails to pay such fees as may be payable to keep his right in force; or

   iii) the breeder does not propose, when the denomination of the variety is cancelled after grant of the right, another suitable denomination.

No breeder's right shall be cancelled for reasons other than those listed above.

Restrictions on the Exercise of Breeder's Rights

A country may restrict the free exercise of a breeder's right in the public interest in exceptional cases. When any such restriction has the effect of authorizing a third party to perform any act for which the breeder's authorization is required, the country concerned shall take all measures necessary to ensure that the breeder receives equitable remuneration.

Duration of the Breeder's Right

The minimum period of protection is 25 years for trees and vines and 20 years for other plants, from the date of application.

Technical Criteria for Plant-variety Protection

Protection (breeder's right) is granted after ascertaining on the basis of UPOV guidelines that the variety (candidate variety, proposed variety) is new, distinct, uniform and stable, and further that it has been appropriately denominated. This is done in the following manner.

1. Establishment of Commercial Novelty

The variety is deemed to be new if at the date of filing application for breeder's right propagated or harvested material of the variety has not been sold or otherwise disposed to others for purposes of exploitation of the variety earlier than one year in the territory of the country where the application has been filed; or in other countries earlier than four years or six years (for trees, vines, etc.).

The applicant breeder is responsible for providing necessary evidence that the variety had not been commercialized by him or with his consent within the time limits stated above. The Competent Authority (Authority established to grant protection in a country under Plant-Variety Protection Law) may examine the novelty (newness) by seeking out information from interested circles. Should negative information be received by the Competent Authority the applicant is given an opportunity to prove to the contrary. If the applicant fails to do so within a time-frame, the application is rejected for lack of novelty.

2. Test for Distinctness

The variety is deemed to be distinct if it is clearly distinguishable from any other variety whose existence is a matter of common knowledge at the time of filing an application.

The distinctness of a new variety is determined on the basis of field examination (comparative field trials) over a period of two to three growing seasons. The variety is grown along with other varieties whose existence is a matter of common knowledge. The variety's distinguishing characters (as per UPOV guidelines) are compared and recorded. The first comparison is normally with those varieties which are considered to be rather similar.[5]

---

[5] Maintenance of an up-to-date reference collection and database of variety distinguishing characters greatly facilitates evaluation with respect to distinctness.

A new variety is considered distinct if the difference in respect of at least one morphological, chemical or physiological characteristic is clear and consistent. In the case of qualitative characters, the difference between two characters is considered clear if the respective characteristics show expressions which fall into two different states of expression. In all other instances, an eventual fluctuation is taken into account while establishing the distinctness. When the distinctness depends upon quantitative characters, the difference is considered clear if it is within one per cent probability of error.

The differences observed in a new variety are considered consistent, if their expression is the same in two consecutive seasons, or in two out of the three growing seasons.

3. Test for Uniformity

The variety is deemed to be uniform if the observed variation is within usually expected limits, depending its mode of reproduction. Varieties of vegetatively propagated and truly self-pollinated crops are not expected to show much variation. Varieties of often-cross-pollinated crops may show somewhat more variation. In varieties of some of the cross-pollinated crops it may be difficult to distinguish off-types and the comparison in respect to variation should be with the variation usually observed in comparable varieties. Single crosses are expected to be more uniform, but some allowance for selfed plants might be necessary in some crops. In other hybrids the segregation of certain characteristics should be in agreement with the formula of variety and comparison done with that variation observed in comparable hybrid varieties already known. Obviously, the observed variation in a variety should not be such as to prevent accurate description and assessment of distinctness and stability of the new variety. The number of off-types should not exceed prescribed tolerance limits.

4. Test for Stability

The variety is deemed to be stable if its relevant characteristics remain unchanged after repeated propagation or, in the case of a particular cycle of propagation, at the end of each such cycle. Usually, when a submitted sample during the period of 2-3 growing seasons has been shown to be uniform, the variety is considered to be stable. When necessary, however, a further generation (or new seed stock) may have to be grown to verify that the material exhibits the same characteristics.

5. Variety Denomination

The variety is required to be designated by a denomination which will be its generic designation. The denomination must enable the variety to be identified.

(i) It may not consist solely of figures and must not be liable to mislead or to cause confusion concerning the characteristics, value or identity of the variety or the identify of the breeder. In particular, it must be different from every denomination which designates an existing variety of the same plant species or of a closely related species.

(ii) Prior rights of third persons shall not be affected.

(iii) A variety must be submitted under the same denomination in all the countries where the breeder is seeking protection.

(iv) A trademark may be associated with a denomination.

Examination of variety denomination consists mainly of searching for the same or similar denominations currently in use in relation to varieties belonging to the same species or closely related species. Maintenance of a database by the Competent Authority greatly facilitates this examination. To further supplement this information, all interested circles are invited to lodge any objection to a proposed variety denomination with the Competent Authority prior to the formal approval of the proposed denomination. If a proposed variety denomination is rejected by the Competent Authority, the applicant is given an opportunity to provide another denomination within a time-frame. If he fails to respond, the application is deemed withdrawn.

32

Hybrid Seed Production

PREREQUISITES FOR HYBRID SEED PRODUCTION

The use of hybrid seed for commercial production is restricted to certain species, for the following reasons.

1. Heterosis (hybrid vigour) must be present. The hybrids must be significantly superior in yielding ability over prevailing varieties (non-hybrids). Besides superiority in yielding ability, other factors, such as uniformity, quality and disease and insect resistance of hybrids may be of considerable economic importance.

2. A system—male sterility, self-incompatibility, or chemical gametocides which eliminate fertile pollen in the female line—should have already been developed to facilitate commercial seed production. Or else the floral structure, reproductive behaviour (dioecism, monoecism, sex reversal, etc.) and/or hand-emasculation procedures for elimination of male organs should be such that these can readily be adopted in hybrid seed production fields.

3. Pollination and fertility restoration by the male parent should be adequate.

4. Increased cost of seed (hybrid seed) is offset by superior yields over non-hybrids.

If these prerequisites are met, hybrid seed production can profitably be organized on the requisite scale.

HYBRID SEED PRODUCTION

The principles and salient features of hybrid seed production for various crops are described below. The usual agronomic practices and handling of the seeds produced are not included.

Corn

Types of corn hybrids have been described in Chapter 20. Production of hybrid corn seed involves the following three steps (Fig. 32.1):

1) maintenance of parental lines (inbred lines);
2) production of single cross;
3) production of commercial hybrid, namely, three-way cross, double-cross hybrid or a double topcross hybrid.

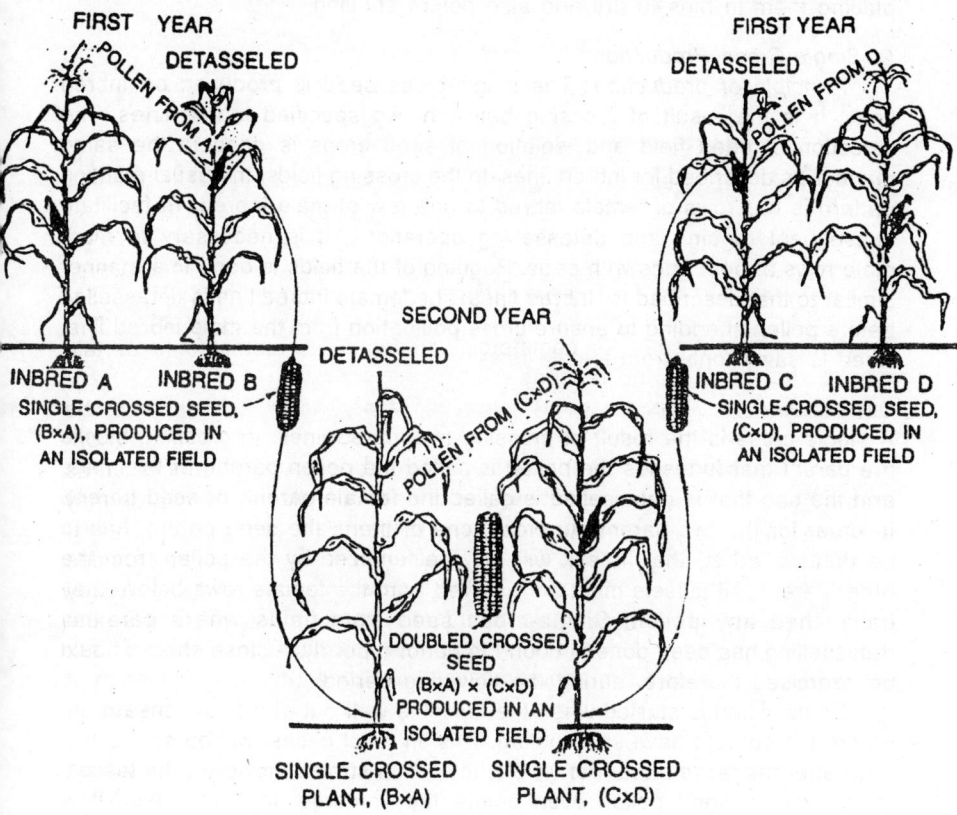

Fig. 32.1. Hybrid corn seed production (from Poehlman and Borthakur, 1969)

1. Maintenance of Inbred Lines

The seed of parental lines is produced in isolated fields. Selected fields should be free of volunteer corn plants. Seed fields must be isolated by not less than 200 metres from any other kind of corn. For hand pollinated seed no isolation is required. In case of fertile inbred lines differential blooming dates are permitted for modifying isolation distances, provided there are no receptive silks at the time pollen is being shed in other corn fields within the prescribed isolation distance.

Rigorous roguing for removal of off-type plants before pollen shedding is essential. At the preflowering stage, rogue out off-type plants which are easily distinguishable on the basis of such plant characteristics as leaf shape, size, plant height, etc. Continue roguing during the flowering stage to remove plants differing in tassel or silk before pollen shedding.

An early harvest prevents losses in the field due to bird damage, stalk breakage, ear rots etc. After harvest off-type ears, particularly those showing different colours and texture, and the diseased ears are sorted out, before placing them in bins to dry and also before shelling.

2. Single-Cross Production

Principle of production: The single-cross seed is produced on inbred lines. It is the result of crossing between two specified inbred lines. The selection of seed field and isolation of seed crops is done in the same manner as described for inbred lines. In the crossing fields, the usual planting pattern is two rows of female inbred to one row of male inbred. To facilitate subsequent roguing and detasselling operations, it is necessary to mark male rows at both ends with pegs. Roguing of the fields is done in a manner similar to that described for inbred lines. The female inbred line is detasselled before pollen shedding to ensure cross-pollination from the male inbred line. Seed is saved only from female rows.

DETASSELLING

A single cross is the result of crossing two inbred lines. In crossing blocks the parent that furnishes the pollen is called the pollen parent, or the male, and the one that is detasselled is called the female parent, or seed parent. In order for the two parents to cross, one of them, the seed parent, has to be detasselled so that its silk will only be fertilized by the pollen from the other parent. All tassels must be removed from the female rows before they have shed any pollen. Single-cross seed from fields where careless detasselling has been done is bound to perform poorly. A close check should be exercised therefore, during the flowering period.

Detasselling is started when the tassel is well out of the leaf sheath but before the anthers have shed pollen. This, in most cases, will be one to two days after the tassels are first visible. In certain strains, however, the tassels may begin to shed pollen even before they emerge. In such cases it is necessary to open out the leaf whorl and remove the tassel. Holding the stalk with the left hand, a little below the tassel, take a firm hold of the entire tassel in the right hand. Remove the tassel by a steady upward pull and throw it on the ground. A loose or imperfect grasp of the tassel may cause a portion of it to be left on the stalk. These portions are difficult to see and will produce contaminating pollen if left on the plant. Do not break or remove leaves, as removal will reduce yields and will also result in lower quality of seed produced.

Precautions in detasselling

(1) Grasp the entire tassel so that all pollen bearing parts are fully removed.
(2) Immature detasselling should be avoided. It may cause a few spikelets being left, which may emerge and shed pollen. Also, the top leaves are likely to be pulled out, leading to reduction in yield or disease attack.
(3) Do not hold the tassel too low on the stalk as otherwise plant tops may be pulled out.

(4) Once detasselling starts in a field it must be repeated daily in all weather. A fixed time should be observed every day. Be particular to start detasselling from the same side every day in the case of a large field.

(5) Mark all male rows at both ends.

(6) Look out for suckers (tillers) on female plants and also for lodged or damaged plants in female rows, as they are likely to pass unnoticed during detasselling.

(7) Plants on the verge of shedding frequently have leaves surrounding the tassel. A gentle shaking of the plant will reveal the tassel and enable the detasseller to grasp the tassel.

(8) Instruct the detasseller to drop the tassels on the ground after removing them and not to carry them in hand, as this may involve the danger of contaminating receptive silks.

(9) Put an experienced detasseller in charge of this operation. He should follow behind the other detassellers and check that no tassels are left.

Male rows are harvested first and kept aside for commercial use in order to avoid mixing male ears with female ears. The harvesting of female rows and further handling of harvested ears is done in the same manner as described for inbred lines.

3. Seed Production of Commercial Hybrids

The hybrid seed (three-way cross, double cross of double topcross) is produced on a high-yielding single cross used as the female parent. The usual planting pattern is to produce hybrid seed on six rows of female single cross to two rows of the pollinator parent (single cross, open-pollinated variety, inbred line, etc., used as the male parent). Female single cross is detasselled before pollen shedding to ensure cross-pollination from the male parent (see Fig. 32.1). The seed is saved only from female rows.

Seed fields should be free from volunteer corn plants. A specific hybrid must be so located that the seed parent is not less than 200 metres from any other corn of different color and texture. However, in the case of same kernel colour and texture as that of the seed parent, the distance may be reduced to 125 metres (410 feet) (sweet corn excepted) and further modified by planting border rows of the male parent. The number of border rows is determined by the acreage of the specific cross and by considering the distance by which the isolation distance falls short.

Differential blooming dates are permitted for modifying isolation distances, provided there are no receptive silks in the seed parent at the time pollen is being shed in other corn fields within the prescribed isolation distance.

Roguing, detasselling and harvesting are done in a manner similar to that described for single-cross seed.

USE OF MALE STERILITY

Hybrid maize seed may be produced without detasselling by utilizing cytoplasmic/cytoplasmic-genetic male sterility in any in the following ways.

1. *One inbred male sterile, no dominant restorer genes*

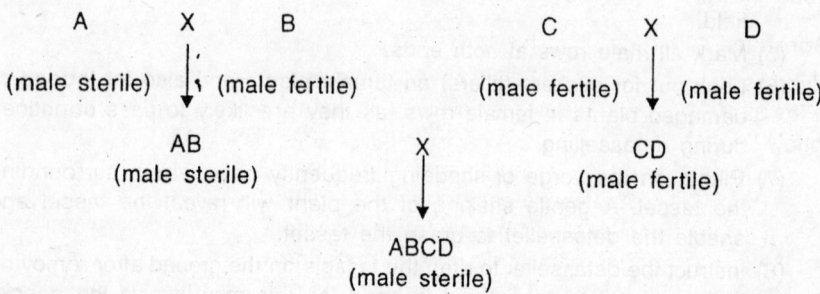

Since none of the inbreds contain pollen fertility-restoring genes, the double-cross ABCD shall also be male sterile. A practical way to ensure adequate pollination in the farmer's field is to make an identical hybrid with male fertile lines throughout and blend the male fertile seed with the male sterile seed in a ratio of 1:2 or 1:3.

2. *One inbred male sterile (A), either one or two inbreds (C x D) with dominant restorer genes*

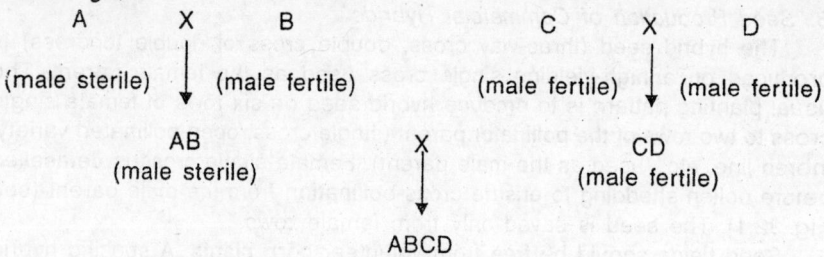

(100% male fertile if both lines carry pollen-restoring ganes and 50% fertile if only one line carries pollen-restoring genes).

3. *Two inbred male steriles, one inbred with dominant restorer genes.*

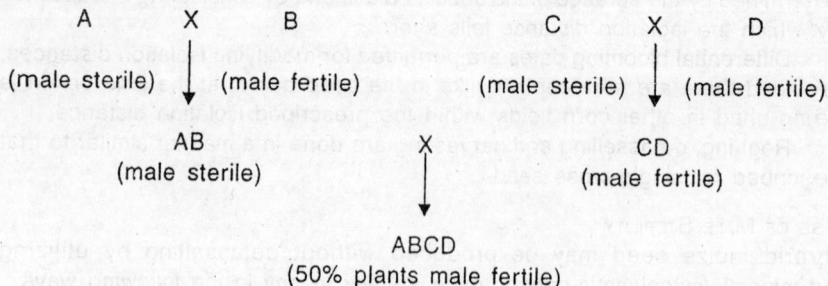

In each scheme the male sterile line is maintained by pollination from a fertile counterpart. Detasselling is eliminated when a male sterile line is used as seed parent.

Sorghum

Hybrid sorghum seed is produced by utilizing cytoplasmic-genetic male sterility (Fig. 32.2). The various steps involved in hybrid seed production are as follow.

1. Maintenance of parental lines, namely, male sterile line (Line A) carrying cytoplasmic-genetic male sterility; maintainer line (Line B) (sister strain of Line A), male fertile, non-pollen restoring; and restorer line (Line R), male fertile, pollen-restoring line.
2. Production of hybrid seed: This involves crossing of male sterile line (Line A) with restorer line (Line R).

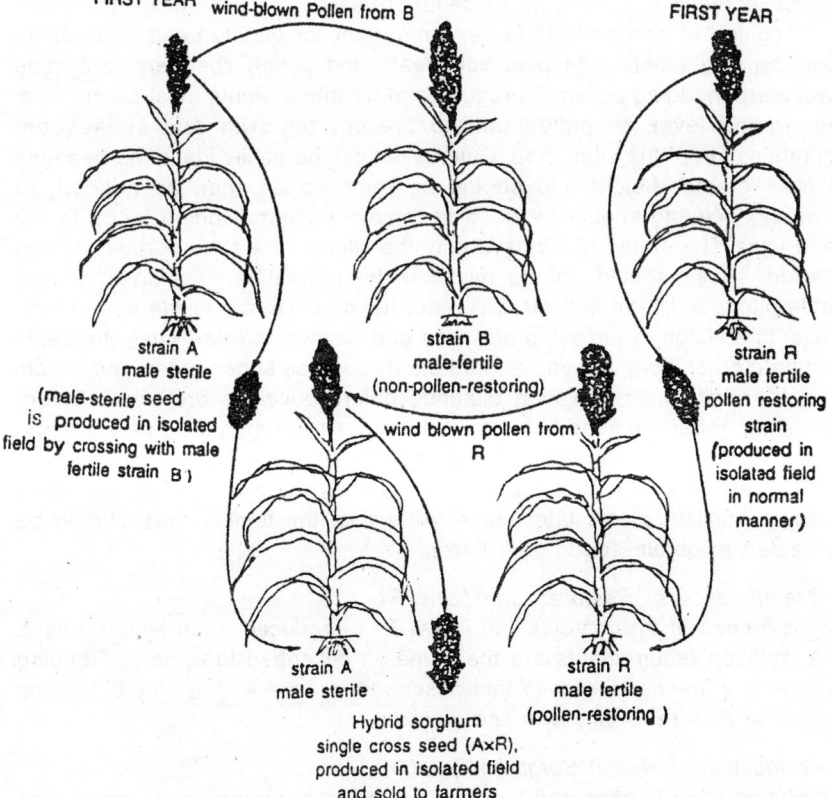

Fig. 32.2. Hybrid sorghum seed production (from Poehlman and Borthakur, 1969)

1. Maintenance of Male Sterile Line (Line A)

Principle of maintenance of male sterile line: The male sterile line (Line A) carries male sterility due to cytoplasmic-genetic factors. It is maintained by crossing with male fertile, non-pollen restoring strain (Line B), which is a sister strain of Line A in an isolated plot.

Line B is essentially similar to Line A in all respect except that Line B is pollen fertile, whereas Line A is pollen sterile.

The seed crop should be raised on land on which the previous crop was of another kind, including sudan grass and broom corn. Seed fields must be isolated by not less than 300 metres from other sorghum fields, including fields of the same line not conforming to varietal purity requirements, and by 400 metres from grass sorghum or broom corn with the same chromosome number (including off-type plants). In a crossing field, the usual planting ratio of Line A and B is 4:2. Four to six border rows with Line B seed are planted all around the seed plot. The seed harvested from Line A is male sterile and is used for hybrid seed production and in future increases of Line A. The seed harvested from Line B is pollen fertile and can be used in further increase of Line A in subsequent years.

Roguing of a seed field is very important for quality seed production. Start roguing before off-types, volunteers and pollen shedders in female rows start shedding pollen. All rogues and volunteer plants must be cut from the ground level, or pulled out, to prevent regrowth and subsequent contamination of the seed crop. Outcrosses can be easily identified because of their greater height and should be removed as soon as noticed. At flowering, roguing should be done to remove pollen-shedding types in the seed rows. The sterile types have only the stigma or a few abortive anthers exerted. These should not be mistaken for normal fertile plants. Normal fertile plants will have anthers which are full of pollen out to the tips of both lobes. In addition to removing off-types and volunteers from within the fields eliminate other *Sorghum* and related plants such as sudan grass and broom corn from within the isolation distance before pollen is produced. Before harvesting (after the seed has matured to the stage when the true plant and seed characters are apparent) a roguing may be done to remove off-types.

Harvest male rows first and keep their heads separate. This is necessary to avoid admixture at a later stage. After this, the female rows should be harvested to obtain seeds of A line.

2. Maintenance of Restorer Line (Line R)

The seed of the restorer line (Line R) is produced in an isolated field. The isolation requirements are the same as described for Line A. Roguing is done in a manner similar to that described for Line A. The only difference is that here there is only one line involved.

3. Production of Hybrid Sorghum Seed

Hybrid seed is produced by crossing the male sterile line (Line A) with the specified restorer line (Line R) in an isolated field. This is the hybrid seed that is sold to farmers.

Land requirements for seed production are the same as described for the male sterile line (A Line). Seed fields should be isolated at least by 200 metres from any variety of the same genetic height as the pollinator parent or from sorghum species with different chromosome number; and by 300 metres from any forage variety or from varieties with a different genetic constitution for height but with the same chromosome number; and also atleast by 400 metres from grass sorghum or broom corn with the same chromosome number (including off-type plants).

Modification of isolation on distance: The modification of isolation distance from 200 metres to 100 metres may be allowed with the planting of additional pollinator rows only in cases in which the hybrid produced by the contaminating pollen would not differ greatly in color, maturity, height, type, or any other important characteristics from the hybrid being produced. Two additional border rows of pollinator[1] to all sides of seed field exposed to contaminating (whether located directly opposite or diagonally) are planted for each 20 metres of reduction in isolation distance.

Differential blooming dates for modifying isolation distances are permitted, provided that the sum of the percentages of plants in bloom in seed rows and in the contaminating field shall not exceed 5% when more than 1% of the plants in either field are in bloom.

The recommended planting ratio is two male rows for every four female rows. At least four border rows should be planted on all sides of the field. This is mainly, to ensure an abundance of pollen on the edges of the field, especially at the ends of rows.

SYNCHRONIZATION OF MALE AND FEMALE PARENTS

In hybrid seed production, perfect synchronization in flowering of the parental lines is most important. The differential behaviour of the parental lines in flowering habit result in non-synchronization of the parents, giving rise to a poor seedset. Therefore, a knowledge of the behaviour of parental lines for the flowering habit is not only essential, but also very useful for careful planning of suitable staggering to ensure nicking and thereby maximum seedset. If before flowering it is known that the parents differ in days to 50% flowering by more than 5 days, staggered sowing is recommended to achieve effective nick. If there is only a slight difference in number of days to 50% flowering, between the 35th and 40th day after sowing, observe primordial initiation by taking a longitudinal section of the internodes in both the parents. Based on primordial initiation, disparity of 4-5 days can be offset by cultural manipulations, namely spraying 1% urea two or three times at the interval of 2-3 days or additional irrigation to the lagging parent. Blowing air by operating an empty duster with the mouth directed horizontally to the male ears, will help disseminate pollen.

---

[1] To be considered a pollinator row the pollinator line must be producing pollen during the entire time 5% or more of the female flowers are receptive.

Under exceptionally difficult situations hand-pollination by collecting and dusting pollen may have to be resorted to. However, by a combination of staggered planting and selective irrigation, fertilization one can meet the requirements of most situations. It should be recognized that achieving a good nick is primarily dependent on personal initiative and experience.

Roguing and harvesting should be done in a manner similar to that described earlier for maintenance of male sterile line (Line A).

Pearl millet

The hybrid pearl millet is produced by utilizing cytoplasmic-genetic male sterility in a manner similar to sorghum.

1. Maintenance of Male Sterile Line

Land to be used for seed production shall be free of volunteer plants. Seed crop may be grown on land on which the preceding crop was of another kind. In a crossing field the usual planting ratio of Line 'A' and Line 'B' is 4:2. In addition, eight border rows are planted with the seed of Line 'B' all around the field. The seed harvested from Line 'A' is male sterile and is used for hybrid seed production.

The male sterile line increase plots of pearl millet should be isolated from other pearl millet fields or fields of the same line increase not conforming to varietal purity requirements of certification at least by 400 metres.

Roguing is very important for high-quality seed production. Roguing should be started before flowering to avoid contamination from foreign pollen. All rogues and volunteers must be cut from the ground level, or pulled out, to prevent regrowth. Outcrosses become noticeable because of their greater height. At flowering, roguing should be done to remove pollen-shedding types in seed rows. The sterile types have only the stigma, or a few shrivelled anthers exerted. Normal fertile plants will have plump anthers which are full of pollen out to the tips of both the lobes. On shedding, they rupture longitudinally along one side and discharge pollen. The field may also be rogued thoroughly before harvesting, but after the seed has been matured to the stage that the true plant and seed characters are apparent.

At maturity, harvest male rows first and keep their heads separate. This is necessary to avoid admixture at a later stage. After this, the female rows should be harvested. After harvesting, sort out the undesirable heads and reject them. The remanent seed should be dried, threshed and cleaned before storage.

2. Maintenance of Maintainer Line (B Line) and Restorer Line (R Line)

Seeds of these lines are produced in an isolated field. The land requirements, isolation requirements, cultural operation, roguing techniques, etc. are similar to those described for the maintenance of male sterile lines; the only difference is that only one parent is handled in contrast to the two parents required for maintenance of male sterile lines.

3. Production of Hybrid Pearl Millet Seed

Principle of production. The seed of hybrid pearl millet is produced by crossing a male sterile line (Line A) with a specified restorer line (Line R) in an isolated field. This is the hybrid seed that is sold to farmers.

Land requirements are the same as described earlier. Hybrid pearl millet seed fields should be isolated at least by 200 metres from fields of other varieties, and from fields of the same variety not conforming to varietal purity requirements for certification.

Where difference in flowering days is observed, the sowing of male and female parents can be suitably adjusted. A moderate variation of blooming period, however, may not cause serious difficulties as the pollinator parent generally has a tendency to develop secondary nodal tillers which shed pollen over a long period. Further, flowering in the early parent can be delayed by a week or so by jerking the earheads in the first stage. This may be continued until the other parent starts booting. The parent which shows slow emergence can be top-dressed by a light urea-sugar solution which enhances initiation of early flowering.

Roguing, harvesting and threshing techniques are similar to those described earlier for maintenance of the male sterile line of pearl millet.

Rice

Hybrid seed production involves multiplication of cytoplasmic-genetic male sterile line (A line), maintainer line (B line) and a restorer line (R line), and production of F_1 hybrid seed (A x R).

The areas of seed production should be so chosen as to provide the best possible conditions at flowering and the pollen-shedding period. The most suitable conditions are 24-28°C daylight average temperature, relative humidity 70-80%, temperature difference between day and night 8-10°C and good sunshine. An average day temperature of more than 30°C or less than 23°C, continuous rains, or strong wind are generally harmful to flowering, pollination and cross-fertilization. As a rule, in high temperature with low humidity or in low temperature with high humidity some glumes will not open. This lowers seed yield.

Growing the hybrid seed crop should be so adjusted that flowering takes place after the high-temperature period ends but before the start of low-temperature period. Selection of prime field plots is necessary. The seed fields should be free of volunteer plants, well levelled, have fertile soil with good physical and chemical characteristics, and be well drained.

Hybrid rice seed-production fields should be isolated from other rice fields, including commercial hybrids of the same variety, and the same hybrid not conforming to varietal purity requirements for certification by at least 200 metres for foundation seed class (A, B and R line production) and by 100 metres for hybrid seed production (A x R production).

Raising of vigorous seedlings is an important factor for obtaining high seed yields. The root system of vigorous seedlings flourish and leaf sheaths have high carbon content, all of which contribute to produce green growth and tillering at the lower nodes so that more dry matter is accumulated, leading to more panicles and a high-seed-setting rate per panicle. It has been observed that tillering at the lower nodes gives more and bigger panicles, which helps to achieve the goal of 100 kernels per panicle.

Seedlings with healthy tillers are the basis for increased panicle size. For hybrid seed production, the seedlings of both parents should be standardized. Seedlings of the male parent for short-duration varieties should be 20-30 days old with 5.5-7.0 leaves and 2-3 tillers and for long-duration varieties 30-35 days old with 5.5-7.0 leaves and 2-3 tillers. The ratio of female and male lines is generally kept at 2:10 and row spacing 10 x 10 cm for the male parent and 20 x 15 cm for the female parent. Two seedlings are planted per hill. Both parents should receive good aeration and equal amounts of sunlight. Row direction should be nearly perpendicular to prevailing winds at flowering to ensure more cross-pollination.

1. Synchronization of Flowering

Synchronizing the flowering of both parents is the key to increased yields. Technical measures such as staggering seeding dates of the male and female parents, sowing the male parents three times to extend the time pollen is available, and predicting and adjusting flowering dates may be adopted. Actual practices would need to be standardized for each hybrid and the locations selected for hybrid seed production.

(i) *Staggered sowing of male parents.* Seeding date is usually determined by leaf age, effective accumulated temperature (EAT) and growth duration. In general, the period from initial to full heading of a CMS line is 4-6 days longer than for a restorer line.

(ii) *Fertilizer application.* Beginning about 30 days before heading, 3 or 4 random samples of the main culm of both parents are taken every 3 days. Young panicle development is compared under magnification. During the first three stages of panicle differentiation, treat the earlier developing parent with quick releasing N-fertilizer and spray the later developing parent with potassium dihydrogen phosphate. This adjusts development differences of 4-5 days.

(iii) *Water management.* During later stages of panicle differentiation, draining water from the field will delay male parent panicle development; higher standing water will accelerate panicle development.

2. Methods of Improving Seed Setting

(i) *Supplementary pollination (rope-pulling).* On calm days during anthesis, supplementary pollination can be carried out. Panicles of the restorer lines are shaken by pulling a long nylon rope (5 mm diameter) back and forth every 30 minutes until no pollen remains on the restorer line. The rope or pole is positioned a little above the flag leaf of plants so that when it is moved back and forth it touches the entire panicle and shakes it. This method is often used on even topography and regularly shaped plots. In hilly, uneven topography with small, irregular plots, a bamboo pole may be used.

(ii) *Leaf clipping.* Leaves taller than the panicles are the main obstacles to cross-pollination. Clipping leaves 1-2 days before initial heading increases the probability of pollination and outcrossing rate. The blade of the flag leaf is cut back 1/2 to 1/3 from the top.

(iii) *GA₃ spray.* Spraying the seed parent with GA₃ (2 or 3 times) increases panicle exsertion and helps increase seed setting.

The seed field should be free of rogues. Remove off-type plants in both the parents first before the onset of the flowering stage and then soon after emergence of the panicle. Rogue out plants of the maintainer line, if any, and the semi-sterile plants in the seed parent as often as necessary. Harvest male rows first to avoid chances of mechanical admixture.

Wheat[2]

The main systems for producing hybrid wheat are described below:

Cytoplasmic Male Sterility/Restorer Systems

A cytoplasmic male sterility and restorer system is available in wheat, the *Triticum timopheevi* (Zhuk.) Zhuk. var. *timopheevi*[3]. Adapted cultivars are converted into female cytoplasmic male steriles by repeated backcrossing. The restorer lines serve as pollinators possessing traits that complement those of the male-sterile females. In the production of F_1 hybrids, however, the difficulties have been encountered in the maintenance of male sterility, in the restoration of full fertility, in synchronizing the times of stigma receptivity and pollen shedding in the parent, and in ensuring adequate cross-pollination, especially under unfavourable weather conditions.

XYZ SYSTEM

The XYZ system of producing hybrid wheat (Driscoll, 1972) involves three lines:

1. The Z line is homozygous for a recessive male sterile mutant on one of the chromosomes. It has the normal 21 pairs of wheat chromosome.

2. The Y line is the same as the Z line except that it contains an additional 'alien' chromosome which bears a compensating gene for male fertility.

3. The X line is the same as the Y line except that it contains a pair of alien chromosomes. The alien chromosomes, for example, chromosomes of rye (*Secale cereale* L.), do not pair with wheat chromosomes and two types of pollen are produced on the Y line, the 21- paired wheat chromosomal type is largely favoured in fertilization. Seed production is done in three steps:

---

[2] Development of hybrid varieties in wheat is still in the experimental stage.

[3] Male sterile: *T. timopheevi* (Zhuk). Zhuk cytoplasm. Male fertility restorer: R5 (*Triticum zhukovskyi* Men & Er./3*)

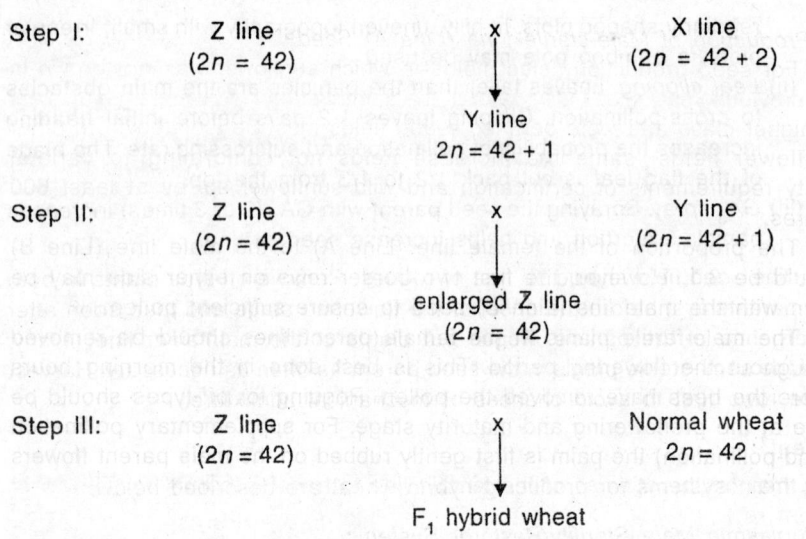

Step I: Z line   x   X line
          (2n = 42) | (2n = 42 + 2)
                                      ↓
                                   Y line
                                 2n = 42 + 1

Step II: Z line   x   Y line
          (2n = 42) | (2n = 42 + 1)
                                      ↓
                              enlarged Z line
                                 (2n = 42)

Step III: Z line   x   Normal wheat
          (2n = 42) | 2n = 42
                                      ↓
                              F₁ hybrid wheat

YZ System (Driscoll, 1985)

The modified XYZ system differs from the original system in which no X line is involved and progeny of the selfed Y line is used instead of the Y line.

Genetic Male Sterility System

Genetic male sterility also occurs in wheat. Various schemes have been proposed to use this form of sterility. This system, however, is not fit for hybrid wheat production due to instability in gene expression for various reasons, such as higher ploidy level of cultivated wheat, genetic interaction effects and influence of environment on gene action.

Chemical Hybridizing Agents

Use of chemical hybridizing agents (CHAs) may eventually offer a rapid and flexible way to induce male sterility in wheat. Using granular ethephon (2-chloroethyl phosphonic acid), sterility levels up to 100 per cent, with no impairment of female fertility or apparent morphological or physiological abnormalities, have been reported. However, successful hybrid seed production is still far away.

Sunflower

Hybrid sunflower is produced by means of the cytoplasmic male sterility and genetic fertility restoration system. The male sterile line (A line) contains sterile cytoplasm and recessive genes for fertility restoration. This is maintained by a male fertile counterpart (Line B), which also contains recessive genes, but has fertile cytoplasm. For production of hybrid seed the male sterile line (Line A) is crossed with a fertility restoring line (Line R) which has the dominant genes for fertility restoration, but may have either sterile or fertile cytoplasm. The restorer line (Line R) should nick well with Line A to produce F₁ hybrid seed.

1. Production of Male-sterile Line (Line A) Seed

For seed production select fields in which sunflower was not grown in the previous year unless they were of the same variety and were of equivalent or higher class and were certified. Seed fields must be isolated from other sunflower fields, same-line increase fields not conforming to varietal purity requirements of certification and wild sunflower sp. by at least 600 metres.

The proportion of the female line (Line A) to the male line (Line B) should be 3:1. However, the first two border rows on either side may be sown with the male line (Line B) seed to ensure sufficient pollen.

The male-fertile plants in the female parent lines should be removed throughout the flowering period. This is best done in the morning hours before the bees have removed the pollen. Roguing for off-types should be done at the preflowering and maturity stage. For supplementary pollination (hand-pollination) the palm is first gently rubbed on the male parent flowers and then on the stigmas of the female line to transfer the pollen. The male parent rows should be harvested prior to harvesting female rows to avoid contamination. No male parent heads should be left intermingled with the female parent rows.

2. Production of Maintainer Line (Line B)/and Restorer Line (Line R) Seed

Seed is produced in an isolated field in a manner similar to that for open-pollinated varieties. The isolation requirements, are higher however, and must be the same as given for production of Line A seed above.

3. Production of Hybrid Sunflower Seed

Seed fields must be isolated by at least 400 metres from field of other varieties, commercial hybrids of the same variety, and fields of the same hybrid seed production not conforming to varietal purity requirements of certification. The proportion of female parent (Line A)/male line (Line R) should be kept at 3:1. However, the first two border rows on either side may be sown with the male parent seed to supply sufficient pollen. Roguing, harvesting and supplementary pollination techniques are the same as described for A line.

Castor

Hybrid castor seed production is done using pistillate lines. (Ramachandran and Prasad, 1995).

1. Sex Expression

Inflorescences are borne terminally on the main and lateral branches. The main stem ends in a raceme which is the first (or primary raceme). After the first raceme appears, two to three branches arise at the nodes immediately below it. Each of these branches terminates in racemes after four or more nodes have formed which are known as secondary racemes. Branches arise from the nodes just beneath secondary racemes, ultimately terminating in tertiary racemes. This sequence of development (indeterminate growth habit) continues.

The racemes of castor are usually monoecious, with the pistillate flowers on the upper 30-50% and staminate flowers on the lower part of the inflorescence. The proportion of pistillate and staminate flowers among the racemes varies a great deal both within and among genotypes. It is influenced by the environment, age of the plant, genotypes and nutrition. Female tendency is highest in winter, while male tendency predominates in summer and rainy seasons. Also, the femaleness in young plants with high levels of nutrition is stronger than in old plants with low levels of nutrition.

2. Genetics of Sex Expression

Shifriss (1960) identified two groups of genes for sex expression. The first group consists of qualitative genes, which determines the kind of flowers in the inflorescence. The second group consists of polygenes, which are responsible for gradient differentiation and for racial differences in sex tendency. In addition to these two major groups, there exist gene modifiers which change the pattern of sex differentiation.

3. Pistillate Mechanism

In addition to monoecism a subform of dioecism exists in castor, which has led to the identification of three different pistillate mechanisms:

1. *N*-type or conventional mechanism: This is governed by a recessive sex-switching gene (ff). Staminate flowers are absent. In the production of F_1 hybrid seed using the *N*-pistillate line, the producer is required to rogue out normal monoecious plants before anthesis to obtain 100 per cent production of pistillate plants in the female rows. This has proved difficult to do for three reasons: (i) uneven emergence, (ii) variation in time of flowering and (iii) higher percentage of monecious plants than expected (50%).

2. *S*-type or non-conventional mechanism: This mechanism is essentially derived from sex reversals, which start out as female and then revert to normal monoecism any time after the first raceme. Continuous inbreeding of sex reversals perpetually gives a spectrum of sex reversals, each reverting at a different time in development, and the some monoecists. If a given sex reversal stock is allowed to reproduce freely in isolation a rapid regression to monoecism will occur in subsequent open-pollinated generations.

3. *NES* System: This system is homozygous for the N-pistillate gene and also involves environmentally sensitive genes for interspersed staminate flowers. Such a line is normally pistillate under moderate temperatures, but produces interspersed staminate flowers under high temperatures. In crossing fields (hybrid seed production fields), usually one or two roguings of the female line are sufficient to ensure that all flowering plants are pistillate and to remove off-types that appear.

In practice, female lines are obtained by selecting all female plants in backcross populations under warmer and longer growing season, roguing before flowering the plants that are monoecious and harvesting bulked seed from both those plants that remain pistillate and those that produce pollen from interspersed staminate flowers. Under proper environmental conditions, this bulk seed produces 100 per cent pistillate plants on the first and second set of racemes. Constant selection pressure is maintained in the foundation seed fields to prevent excessive expression of genes for interespersed staminate flowers, which can result in the selfing of the female parent.

Planting is usually done in an isolated field in a ratio of six lines of females and one line of males. The isolation distance is kept at 300 metres and 150 metres from other castor fields for foundation and certified seed class respectively. The direction of rows is so oriented that prevailing winds blow across the rows for best pollen distribution. The male parent should be earlier flowering than the female to avoid necessity of planting at different times. Both male and female parents should be rogued for off-type plants prior to blooming and the female parent should be rogued for any pollen-shedding plants at the early bloom stage of the primary raceme.

Cotton

Intraspecific (*Gossypium hirsutum*) and interspecific (*G. hirsutum* × *G. barbadense* and *G. herbaceum* × *G. arboreum*) hybrids of cotton are under commercial cultivation. Hybrid seed production is presently being done by hand-emasculation and pollination. The genetic and cytoplasmic-genetic male sterility system is being researched and has been in use on a limited scale. (Basu, Narayanan and Bhat, 1995).

Hand-emasculation and Pollination Method of Hybrid Seed Production

The Doak's method with minor modifications is being used for hybrid seed production. The female and male parents of a hybrid are sown in separate blocks within a seed field. Usually the area under female and male parent is kept in a ratio of 4:1 or 5:1. Sowing of the male parent is done in two to three instalments at a week's interval. Off-types are rogued before commencement of the flowering season. Hand-emasculation and pollination work is undertaken soon after commencement of the flowering season and continued for seven to ten weeks depending on the duration of flowering. Seed-set and development of bolls is better in the early and medium flushes. Late flush is usually avoided. After the crossing programme is discontinued, all the buds and flowers that appear subsequently are removed and the top end of the shoot is nipped to stop further growth of branches to facilitate better development of bolls. During the entire crossing programme any unemasculated bud that flowers (opens) is removed without fail. Ripe and completely open bolls are picked.

Method of Emasculation and Pollination

(a) Floral biology: Cotton is both a self-pollinated and cross-pollinated crop. The extent of natural outcrossing varies a great deal, from 10-50% or more in *G. hirsutum*, 1-2% in *G. arboreum* and 5-10% in *G. barbadense*. Effective flowering duration varies from 45-90 days depending on the duration of parents involved.

The stigma tip is above the staminal column but the stigma is not exserted clear of the uppermost anthers. The pollen is heavy, warty and sticky, and is not windblown. Natural outcrossing results from insect-pollination. Insects are attracted by large showy flowers and intra- and extrafloral nectaries. Anthesis usually takes place between 9-12 a.m. Once in bloom the cotton flower is receptive to pollination for about 8 hours or less. The ovary after fertilization develops into a boll and takes about 40-80 days to mature.

(b) Emasculation: Flower buds are ready for emasculation 12 hours before the day of anthesis (white bud stage). Too young or too old buds are avoided. Emasculation is done in the afternoon. The petals and anthers are removed by the thumbnail method in a deft operation, or preferably the petals should be cut and the anthers stripped gently using a small knife and forceps. Some strains of G. barbadense may not tolerate loss of petals before pollination. In this case, anthers may be removed through a slit in the corolla. Stigmas of emasculated flowers are protected by covering with a length of soda straw. Emasculation should be complete and perfect.

(c) Pollination: Emasculated flowers are pollinated the next day during the anthesis hours (9-12 a.m.). The male buds[4] selected for pollination should be shedding abundant pollen at the time of use. Pollen is collected in straw tubes and applied to the stigma within minutes after the male flower is collected. Stigmas are covered with the soda-straw tube used in pollinating the emasculated bud. A piece of thread is tied to the pedicel of the bud immediately after pollination.

(d) Picking of hand-pollinated bolls: The ripe and completely opened bolls (with their threads intact) are picked and collected in baskets. After sun-drying for a day or two they are supplied to seed-processing plants for further handling.

Use of Male Sterility in Hybrid Seed Production

1. GENETIC MALE STERILITY SYSTEM (GMS)
Genetic male sterility has been investigated and 11 loci (ten in G. hirsutum and one in G. barbadense) have been identified. Four of these loci-MS 4, MS 7, MS 10 and MS 11 behave as dominant, three loci-ms 1, ms 2 and ms 3-behave as recessives and the remaining four-ms 5, ms 6, ms 8 and ms 9-behave as duplicate recessives. Gregg, with ms 5, ms 6 genes, is considered a good basic source for developing stable GM 5 female parents of a new hybrid. Four to five backcrosses are required for converting the potential female parent into a GM 5 line and is achieved by alternate selfing and selection, each time choosing lines giving 1F:1S. GM 5 line is maintained by sib-mating between fertile and sterile. Fertility-restoring genes MS 5 and MS 6 are present in all the American varieties.

---

[4] The practice to cover male buds with paper bags the previous evening has been abandoned now with no adverse effect on genetic purity of the seeds produced.

2. Cytoplasmic-genetic Male Sterility

(a) Sources of sterility and fertility restoration: Among the various sources, the cytoplasm of *G. harknessii* and the genome of *G. hirsutum* interaction produce stable and dependable cytoplasmic male sterility in cotton. A single dominant gene *Rf* from *G. harknessii* is essential for fertility restoration. *G. barbadense* possesses modifier gene or fertility enhancement factor 'E' for fertility in the male-sterile restorer system involving *G. harknessii*. This system also contributes to excellent agronomic properties. However, some adverse effects on ginning outturn and bacterial susceptibility have been reported.

Some other cytoplasms, especially *G. longicalyx* and *G. aridum*, are also being experimented with to develop alternate systems. The male sterility resulting from the system contributes to degradation of tapetal cells and coalescence of pollen mother cells.

(b) CMS Conversion: The original *G. harknessii* conversion in cross with recurrent parent *G. hirsutum* takes 10-12 backcrosses to obtain the CMS-A line by eliminating the *Rf* factor. The proposed female parent of a hybrid in 5-6 backcrosses converted with CMS-A results in a converted 'A' line, and the original female when selfed maintains the 'B' line.

The F_1 of CMS 'A' line x *G. hirsutum* 'R' line (with enhancement factor 'E') may be used as the female and crossed with the P_2 of the proposed new hybrid as the recurrent parent. The converted restorer line with '*Rf*' factor can be obtained in 5-6 backcrosses in the desired combiner male parent. Like 'B', the 'R' line is also maintained by selfing.

Using this system experimental hybrids have been identified.

(c) It has been found convenient to create 'R' lines by keeping the restorer gene(s) together with *G. harknessii* cytoplasm as the female line in the backcross cycle. The donor line is exclusively used as the pollen parent. Since the restorer genes are fully dominant, segregates carrying the restorer factor can be identified in the F_1 generation of each cross.

It is thus possible to make two backcrosses per year in creating 'R' lines as well as 'A' lines.

Practical Applicability

Since the pollination problem is not yet solved, mass-scale commercial hybrid seed production in cotton is still being done by hand-emasculation and pollination. Also, the presence of strong cytoplasm in CMS hybrids may act to pull down the yield.

Sugar-beet

1. Diploid Hybrid Varieties

The hybrid seed of sugar-beet is produced by utilizing cytoplasmic and genetic factors that condition sterility (Fig. 32.3).

This is called a four-way hybrid, or a double-cross hybrid, which is comparable to that used in hybrid corn production. As can be seen from the diagram, four inbred lines are required for production of the four-way hybrids.

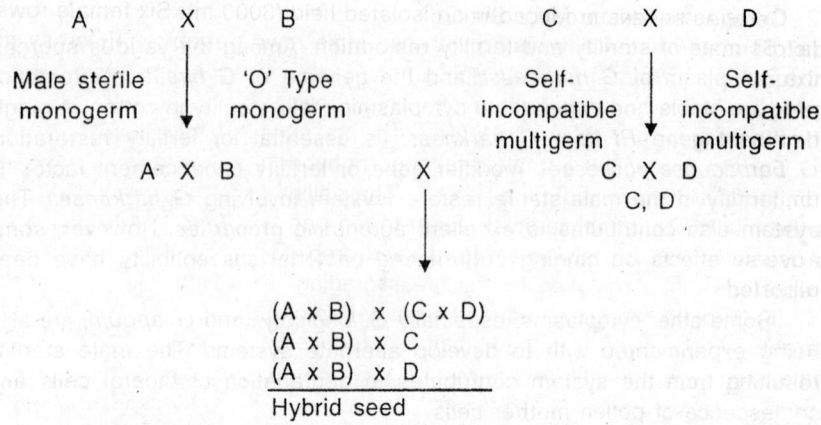

Fig. 32.3. Hybrid seed production of sugar-beet

Lines A and B are monogerm, 'O' type, with the A being the male-sterile equivalent.

The hybrid between these two is male sterile and serves as a seed parent in commercial production; the hybrid between lines C and D is produced by planting the two lines in alternate rows. The hybrid between C and D which is multigerm, is mixed with the monogerm A x B parent to constitute the stock seed for commercial production. Ten per cent seed of the pollinator is usually mixed for raising a hybrid seed crop.

The hybrid seed production fields are isolated from other beet fields. The isolation distance is kept at 3000 metres.

All the A x B seed, which is monogerm, when harvested in the commercial field will be hybrid and can be easily separated from the multigerm pollinator seed by using thickness grading with a cylindrical slotted screen. Production of hybrid seed is a rather simple process after the inbred lines have been selected.

2. Triploid Hybrids

Triploid hybrids may be produced in two ways:

(i) Triploid hybrids produced from diploid male steriles : The prebasic material consists of a diploid male sterile line, a diploid pollinator ('O' type) and a tetraploid F_1 cross or mixture of F_1 crosses.

The basic seed is produced in isolated fields (3000 m). The male sterile seed is produced by planting out rows of the male sterile line and pollinator line ('O' type) in the ratio of 4:2 or 6:2. The pollinator rows are removed before the seed is harvested from the female rows. The entire field is checked plant by plant for sterility and genetic purity in the female rows. The seed of tetraploid F_1 cross or mixture of F_1 crosses is produced in an isolated field. Strict roguing and inspection is necessary for ensuring genetic purity.

The basic seed produced in the manner outlined above is tested for varietal purity, sterility and ploidy prior to its use for certified seed production.

Certified seed is produced in an isolated field (3000 m). Six female rows (diploid male sterile) alternating with two male rows (tetraploid F_1 cross or mixture of F_1 crosses) are planted out. The male rows are eliminated before the hybrid seed is harvested from the female rows. Strict roguing for removal of off-types is done to ensure purity.

For the production of triploid monogerm hybrids, it is possible to use an admixture of female and male parents. The pollinator in this case, however, is the multigerm tetraploid stock. Eight to ten per cent seed of the pollinator is usually mixed for raising a hybrid seed crop. The multigerm seed of the pollinator is later removed by thickness grading.

(ii) Triploid hybrids produced from tetraploid male sterile monogerm parents : These hybrids are now being introduced but handling of tetraploid male steriles is difficult.

Tomato

Tomato is a highly self-fertilized crop. The varieties are pure lines and quite homozygous so that seeds of uniform F_1 hybrids can be obtained straightaway from pair crosses. Commercial production of tomato seed is undertaken when particular pair cross combinations have produced a high value hybrid. The parent which is better both for pollen production and number of seeds produced is chosen as a male parent and the other one as a female parent. Also, the disease-resistant parent should be used as a female because an occasional self-pollination would still produce resistant plants in the general F_1 hybrid population.

F_1 hybrid seed production of tomato is done by hand-pollination more generally in glasshouses than in seed fields. Making controlled pollinations under field conditions may be less efficient than under greenhouse environments because hot, drying winds may cause rapid desiccation of the exposed pistil before fertilization is achieved. Cool, dry, and relatively wind-free weather is preferred for high success rates. The hand-pollination technique is described below.

Hand-pollination

A) FLORAL TRAITS
The tomato flower is normally perfect, having functional male (anthers) and female (pistils) parts. Several (usually four to eight) flowers are borne in each compound inflorescence and a single plant may produce as many as 20 inflorescences during its life-cycle. This feature facilitates crosses between varieties that represent extremes in variation for maturity since flowering occurs over a long period of time. A tight protective anther cone surrounds the stigma, which greatly reduces the possibility for natural cross-fertilization. Outdoor (fields) flower movement aided by wind is sufficient to release pollen, but under greenhouse conditions, manual vibration of open flowers is required to effect pollination and fruit set. Crosses to produce F_1 hybrid seed are normally made in one direction only, using four or five times as many plants as females than as males. The parental lines are maintained by selfing.

b) EMASCULATION

In the female parent flower buds about to open in the next two to three days are selected for emasculation to avoid accidental self-pollination. At this time, sepals have begun to separate and the anthers and corolla are beginning to change from light to dark yellow, characteristic of fully opened flowers. The stigma appears to be fully receptive at this stage, thus allowing for pollination immediately after emasculation. Use sharp-pointed forceps to force open the selected buds. Disinfect forceps, scissors, hands or gloves before starting emasculation. After opening the buds split open the anther cone and carefully pull it out of the bud. To help identify the hybrid fruits from selfed fruits at the time or harvest cut the corolla and the calyx. Generally, under greenhouse conditions no protection is required following emasculation to prevent uncontrolled crossing. Under field conditions protection of emasculated buds with glassine bags is necessary to avoid chance crosses.

c) POLLINATION

When the corolla of the emasculated flower starts to turn bright yellow, the stigma is ready for pollination. Because of large-scale production of F_1 hybrid tomatoes, mechanical aids are used for pollen collection and pollination whenever possible.

Pollen Collection

Collect flowers from the male parent to extract pollen. The best time for pollen collection is during the early morning hours when anthers have dehisced since most of the pollen sheds if done later.

Pollen may be collected with a battery-powered electric 'bee' or into a 'thumbnail' receptacle whose effect is to shake pollen from the anthers into a receptacle borne on the operator's third finger.

Or

Remove the anther cones from the flowers and put them in suitable containers, such as, cellophane or paper bags or plastic pans. Dry the anther containers under a 100-watt lamp (above 30 cm above the bag) for 24 hours. The optimum temperature for drying is 30°C. Put the dried anther cone into a plastic pan, cover it with a fine mesh screen, then seal it with a similar tight-fitting cup (serving as a lid). Shake the cup so that pollen is collected in the 'lid' cup. Transfer the pollen into a small container for pollination. Fresh pollen is best for good fruit set. Pollen is applied to the female flower either by brush from the pollen supply in the glass tube or directly by inverting the emasculated flower over the 'thumbnail' receptacle. Generally no protection is required under greenhouse conditions but under field conditions it is necessary to place glassine bags.

Precautions: Remove naturally pollinated flowers and developing selfed fruits of the female plant to prevent admixture of selfed seed with the hybrid seeds. The number of hybrid fruits produced per plant may be around 20, 30 or more for large- medium- and small-fruited varieties.

Making controlled pollinations under field conditions may be less efficient.

Brinjal (Egg-plant)

In brinjal, hybrid seed can be produced by the hand-emasculation and pollination technique, as its floral morphology favours rapid emasculation and its pollination. Also, a large number of seeds are formed in a single fruit (800 to 1000 seeds in long brinjal, and 1000 to 1500 in round brinjal). Therefore, a very limited number of fruits produce a sizeable amount of seed.

Technique of Hybrid Seed Production

In producing the hybrid seed, the variety setting a large number of seeds in a single fruit should be taken as the female parent, so that a large amount of seed can be obtained in a single attempt. The flowers are borne singly or in clusters. There are four types of flowers, namely long styled, medium styled, short styled and pseudo-short styled. The former two types with normal ovary are the flowers that set fruits. Flowers with rudimentary ovary do not set fruits.

EMASCULATION

Flower buds which are expected to open the next day are selected on the female parent. Using a forceps the flower buds are opened and the stamens, the number of which varies from five to seven, are removed one by one. The emasculated buds are then bagged in butter-paper or muslin bags to prevent pollination with the undesirable pollens. While emasculating the flower, care should be exercised that no anther is ruptured or crushed. If this happens, such flowers should be rejected and the forceps should be sterilized with spirit or alcohol. Emasculation is usually done in the afternoon. Stigma receptivity is maximum on the day of anthesis.

POLLINATION

The flower buds of the male parent should also be bagged to avoid contamination. Anthesis usually starts from 7:30 a.m. and continues up to 11:00 a.m. Peak time for anthesis is 8:30-10:30 a.m. Pollen dehiscence usually starts from 9:30 a.m. The flowers which were bagged for taking the pollen grains are plucked and collected in a petri dish. The female buds are then uncovered. The anther from the male flower is removed and held in the forceps. As the pollen grains in the anthers of brinjal are released through apical pores, the anther is held perpendicular to the stigmatal surface, keeping the apical pores of the anther opposite to this surface. The forceps are tapped and the yellow-coloured powder of pollen mass is dusted on the stigma. The pollinated buds are again bagged to prevent cross-pollination.

Emasculation and pollination can be done simultaneously. However, success in fruit setting when this method is followed is marginally reduced, but the labour and time required for bagging the emasculated buds and unbagging them the next day for pollination is effectively saved. Emasculated and unpollinated buds and male buds are tagged with tags of different colours for ready identification.

Okra

Hybrid seeds of okra may be produced by hand emasculation and pollination much in the same way as cotton. Joshi et al. (1958) reported that opening of okra flowers varies with the season. Dehiscence of anthers takes place only after flower opening. Insects start to visit flowers soon after opening. The flower structure of okra is such that without the aid of insects only the lower surface of the stigmatal lobes can receive pollen after dehiscence; the top and sides of the stigma usually receive pollen when insects visit the flowers. Joshi et al. (1958) studied four methods of hybrid seed production in okra without resorting to actual emasculation. They recommended that for hybrid seed production anthers should be scraped soon after the flower opens and flower then be hand pollinated. The flowers after hand pollination could be left as such. They obtained 85-90% hybrid seed in this manner. This study is significant from the viewpoint of hybrid seed production in okra. If the okra crop is grown in insect-proof conditions (wiremesh cages, glasshouse, etc.) a hybrid seed of still higher purity by this method can be obtained.

Pepper

1. Use of Genetic Male Sterility

Genetic male sterility to produce F_1 hybrid seed has been used on a limited scale in some European countries. Abnormal anther development in ms/ms^+ hybrids at low temperatures and consequently poor pollination have been reported.

Male steriles constitute only 50% of the plants ($ms\ ms \times ms^+\ ms$); one-half therefore must be eliminated in the seedling stage before transplanting to the field. This has not been feasible because closely linked marker genes have not been found and the ms genes have no obvious phenotypic effect prior to flowering. However, male sterile plants are easily identified at anthesis and hand-pollinated in an insect-screened greenhouse.

2. Cytoplasmic-genetic Male Sterility

This type of male sterility is due to the interaction of sterility inducing S-type cytoplasm with a recessive nuclear male sterility inducing ms gene. The ms gene is only expressed in S-cytoplasm. The only plasmon-genome combination that induces male sterility is $S\ ms/ms$. The other combinations $S\ ms^+/ms$, Sms^+/ms^+, $N\ ms/ms$, $N\ ms^+/ms$ and $N\ ms^+/ms^+$ all produce fertile pollen. The CMS system for producing F_1 hybrids has an advantage over the genic system for mass hybridization, because all of the female parent plants are male sterile compared to only 50% with the ms-gene method. (Greenleaf, 1986).

Three parent lines are required to produce F_1 hybrid seed by the CMS system:

The male sterile line, or A line ($S\ ms/ms$)

The maintainer line or B line ($N\ ms/ms$)

Fertility restorer or R line ($N\ ms^+/ms^+$) or ($S\ ms^+/ms^+$)

The male sterile line is maintained by hybridization in isolation with a maintainer line of the same cultivar. In the field, crossing by bees would need to be supplemented by hand-pollination during periods of maximum flowering. The commercial hybrid is produced by crossing the S ms/ms line with a good combiner line.

The major problems associated with the CMS system are instability of S ms/ms lines in fluctuating environments and deleterious effects on growth and fruit setting. Further improvement of CMS lines (now variable in pollen sterility), should be sought by (i) use of genetically different maintainer lines, (ii) incorportion of recessive marker genes to permit elimination of selfed plants in the seedling stage and (iii) production of F_1 hybrid seed under more constant high-temperature environments.

3. Hand-pollination

While crosses can be made any time during daylight hours, the best times are early morning or late afternoon when the flowers are in the mature bud stage and have not been disturbed by insects.

a) EMASCULATION

Select flowers that are at the mature bud stage and have not been disturbed by insects. Several flowers are usually emasculated and prepared for pollination at one time to speed up the process. The stamens alternate with the petals and correspond with them in number. The stamens are carefully removed with forceps. The hands and tools must be disinfected.

b) POLLINATION

The stigma of the emasculated flower is first carefully examined with a hand lens and, if found contaminated with pollen is pinched off. If the flower has already been visited by bees, the pollen will appear evenly distributed over the stigma and the flower must be discarded. Earlier pollen penetration, if any, is evident by swelling and puffiness. After checking that the flower is free from either pollen contamination or pollen penetration, the flower may be pollinated. Pollen from the male parent is gently transferred to the stigma either from matured undehisced anthers by scooping it out through the lateral sutures with a spear needle, or by touching a freshly dehisced anther to the stigma with forceps. Pollinated flowers are marked by loosely tying coloured thread around the delicate pedicels, preferably enclosing a leaf petiole for protection. Pollinated flowers must be protected from bees by a double layer of cheesecloth, loosely wrapped around the branch, enclosing leaves and flowers, and securely fastened. Pollinated flowers should be periodically checked and the cheesecloth removed in 4-6 days. Fruits should mature in about 45 days.

Cucumber

Monoecism and Sex Reversal

Cucumber is a monoecious plant. Staminate, pistillate, and hermaphroditic flowers occur in various arrangements, yielding several types of sex expression. Furthermore, these types are greatly influenced by environmental conditions, producing a virtually continuous spectrum of sex expression.

In monoecious forms, the succession of flower types may vary with location on the plants. The main stem of monoecious varieties is typically characterized by three phases of sex expression, each of variable duration. Only the staminate flower occurs in the first phase. This is followed by a phase of irregularly alternating female, male or mixed nodes, and finally a phase of only pistillate nodes. Lateral shoots usually have stronger female tendencies, possibly even to the degree of bearing only pistillate flowers, while the main stem produces only staminate flowers.

The principle of sex reversal of lines, whether gynoecious, monoecious of androecious by exogenous application of growth regulators, is of tremendous advantage both in breeding and commercial hybrid seed production. Production of gynoecious x monoecious and gynoecious x gynoecious hybrids is contingent upon the ability to maintain and increase gynoecious parental stock. Only with the advent of growth regulators, such as silver nitrate, that promote male flowering has it become possible to self-pollinate genetically gynoecious plants and subsequently develop gynoecious inbred lines and maintain them, and to produce all female hybrids (gynoecious x gynoecious).

Methods of Hybrid Seed Production

(1) Hybrid cucumber is mostly produced by using a gynoecious[5] line as the female parent and a monoecious line as the male parent. (Lower and Edwards 1986). The resultant hybrid is heterozygous at the F locus and is not totally gynoecious but is predominantly female. A major advantage of gynoecious cucumbers is the potential for earlier harvest, which is important to assure a premium value for the crop.

$$m^+m^+ FF \qquad\qquad X \qquad\qquad m^+m^+ F^+ F^+$$
$$\text{gynoecious} \qquad\qquad\qquad\qquad \text{monoecious}$$

(2) Gynoecious sex expression, however, is greatly influenced by the environment. Because of this some hybrids have been developed from (gynoecious x hermaphrodite) x monoecious three-way cross. The gynoecious x hermaphrodite hybrid confers stability to the gynoecious sex expression

---

[5] The inheritance of female tendency is under genetic control. At least three major loci are involved. The F allele is partially dominant and intensifies female expression, but is subject to strong environmental influence and background genetic modification. The allele a intensifies male tendency, but is hyposatic to the F locus. The allele m exhibits non-specific development of staminate and pistillate flower parts resulting in hermaphrodite flowers. The genetic make-up of lines is gynoecious m^+m^+F/F, androecious F^+F^+aa, hermaphrodite $mmFF$ and monoecious $m^+m^+F^+F^+$. With the advent of growth regulators that promote male flowering it has become possible to self-pollinate genetically gynoecious plants and subsequently developing gynoecious inbred lines.

$$m^+m^+\ FF \quad \times \quad mm\ FF \qquad X \qquad m^+m^+\ F^+F^+$$

gynoecious	hermaphrodite		monoecious

of the seed parent and simplifies increase of the seed parent stock. Three-way crosses, however, may be less uniform for flowering and fruit production.

(3) Another scheme used to ensure all-female hybrids involves releasing gynoecious × hermaphroditic hybrids.

$$m^+m^+\ FF \qquad\qquad X \qquad\qquad mm\ FF$$

gynoecious	hermaphrodite

Such hybrid varieties have expressed extremely stable sex expression but require blending seed of the monoecious pollinator with the hybrid seed to ensure adequate fruit set in commercial production.

(4) The present trend is to produce all-female hybrids involving two gynoecious lines. This has become possible with the advent of silver compounds. Spraying

$$FF \quad \times \quad FF$$

rows of the gynoecious line with silver nitrate ($AgNO_3$) induces sufficient pollen production of the gynoecious line to serve as male parents in gynoecious × gynoecious hybrids.

(5) Alteration of sex expression in field-scale hybrid production involving monoecious × monoecious lines is also tenable. The hybrids may be produced without contamination due to undesired self- or sib-pollination, by spraying the designated female inbred regularly with ethephon[6] beginning at an early developmental stage.

Hand-pollination

The cucumber is an insect-pollinated crop. If natural pollen vectors can be excluded, for example in greenhouses, one need not worry about measures to prevent selfing or other pollen contamination because of diclinous nature of cucumbers and the stickiness of pollen to its source flower. There is no wind dissemination of pollen. Pistillate flowers are receptive in the morning or up to midday on the day they open. Hot, dry conditions reduce pistil receptivity and pollen viability. Development of fruit generally proceeds according to the sequence of pollination and the first fertilized flower inhibits development of subsequent fruit. For this reason controlled pollinations should be done as soon as possible soon after flowering begins.

In outdoor plantings the male and female flowers to be used in pollination must be selected and identified the day before they open and covered with half of a size 000 gelatin capsule or a section of wire tie to prevent them from opening. The following day these flowers are relocated and pollination is conducted.

---

[6] The synthetic ethylene-releasing compound 2-chloroethyl phosphonic acid (ethephon) has been found to provide an effective means of female flower promotion in both monoecious and andromonoecious varieties.

POLLINATION

The newly opened male parent flower (greenhouse conditions) or identified male flower (outdoor plantings) should be removed, keeping a portion of the pedicel attached for easy handling. The corolla may be gently torn from the point of attachment with the sepals by grasping it between the thumb and forefinger of each hand. By using one hand to grasp the calyx, the corolla may then be removed by tearing it from its attachment all the way around the calyx. This leaves only the pedicel and the calyx with the stamens protruding. The pedicel may then be used as a handle to transfer pollen from the staminate flower, stamens first, onto the pistil of the open female flower. When in this position, gently rotating the male flower helps to dislodge pollen onto the pistil to assure thorough pollination. The male flower may be left in contact with the pistil. The female flower must be covered or closed after pollination to prevent subsequent contamination. This can be accomplished with a wire tie. Tagging of the female flowers completes the procedure.

Water-melon

1. Seedless Triploid Hybrids

Seedless triploids, though a notable plant-breeding achievement, are presently considered to be non-feasible on economic considerations. Also, production of seedless fruit by triploid plants in the grower's field is dependent upon the pollination of flowers of the triploids with pollen from diploid plants. The diploid pollen stimulates parthenocarpy, but the ovule fails to develop because of sterility accompanying the triploid condition. Thus a diploid cultivar would have to be interplanted with the seedless hybrid. An additional disadvantage to growers of seedless melons is the difficulty in germinating the seeds.

METHOD OF SEED PRODUCTION OF TRIPLOID HYBRIDS

The method of producing seedless triploid hybrids is shown in Fig. 32.4.

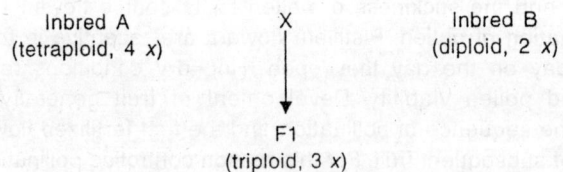

Fig. 32.4. Method of producing seedless triploid hybrids.

To produce triploid seeds a tetraploid line is first established by giving colchicine treatment to induce polyploidy in a diploid line. The tetraploid line must be used as a female parent because the reciprocal cross ($2x$ x $4x$) using the diploid as a female results in empty seeds.

Both the tetraploid and diploid lines are maintained by selfing. The tetraploid lines produce a much smaller number of seeds than diploids. This also makes it expensive to maintain the tetraploid lines.

2. Production of Diploid Hybrid Water-melon Seeds by Hand-pollination

F_1 hybrid seed may be produced by hand-pollination and by hand-removal of male blossoms. It has been experimentally established that the cost of producing F_1 hybrid seed in this manner is not prohibitive since each fruit produces a very large number of seeds (approx. 225 seeds). F_1 hybrid is produced by crossing two selected inbred lines, one as a female parent and the other as a male parent (pollen parent) (Mohr, 1986).

HAND-POLLINATION TECHNIQUE

1) The female parent and male parent are planted in separate fields.
2) At flowering, in the female parent field the unopened buds of selected pistillate flowers which normally occur on every seventh leaf axil are protected from insect visitation. This is done by placing small screen cages over the selected flowers or by preventing the buds from opening with paper clips or scotch tape. The ovary, which is inferior, makes pistillate flowers readily apparent. The relative size of the ovary is an important factor in fruit setting and hence flowers with large ovaries are selected. These are generally found near the tips of the most vigorous branches of a plant. Staminate flowers in the intervening six-leaf axils are removed.
3) For hand-pollination, the staminate flowers from the male parent are plucked. Usually the anthers have dehisced when the corolla expands and the pollen is visually evident in sticky masses adhering to the anther.
4) Though the stigma is receptive throughout the day, fruit setting is reported to be much higher if hand-pollinations are done between 6:00 and 9:00 a.m. rather than later in the day. High atmospheric humidity favours fruit setting.
5) Petals of the staminate flowers from the male parent are bent until they break, leaving the stamens protruding prominently. Holding this flower by the pedicel, the anthers with their masses of sticky pollen are lightly brushed against the stigmatic surface of the pistillate flower.
6) After hand-pollination the protection must be replaced and maintained for at least one day.
7) Until such time that hand-pollinations desired on a plant have not been completed and the resultant fruits are set, all the open-pollinated fruit should be continuously removed. This is necessary because there is an inhibitory influence produced by fruit already set that reduces further fruit setting.

A helpful procedure for locating flowers to be used in controlled pollinations is the placement of plastic flags at each flower to be used. Different colours can be used to designate the location of developing fruits, flowers to provide pollen, and pistillate buds not yet opened. Tags should be attached to the pedicel of the hand-pollinated pistillate flowers indicating the date of hand-pollination.

Squash

1. Hand-removal of Male Flowers (Manual Defloration)

This method is in common use today. Several rows of female parent inbred for each row of male parent inbred are grown in isolated fields. Seed fields are isolated from contaminants at least by 1500 and 1000 m for foundation and certified seed classes respectively. The male flowers are not allowed to open on the female parent. Pollination is done by bees bringing pollen from the male inbred grown in adjoining rows. Hives of honey-bees are often placed in seed production fields to ensure pollination.

Plant breeders have succeeded in developing inbred lines with a high degree of female sex expression, that is, producing a high ratio of female to male blossoms. Use of such lines as the female parent when available reduces the labour of removing male flower buds from the female parent (Whitaker and Robinson, 1986). Rows of the pollen parent are cut before seed harvest, to avoid seed admixture.

2. By Using Sex Suppression through Endogenous Chemicals

In this system one row of staminate inbred line is planted earlier than the intervening two rows of the pistillate inbred parent in isolated fields. Young seedlings of the female inbred parent are treated with ethephon (250 ppm)[7] to inhibit staminate flower formation. Pollination is done by bees. Hives of honey-bees are placed in seed production fields to ensure pollination. (Robinson et al., 1970; Rudich et al., 1970).

3. By Using Genetic Male Sterility

Two different single recessive genes for male sterility in *Cucurbita pepo* have been reported. Male sterility eliminates the need for emasculation or defloration in the production of hybrid seed, but it is necessary to remove 50% male fertile plants. Despite this, genetic male sterility is now being used to a limited extent for producing F_1 hybrid seed of *C. maxima*.

4. Hand-pollination

The male and female parent lines are planted separately. In the female parent line the pistillate flowers are located the afternoon prior to anthesis by the appearance of a slight touch of yellow at the apex of the corolla tube. They are prevented from opening by tying the tips of the corolla tube. Similarly, in the male parent inbred line staminate flowers are secured. The following morning, as soon as the pollen sacs dehisce, transfer of the pollen from anther to stigma can commence. Pollinations can be made from anthesis until about noon.

It is desirable to pollinate the first pistillate flowers to develop on a plant, since fruit set is usually better if the plant has not previously produced fruit. Open pollinated fruits should be removed. After pollination the pistillate flowers are labelled and tied shut to prevent insect visitations.

---

[7] Application of 250 ppm prevents the development of male flowers but has no effect on the production of pistillate flowers. Some workers have recommended the use of two applications of 400-600 ppm of ethephon for best results.

Brassicas

Uniformity in all respects, and especially for time of harvest, is the major advantage of hybrid varieties. Hybrid seed production is done in the following manner.

1. Use of Self-incompatibility

The self-incompatibility character is used for producing hybrids seeds of cabbage, cauliflower, broccoli, Brussels sprouts and kale. The incompatibility specificities are controlled by the S-gene. The homozygous S-allele genotype, for example S_1S_1; S_2S_2; S_3S_3 etc. have incompatibility specificities of S_1S_2 etc. for both their stigma (female) and pollen (male). Control of pollen specificity is by the sporohyte (sporophytic control) (Dickson and Wallace, 1986). Seed fields are isolated from the contaminants by at least 1600 metres.

a) SINGLE-CROSS HYBRIDS

Single plants of inbred lines that have been selected for homozygosity of an S-allele followed by selection for strong expression by this S-allele for self-incompatibility do not cross-fertilize each other. As a result there is little selfed seed. On the other hand, such an inbred S_1S_1 when planted in rows alternating with another inbred S_2S_2 in every other or every second or third row, or with both inbreds in alternating blocks of three or four rows, will be readily cross-fertilized. Fertilization is mostly done by bees. The cross-compatibility between inbreds S_1S_1 and S_2S_2 assures production of F_1 hybrid seed and also that the seed produced on both the inbreds will be identical. Thus commercial seed of the same F_1 hybrid can be harvested from both inbreds (Fig. 32.5).

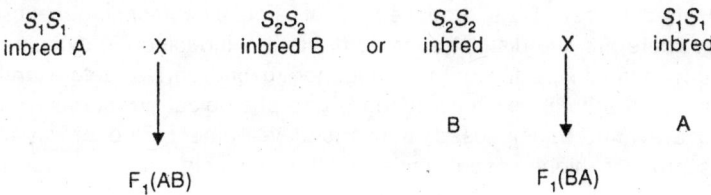

Fig. 32.5. Seeds from both the crosses A x B or B x A constitute a standard A x B single-cross F_1 hybrid

b) TOPCROSS HYBRIDS

For topcrosses an open-pollinated variety is used as the pollen parent and crossed with a single self-incompatible inbred line that is used as the female parent.

c) THREE-WAY AND FOUR-WAY CROSS

A single-cross F_1 such as A x B derived above is used as the seed parent. A three-way cross may use an inbred line as a pollen parent only, or if it is self-incompatible, the inbred may simultaneously be used as a seed parent and seed can be harvested from all the plants.

When a highly self-incompatible single cross C x D is used in conjunction with A x B, the resultant four-way hybrid seed can also be harvested from every plant in the field. This plus the larger seed production per plant due to hybrid vigour of both single-cross parents will give the lowest production cost for the hybrid seed.

d) THREE-WAY TOPCROSS HYBRIDS
An open-pollinated variety used as a pollen parent for crossing with a single-cross A x B will give three-way hybrids.

2. Use of Genetic Male Sterility

Genetic male sterility may be used in the commercial production of hybrids on a rather small scale. The male sterility system has certain advantages over self-incompatibility as far as inconsistent and variable proportion of sibs within a so-called F_1 hybrid variety is concerned. No sibs will be present when the hybrids are produced from male sterile plants.

So far only recessive genes for male sterility have been widely recorded. The drawback to hybrid production by this method is apparent at the flowering stage, since one-half of the plants of the female line will be male fertile (*Ms ms*) and must be removed as soon as they can be identified through the appearance of first flowers, which is a rather tedious job. The one way to overcome this difficulty is through the use of a marker gene which is linked to one or the other genotype. If present at the juvenile plant stage, removal of the unwanted types can be done early without difficulty but unfortunately no such marker gene (linked to that for male sterility) has yet been reported.

Carrot

Hybrid seed of carrots is produced by utilising cytoplasmic-genetic male sterility. There are two distinct sources of sterility inducing (*S*) cytoplasm. In the *brown anther type* the anthers degenerate and shrivel before anthesis while *petaloid steriles* exhibit a range of morphological structures. Petaloid steriles are more widely used in hybrid development. The brown anther steriles produce higher seed yields but the available ones are not stable compared to the petaloid type. At least two and probably three duplicate dominant maintainer genes and an epistatic restorer gene are involved. The useful maintainer line would therefore have to be free of the restorer and homozygous dominant at one of the *MS* loci.

The characteristically low seed yields from many cytosterile inbreds used as female parents in single-cross hybrids have encouraged breeders and seedsmen to adopt three-way hybrids, in which male sterile F_1 hybrids are used for seed parents. An obvious but essential requirement for hybrid seed parents is that roots of the inbred components must be as nearly identical as possible in visible characteristics, such as shape, length and colour, to minimize segregation in the three-way hybrid. (Peterson and Simon, 1986).

Commercial Seed Production

Most commercial production of hybrid seed is done by the seed-to-seed method. However, the parent stocks are increased by root-to-seed method for reasons of ensuring higher genetic purity of seeds. Elite seed stock is increased from carefully selected roots under screen isolation. This cage-grown seed is adequate for outdoor root-to-seed production of the stock seed for commercial seed fields. Approximately 100 plants grown under screen from carefully selected roots produce enough seed for 2 ha of root-to-seed increase. Only one outdoor root-to-seed cycle should be used for increasing the component lines of the hybrid variety. To ensure uniformity, the roots for this one cycle should be rogued for off-types that result from admixtures of outcrosses.

In hybrid seed production fields, the female and male parent lines are planted in 4:1 or 8:2 ratio. The seed fields are isolated from contaminants by at least 1000 and 800 metres for foundation and certified seed classes respectively. Roguing for pollen shedders and off-types is done during the flowering period in the usual manner. For good pollination and subsequent seed setting, keeping of beehives may be advantageous.

Onion

Hybrid onion seed is produced by utilizing cytoplasmic-genetic male sterility involving the use of male sterile line (A line), its maintainer line (B line) and a 'C' line, that is, an unrelated inbred used as pollen parent in hybrid seed production. The most obvious reason for making hybrid onions is to obtain uniformity of the onions produced and to control the date of maturity. In some instances seedling vigour and yields may be improved. Other reasons might be disease resistance, insect resistance, or environmental stress resistance.

Asparagus (*Asparagus officinalis*) - All-Male Hybrids

Asparagus (*Asparagus officinalis*) is *dioecious*, with male and female flowers borne on different plants in about equal numbers. A single gene factor controls the sex expression, the dominant expression *M-* is male and recessive *mm* is female. Hermaphrodite plants are occasionally found, however, which have a preponderance of normal male flowers plus a few androgenous flowers with small but functional pistils. Such plants are known as *andromonoecious*.

The male hybrids of asparagus are developed by using *andromonoecism*. The male hybrids are produced by crossing selected female lines (*mm*) with selected male lines (*MM*) which are called supermales. Homogametic (*MM*) supermales are distinguished from heterogametic males (*Mm*) by a sex genotype progeny test. The progeny of a supermale is all male, while the progeny of a heterogametic male segregates into male and female in a 1:1 ratio (Ellison, 1986).

Advantages of Male Asparagus Hybrids

There are many advantages of male hybrids

1. Male asparagus hybrids live longer and yield more spears than dioecious asparagus.
2. They mature relatively early in the spring season.

3. There is no seed to make seedling weeds in the field.
4. There is no seed to compete with the storage roots for photosynthate.
5. There is no opportunity for growers to save seed that may deteriorate the genetic purity and quality of the hybrids.

Method of Developing Male Hybrids

Figure 32.6 illustrates the method of selecting a seed parent among female plants derived from a self-pollinated andromonoecious male plant. Desirable characters that should be considered during selection are increased disease resistance and increased uniformity in the F_1 progeny. The low-vigour S_1 females are discarded. Careful progeny testing of selected S_1 seed parents is done to avoid the berry-bearing character in F_1 male hybrids. The seed parents are not selected in subsequent selfed generations for such reasons, as poor vigour and higher probability of berry-bearing F_1 males than expected from S_1 seed parents. The general and specific combining ability is evaluated through diallel crossing between several selected female and supermale plants. In crosses the tendency to transmit andromonoecism is observed along with other desirable characters.

Parent material is maintained through cloning (tissue culture) while the progenies are being evaluated, which may take 5-10 years. Currently used systems depend on meristem or bud culture and are known to yield genetically stable explants. Callus that may develop is carefully trimmed away. The new explants are field planted in a holding block where they can be kept indefinitely. The male clone can be managed best in pots.

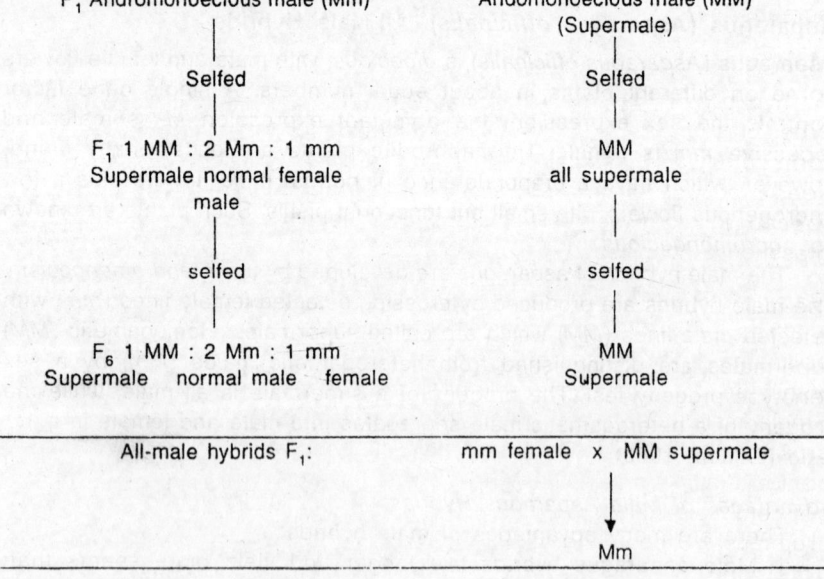

Fig. 32.6. Scheme for inbreeding andromonoecious asparagus plants for production of male hybrids (from Ellison, 1986).

Seed Production

The parents for commercial seed production are cloned through commercial tissue-culture laboratories. Care must be taken that only the parent clones are included in the seed block. A convenient spacing of plants in the seed block is 150 cm between rows and 60 cm between plants of the row. One row of male is needed for every set of four female rows. Planting should start with a male row, then four female rows and so on across the field.

Roguing: Seedling volunteers must be removed before they can flower and contaminate the source of pollen. Besides, take adequate care to control morning-glory, an objectionable weed, and it is necessary to rogue it out.

Isolation: 300 metres

General Precautions
   (i) Do not harvest the spears to obtain maximum seed yields.
   (ii) Provide supplementary pollination by keeping beehives (5-10 hives per ha).
   (iii) Effectively control asparagus beetles and keep them away from seed fields.
   (iv) Harvest only red ripe berries. Do not harvest very late brush.

Harvesting

Asparagus seed is harvested by crushing the berries. Care must be taken not to damage the seeds. The crushed berries are washed with water to separate the berry skins and pulp from the seed and later spread out on screens to dry overnight.

Marigolds

F_1 hybrid marigolds are produced by using male sterility. There are two types of male steriles, namely, *apetalous* form (stamenless), and true double-form (produces no anther in the central part of the flower). Commercial seed production mainly involves use of the apetalous form of male sterility (because of its greater reliability) and ease in hand-pollination under glasshouse conditions.

Raising of Male Sterile Plants

The seed (that will give rise to male sterile plants) is produced by pollinating male steriles of an inbred line with male fertiles which appear within the same line. These male fertiles act as *restorers* of varying efficiency. Single male-fertile plants within a predominantly male-sterile line may, in crosses with male-sterile inbreds give offspring where only one or two per cent are male sterile. Other single plants have given offspring with up to 50 per cent male-sterile plants. For this reason, the commercial seed producing crop must be rogued heavily to remove male fertiles before plants (raised in pots) are transferred to their final seeding positions. This is easily accomplished by noting bud shape.

Production of Hybrid Seed

The optimum ratio of rows of male-sterile line plants to those of the male-fertile inbred is 3:1. The male parent should be sown about two weeks before the females in order to provide an adequate pollen supply. The first male-sterile flowers can be removed since this encourages new vegetative and bud growth. It is most common for the male-fertile line to be more uniform in habit than the male sterile and removal of any off-types should be straight-forward. Any variation within the female (male-sterile) is usually marked in the F_1 generation.

Pollination of male-sterile plants is achieved by taking flowers from male plants, inverting and twisting them lightly over the stigmatic surface of the female. Each male-sterile flower bears up to 300 styles which mature over a period of two to three weeks and pollinations should therefore be repeated at least twice a week. In practice, on a large scale, male flowers would be constantly picked and used as pollinators over all open male-sterile flowers so that repeat pollinations become a matter of course.

If pollen supply runs short, the blooms removed from the male plants can be stood in water until the next pollination period, by which time their immature anthers will again be shedding further pollen. With a good pollen parent, one detached flower should pollinate five or six male-sterile flowers. A more sophisticated method involves collecting pollen from the male parents through a vacuum pump. Pollen, deposited in a glass tube, can then be dusted onto the male steriles. Seed from the male steriles should be harvested in its capitulum as it ripens (usually six to eight weeks after the first pollination). Only the black seeds (compared with grey or light brown) within the capitulum will be viable and all should produce perfectly formed F_1 plants.

Petunias

Hybrid seed is produced in any of the following ways.

Use of Self-incompatibility

Petunias have a straight-forward gametophytic incompatibility system. Thus, an inbred population of plants that would not self- or cross-fertilize each other may be built up. Plants of one inbred line can be pollinated, without emasculation, by plants of another inbred line which carry a different S-allele. A reciprocal cross can also be made and large quantities of seed produced. The choice of this method, however, depends on the reliability of the incompatibility system throughout the pollinating season.

Hand-pollination

Another method (sib mating) which does not require much intensive inbreeding can be used. A group of up to about eight selected plants can be mass pollinated and the process repeated with smaller numbers for three to four generations until a very uniform line is produced with no loss of vigour and no self-incompatibility. From this point, F_1 hybrid seed is produced by crossing two such lines after emasculation of the female plants in the bud stage. Emasculation and pollination of a large number of buds can be carried out rapidly at the rate of at least five to six buds per minute giving over 1000 seeds. The F_1 hybrids of petunia are usually produced commercially by hand-pollination under controlled glass-house conditions.

Production of True Potato Seed (TPS)

Climatic Requirements

Potato usually requires long-day conditions for flowering. High temperature and low relative humidity conditions adversely affect flowering. TPS seed production is therefore done in areas where weather during blooming is cool and humid. Under short day conditions TPS seed production may be successfully done by providing artificial illumination (200 W incandescent bulbs can be used as source of illumination) up to 10:00 p.m., coupled with GA_3 sprays (50 ppm) at 20 and 30 days after planting for induction of flowering.

Seed-production fields should not be infested with wart- or cyst-forming nematodes, brown rot or non-cyst-forming nematodes (within last three years), and common scab. The seed fields should be isolated from other variety fields or fields of the same variety not conforming to varietal purity requirements of certification by at least 50 m. The blocks of parental lines must be isolated at least 5 m. The healthy and pure seed (tubers, clones) of parental lines should be obtained from an approved source for planting seed fields. Female parent and male parent are planted in separate blocks. One week prior to sowing the female block, the male block is planted in three phases at weekly intervals. This is done to ensure abundant pollen availability during the period the female parent is in bloom. Spacing of the female parent is kept at 50 x 15 cm and that of the male parent at 60 x 20 cm. Use of cut tubers (25-30 gm) having atleast one eye is more advantageous than planting whole tubers. Planting may be done in the usual manner. Brick planting gives much higher seed yields, however.

Roguing should be done in male and female blocks to remove off-types and otherwise undesirable plants. Flower bunches in the female parent are trimmed to retain six buds per bunch. This ensures synchronous bud and berry development and reduction in bud abscission; besides, the time to complete artificial pollination is reduced to 48 hours.

Pollination

a) *Pollen collection*: Flowers from the male parent with corolla open and anthers at the verge of dehiscence are collected and stored overnight at room temperature. Next morning, pollen is extracted either by manually shaking the anther cones using forceps, or using a vibratory device. If need be, the collected pollen may be dried on silica gel placed in a desiccator and stored at 6-10°C for subsequent use.

b) *Pollination*: Trimmed bunches in the female parent are hand-pollinated when the stigmas become receptive. Every stigma is pollinated a number of times (usually three) at an interval of 8 or more number of hours. Repeated pollination gives higher seed yields.

Seed Collection

Berries from the female parent are collected 45 days after hand-pollination.

Seed Extraction

Berries are macerated in many different ways. Whatever the method used, the important point is seeds should not be damaged during the process. A hand-operated reverse juice extractor gives satisfactory results. The macerated berries and pulp mass are continuously stirred in 10% hydrochloric acid (HCl) for 20 minutes: thereafter the pulp is decanted and the seed is washed 3-4 times so that there are no traces of acid on the seeds. The washed seeds are dried in thin layers under shade. The usual practice is to thinly spread seeds on muslin stretched over a wooden frame in an airy shed. The seeds are subsequently dried to 5-7% moisture content, cleaned and stored usually in vapour-proof containers for better storage life.

References

Abel, P.P., Nelson, R.S., De, B., Hoffmann, N., Rogers, S.G., Fraley, R.T. and Beachy, R.N. (1986). Delay of disease development in transgenic plants that express the tobacco mosaic virus coat protein gene. *Science*, 232, 738-743.

Agrawal, R.L. (1995). *Seed Technology*, Oxford & IBH Publishing Co., New Delhi (2nd ed.).

Alcala, J., Giovannoni, J.J., Pike, L.M. and Reddy, A.S. (1997). Application of Genetic Bit Analysis (GBA™) for allelic selection in plant breeding. *Molecular Breeding*, 3, 492-502.

Allan, R.E. (1980). Wheat, In: *Hybridization of Crop Plants*, pp. 709-720. W.R. Fehr, H.H. Hadley (eds.). ASA and CSSA, Madison, Wisc., USA.

Allard, R. W. (1960). *Principles of Plant Breeding*. John Wiley, New York.

Allison, J.C.S. and Curnow, R.W. (1966). On the choice of tester parent for the breeding of synthetic varieties of maize (*Zea mays* L.). *Crop Sci.*, 6, 541-544.

Anderson, E. and Dewinton, D. (1931). The genetic analysis of an unusual relationship between self-sterility and self-fertility in *Nicotiana*. *Ann. Missouri Bot. Gard.*, 18, 97-116.

Anonymous (1961). *Agricultural and Horticultural Seeds: Their Production, Control and Distribution*. FAO Agriculture Studies No. 55, Rome.

Asana, R.D. (1957). The problem of assessment of drought resistance in crop plants. *Ind. J. Genet. & Pl. Breed.*, 18, 370-378.

Asana, R.D. (1966). Physiological analysis of yield of wheat in relation to water stress and temperature. *J. Post Graduate School*, I.A.R.I., New Delhi, 17-32.

Ashby, E. (1930). Studies in the inheritance of physiological characters I. A physiological investigation of nature of hybrid vigor in maize. *Ann. Botany*, 44, 457-467.

Association of Official Seed Certifying Agencies, U.S.A. (1971). AOSCA Certification Handbook. Publication No. 23.

Auckland, A.K. and Van Der Maesen, L.J.G. (1980). Chickpea. In: *Hybridization of Crop Plants*, pp. 249-259. W.R. Fehr and H.H. Hadley (eds.). ASA and CSSA, Madison, Wisc., USA.

Barnes, D.K. (1980). Alfalfa. In: *Hybridization of Crop Plants*, pp. 117-187. W.R. Fehr and H.H. Hadley (eds.). ASA and CSSA, Madison, Wisc., USA.

Basu, A.K., Narayanan, S.S. and Bhatt, M G. (1995). Hybrid cottons : Historical development, present status and future perspective in Indian context. In: *Hybrid Research and Development*, pp. 63-80. M. Rai and S. Mauria (eds.). Indian Society of Seed Technology, New Delhi.

*Battaglia, E. (1963). Apomixis, In: *Recent Advances in Embryology of Angiosperms*, pp. 221-264 P. Maheswari (ed.) Intern. Soc. Plant Morphologists, Univ. of Delhi, India.

Beard, B.H. and Comstock, V.E. (1980). Flax. In: *Hybridization of Crop Plants*, pp. 357-366. W.R. Fehr and H.H. Hadley (eds.). ASA and CSSA Madison. Wisc., USA.

Bicar, E.H. and Darvey. N.L. (1997). Development of the components of a cytoplasmic male sterility hybrid system in rye through anther culture. *Euphytica*, 97, 151-160.

*Binns, A.N. (1990). Agrobacterium-mediated gene delivery and the biology of host range limitations. *Physiol. Plant*, 79, 135-139.

Bliss, F.A. and Gates, C.E. (1968). Directional selection in simulated populations of self-pollinated plants. *Aust. J. Biol. Sci.*, 21, 705-719.

Blum, A.(1988). *Plant Breeding for Stress Environments*, CRC Press, Boca Raton.

*Bolanos, J. and Ed Meades, G.O. (1989). *Combios en la poblacion tuxpeno despues de oeho de mejoramiento, para resistencia a sequia*. Meeting at San Pedro Sula, 2-9 April, 1989. PCMMA

*Bos, I. (1981). The relative efficiency of honey-comb selection and other procedures for mass selection in winter rye (*Secale cereale* L.). Dr. Thesis. Agriculture University, Wageningen.

Brewbaker, J.L. (1957). Pollen cytology and self-incompatibility systems in plants. *J. Hered.*, 48, 271-277.

Brewbaker, J.L. (1959). Biology of the angiosperm pollen grain. *Ind. J. Genet. &. Pl. Breed.*, 19,121-133.

Briggs, F.N. and Knowles, P.F. (1967). *Introduction to Plant Breeding*. Reinhold Publishing Corp., New York.

Brim, C.A. (1966), A modified pedigree method of selection in soybeans. *Crop Sci.*, 6, 220.

Brim, C.A. and Cockerham, C.C. (1961). Inheritance of quantitative characters in soybeans. *Crop Sci.*, 1,187-190.

Broertjes, C. and Van Harten, A.M. (1978). *Application of Mutation Breeding Methods in the Improvement of Vegetatively Propagated Crops. Developments in Crop Science*. 2. Elsevier Scientific Publishing Co., Amsterdam.

Brown, J.A.M. and Ellis, S.E. (1976). Breeding for resistance to cereal cyst nematode in wheat. *Euphytica*, 25, 73-82.

Brown, A.H.D., Marshall, D.R., Frankel, O.H. and Wiliams, J.T. (eds.) (1989). *The Use of Plant Genetic Resources*. Cambridge University Press, Cambridge.

Brown, C.M. (1980). Oats. In: *Hybridization of Crop Plants*, pp. 427-441 W.R. Fehr and H.H. Hadley (eds.). ASA and CSSA, Madison, Wisc., USA.

*Bruce, A.B. (1910). The Mendelian theory of heredity and the augmentation of vigor. *Science, N.S.*, 32, 627-628.

Brule-Babel, A.L. and Fowler, D.B. (1989). Use of controlled environments for winter cereal cold hardiness evaluation: Controlled freeze tests and tissue water content as prediction tests. *Can. J. Plant Sci.*, 69, 355-366.

Burton, G.W. and Forbes, J. Jr. (1961). The genetics and manipulation of obligate apomixis in common bahia grass (*Paspalum notatum* Flugge.) pp. 66-71 In: *Proc. VIIIth Int. Grassland Congress*, Reading, C.L. Skidmore (ed.). Alden Press, Oxford.

Burton, G.W. (1980). Pearl millet. In: *Hybridization of Crop Plants*, pp. 457-469. W.R. Fehr and H.H. Hadley (eds.). ASA and CSSA, Madison, Wisc., USA.

Burton, J.W, Penny, L.H., Hallauer, A.R. and Eberhart, S.A. (1971). Evaluation of synthetic populations developed from a maize population (BSK) by two methods of recurrent selection. *Crop Sci.*, 11, 361-367.

*Chen, Y. and Li, L.T. (1978). Investigation and utilization of pollen-derived haploid plants in rice and wheat pp. 199-211. In: *Proceedings of the Symposium on Plant Tissue Culture*, Science Press, Beijing, China.

Coffman, W.R. and Herrera, R.M. (1980). Rice. In: *Hybridization of Crop Plants*, pp. 511-522 W.R. Fehr and H.H. Hadley (eds.) ASA and CSSA, Madison, Wisc., USA.

*Collins, G.N. (1921). Dominance and the vigor of first generation hybrids, *Am. Natur.*, 55,116-133.

Compton, W.A. and Comstock, R.E. (1976). More on modified ear-to-row selection in corn. *Crop Sci.*, 16, 122.

Comstock, R.E. (1964). Selection procedures in corn improvement, *Proc. 19th Annu. Hybrid Corn Ind. Res. Conf.*, Washington, D.C., pp. 1-8.

Comstock, R.E. and Robinson, H.F. (1948). The components of genetic variance in populations of biparental progenies and their use in estimating the average degree of dominance. *Biometrics*, 4, 254-266.

Comstock, R.E. and Robinson, H.F. (1952). Genetic parameters, their estimation and significance. *Proc. Sixth Intern. Grassland Congr.*, pp. 284-291.

Comstock, R.E., Robinson, H.F. and Harvey, P.H. (1949). A breeding procedure designed to make maximum use of both general and specific combining ability. *Agron. Jour.*, 41, 360-367.

Creech, J.L. and Reitz, L.P. (1971). Plant germplasm for now and tomorrow. *Advances in Agronomy*, 23, 1-47.

Davey, M.R., Rech, E.L. and Mulligan, B.J. (1989). Direct DNA transfer to plant cells. *Plant Mol. Biol.*, 13, 273-285.

Dedio, W. and Putt, E.D. (1980). Sunflower. In: *Hybridization of Crop Plants*, pp. 631-644. W.R. Fehr and H.H. Hadley (eds.). ASA and CSSA, Madison, Wisc., USA.

Dewey, D.R. (1980). Some applications and misapplications of induced polyploidy. pp. 445-470. In: *Biological Relevance*, W.H. Lewis (ed.), Plenum Press, New York, London.

Dhawan, N.L. (1965). Breeding methodology for yield and other quantitative traits in maize. *Proc. Second Inter-Asian Maize Conference*, Los Banos, Philippines.

374

Dickson, M.H. and Wallace, D.H. (1986). Cabbage breeding. In: *Breeding Vegetable Crops*, pp. 396-432. W.J. Bassett (ed.) AVI Publishing Company Inc., Westport, Connecticut, USA.

Dixon, R.A. and Gonzales, R.A. (eds.) (1994). *Plant Cell Culture. A Practical Approach*. IRL Press, Oxford (2nd ed.).

Dixon, R.A. and Arntzen (1997). Transgenic plant technology is entering the era of metaboiic engineering. *TIBTECH*, 15, 441-444.

Donald, C.M. (1968). The breeding crop ideotypes. *Euphytica*, 17, 385-403.

Dowker, B.D. (1969). Field methods of assessing winter hardiness in peas. *Euphytica*, 18, 398-402.

Driscoll, C.J. (1972). X Y Z System of Production of hybrid wheat. *Crop Sci.*, 12, 516-517.

Driscoll, C.J. (1985). Modified X Y Z system of producing hybrid wheat. *Crop Sci.*, 25,1115-1116.

Driscoll, C.J. and Jensen, N.F. (1964). Characteristics of leaf-rust resistance transferred from rye to wheat, *Crop Sci.*, 4, 372-374.

Duncan, R.R., Bockholt, A.J. and Miller, F.R. (1981). Descriptive comparison of senescent and non-senescent sorghum genotypes. *Agron. J.*, 73, 849-853.

*East, E.M. (1929). Self-sterility. *Bibliogr. Genet.* 5, 331-368.

*East, E.M. (1936). Heterosis. *Genetics*, 21, 375-397.

*East, E.M. and Yarnell, S.H. (1929). Studies on self-fertility. *Genetics*, 14, 455-487.

Eberhart, S.A. and Russel, W.L. (1966). Stability parameters for comparing varieties. *Crop Sci.*, 6, 36-40.

Ellingboe, A.H. (1984). Genetics of host-parasite relations: an essay. In: *Advances in Plant Pathology*, vol. 2. D.S. Ingram, and P.H. Williams, (eds.). Academic Press, New York.

Elliot, F.C. (1958). *Plant Breeding and Cytogenetics*. McGraw-Hill, New York.

Ellison, J.H. (1986). Asparagus breeding. In: *Breeding Vegetable Crops*, pp. 523-569. M.J. Bassett (ed.) AVI Publishing Company Inc., Westport, Connecticut, USA.

Empig, L.T. and Fehr, W.R. (1971). Evaluation of methods for generation advance in bulk-hybrid soybean populations. *Crop Sci.*, 11, 51-54.

*Fagerlind, F. (1944). Is my terminology of the apomixtic phenomenon of 1940 incorrect and inappropriate? *Hereditas*, 30, 590-596.

Falconer, D.S. (1960). *Introduction to Quantitative Genetics*. Oliver & Boyd, Edinburgh.

*Fasoulas, A. (1973), A new approach to breeding superior yielding varieties. *Publication 4, Department of Genetics & Plant Breeding, Aristotelian University of Thessaloniki.*

Fehr, W.R. (1980). Soybean. In: *Hybridization of crop plants*, pp. 589-599, W.R. Fehr and H.H. Hadley (eds.), ASA and CSSA, Madison, Wisc., USA.

Fehr, W.R. (1980). Artificial hybridization and self-pollination. In: *Hybridization of Crop Plants*, pp.105-131 W.R. Fehr and H.H. Hadley (eds.). ASA and CSSA, Madison, Wisc., USA.

Finlay, K.W. and Wilkinson, G.N. (1963). The analysis of adaptation in a plant breeding program. *Aust. J. Agr. Res.*, 14, 742-754.

Flor, H.H. (1955). Host-parasite interactions in flax rust—its genetics and other implications. *Phytopathology*, 45, 680-685.

Fowler, D.B. and Carles, R.J. (1979). Growth, development and cold tolerance of fall-acclimated cereal grains. *Crop Sci.*, 19, 915-922.

Fowler, D.B. and Gusta, L.V. (1979). Selection for winter hardiness in wheat. I. Identification of genotype variability. *Crop Sci.*, 19, 769-772.

Fowler, D.B., Gusta, L.V. and Tyler, N.J. (1981). Selection for winter hardiness in wheat III. Screening methods. *Crop Sci*, 21, 896-901.

Frankel, O.H. (1957). The biological system of plant introduction. *Ind. Jour. Genet. & Pl. Breed.*, 17: 336-342.

Frankel, O.H. (1958). The dynamics of plant breeding. *J. Aus. Ins. Agri. Sci.*, 24, 112-123.

Frankel, R. and Galun, E. (1977). *Pollination Mechanisms, Reproduction and Plant Breeding.* Springer-Verlag, New York.

Freeman, G.H. and Perkins, J.M. (1971). Environmental and genotype-environmental components of variability VIII. Relation between genotypes grown in different environments. *Heredity*, 27, 15-23.

Fuchs, M., McFerson, J.R., Tricoli, D.M. McMaster, J.R., Deng, R.Z., Boeshore, M.L., Reynolds, J.F., Russel, P.F., Quemada, H.D. and Gonsalves, D. (1997). Cantaloupe line CZW-30 containing coat protein genes of cucumber mosaic virus, and watermelon mosaic virus-2 is resistant to these three virus in the field. *Molecular Breeding*, 3, 279-290.

Funk, C.R. and Han, S.J. (1967). *Recurrent intraspecific hybridization* : a proposed method of breeding Kentucky blue grass, *Poa pratensis*, N.J. Agic. Expt. Sta., Bull. 818, 3-14.

Fyfe, J.L. and Gilbert, N.E. (1963). Partial diallel cross. *Biometrics*, 19, 278-286.

Gardner, C.O. (1961). An evaluation of effects of mass selection and seed irradiation with thermal neutrons on yield of corn. *Crop Sci.*, 1, 241-245.

Gardner, C.O. and Eberhart, S.A. (1966). Analysis and interpretation of the variety cross diallel and related populations. *Biometrics*, 22, 439-452.

Gatehouse, A.M.R., Davison, G.M., Newell, C.A., Merryweather, A., Hamilton, W.D.O., Burgess, E.P.J., Gilbert, R.J.C. and Gatehouse, J.A. (1997). Transgenic potato plants with enhanced resistance to the tomato moth, *Lacanobia oleracea*: growth room trials. *Molecular Breeding*, 3, 49-63.

Genter, C.F. (1973). Comparison of S_1 and test cross evaluation after two cycles of recurrent selection in maize. *Crop Sci.*, 13: 524-527.

Genter, C.F. and Alexander, M.W. (1962). Comparative performance of S_1 progenies and test crosses of corn. *Crop Sci.*, 2, 516-519.

Grafius, J.E. (1963). Vector analysis applied to crop eugenics and genotype-environment interaction, pp. 197-213. In: *Statistical Genetics and Plant Breeding.*, Pub. 982, W.D. Hanson and H.F. Robinson (eds.), NAS-NRC, Washington D.C.

Grafius, J.E. (1965). Short cuts in plant breeding. *Crop Sci.*, 5, 377.

Grant, V. (1971). *Plant Speciation*, Colombia University Press, New York.

Greenleaf, W.H. (1986). Pepper breeding. In: *Breeding Vegetable Crops*, pp. 69-134. M.J. Bassett (ed.). AVI Publishing Company, Inc., Westport, Connecticut, USA.

Griffing, B. (1956). Concept of general and specific combining ability in relation to diallel crossing systems. *Australian Jour. Biol. Sci.*, 9, 463-493.

Gusta, L.V., Burke, M.J. and Kapoor, A.C. (1975). Determination of Unfrozen water in winter cereals at sub-freezing temperatures. *Plant Physiol*, 56, 707-709.

Gusta, L.V., Boyachek, M. and Fowler, D.B. (1978). A system of freezing biological materials. *Hort. Science*, 13, 171-172.

Haddad, N.I. and Muehlbauer, F.J. (1981). Comparison of random bulk population and single-seed-descent methods for lentil breeding. *Euphytica*, 30, 643-651.

Hadley, H.H. and Openshaw, S.J. (1980). Interspecific and intergeneric hybridization. In: *Hybridization of Crop Plants*, pp.133-159. W.R. Fehr and H.H. Hadley (eds.). ASA and CSSA, Madison Wisc., USA.

Hall, O.L. (1954). Hybridization of wheat and rye after embryo transplantation. *Hereditas*, 40, 453-458.

Hallauer, A.R. (1967). Development of single cross hybrids from two-eared maize populations. *Crop Sci.*, 7, 192-95.

Hallauer, A.R. (1970). Genetic variability for yield after four cycles of reciprocal recurrent selection in maize. *Crop Sci.*, 10, 482-485.

Hallauer, A.R. (1975). Relation of gene action and type of tester in maize breeding procedures. *Proc Ann. Corn Sorghum Conf.*, 30, 150-165.

Hallauer, A.R. and Eberhart, S.A. (1970). Reciprocal full-sib selection. *Crop Sci.*, 10, 315-316.

Hanson, W.D. and Johnson, H.W. (1957). Methods for calculating and evaluating a general selection index obtained by pooling information for two or more experiments. *Genetics*, 42, 421-432.

Hanson, W.D. and Robinson, H.F. (1963). Statistical Genetics and Plant Breeding. Pub. 982, NAS-NRC, Washington D.C.

Harrington, J.B. (1952). *Cereal Breeding Procedures*. FAO, paper 28, Rome.

Hawkes, J.G. (1983). *The Diversity of Crop Plants*. Harvard University Press, Cambridge, Mass.

Hayes, H.K., Immer, F.R. and Smith, D.C. (1955). *Methods of Plant Breeding*. McGraw-Hill, New York.

Hayman, B.I. (1954). The theory and analysis of diallel tables. *Biometrics*, 10, 235-244.

Hayman, B.I. (1954). The theory and analysis of diallel crosses. *Genetics*, 39, 789-809.

Hayman, B.I. (1958). The separation of epistatic from additive and dominance variation in generation means. *Heredity*, 12, 371-390.

Hayman, B.I. (1958). The theory of diallel crosses, II. *Genetics*, 43, 63-85.

*Hayman, B.I. (1960). The separation of epistatic from additive and dominance variation in generation means, II. *Genetica*, 31, 133-146.

Hayman, B.I. (1960). The theory of diallel crosses, III. *Genetics*, 45, 155-172.

Hayman, B.I. and Mather, K. (1955). The description of genic interactions in continuous variation. *Biometrics*, 11, 69-82.

Henderson, C.R. (1963). Selection index and expected genetic advance. In: *Statistical Genetics and Plant Breeding*, Pub. 982, W.D. Hanson and H.F. Robinson (eds.)., NAS-NRC, Washington D.C.

Hittle, C.N. (1954). A study of the polycross testing technique as used in the breeding of smooth bromegrass. *Agron. Jour.*, pp. 46, 521-523.

*Höfte, H., De Greve, H., Seurinck, J., Jansens, S., Mahillon, J., Ampe, C., Vandekerckhove, J., Vanderbruggen, H., Vanmontagu, M., Zabeau, M. and Vaeck, M. (1986). Structural and functional analysis of a cloned delta endotoxin of *Bacillus thuringiensis berlinger* 1715. *Eur. J. Biochem.*, 161, 273-280.

*Hopkins, C.G. (1899). Improvement of the chemical composition of corn, kernel pp, 205-240. *Bull. 55, Agr. Exp. St., University of Illinois*, USA.

Horner, E.S.,. Lundy, H.W., Lutrick, M.C. and Chapman, W.H. (1973). Comparison of three methods of recurrent selection in maize. *Crop Sci.*, 13, 485-489.

Horner, T.W., Comstock, R.E. and Robinson, H.F. (1955). Non-allelic gene interactions and the interpretation of quantitative data. N.C. Agr. Exp. Sta. Tech Bull. 118, Raleigh, N.C.

Hubert, C.M. (1986). Watermelon breeding. In: *Breeding Vegetable Crops*, pp. 37-66. M.J. Bassett (ed.) AVI Publishing Company Inc., Westport, Connecticut, USA.

Hull, F.H. (1945). Recurrent selection for specific combining ability in corn. *J. Amer. Soc. Agron.*, 37, 134-145.

Jain, S.K. (1961). Studies on breeding of self-pollinated cereals. The composite cross bulk population method. *Euphytica*, 10, 315-324.

James, N.I. (1980). Sugar-cane. In: *Hybridization of Crop Plants*, pp. 617-629. W.R. Fehr and H.H. Hadley (eds.). ASA and CSSA, Madison, Wisc., USA.

Jenkins, M.T. (1934). Methods of estimating the performance of double crosses in corn, *J. Amer. Soc. Agron.*, 26, 199-204.

Jensen, N.F. (1952). Intravarietal diversification in oat breeding. *Agron. Jour.*, 44, 30-34.

Jinks, J.L. (1954). The analysis of continuous variation in diallel cross of *Nicotiana rustica* varieties. *Genetics*, 39, 767-788.

Jinks, J.L. (1955). A survey of genetic basis of heterosis in a variety of diallel cross. *Heredity*, 9, 223-239.

Jinks, J.L. and Jones, R.M. (1958). Estimation of the components of heterosis. *Genetics*, 43, 223-234.

*Johansen, W.L. (1903). *Ueber Erblichkeit in Populationen und in reinen Lenein, Gustav Fisher*, Jena.

Jones, D.F. (1917). Dominance of linked factors as a means of accounting for heterosis. *Genet.*, 2, 466-479.

Jones, L.P., Compton, W.A. and Gardner, C.O. (1971). Comparison of full-sib and half-sib reciprocal recurrent selection. *Theor. & Appl. Genet.*, 41, 36-39.

Joshi B.S., Singh, H.B. and Gupta, P.S. (1958). Studies in hybrid vigour-III. *Bhindi. Ind. Jour. Genet. & Pl. Breed.*, 18, 57-68.

Jugenheimer, R.W. (1958). *Hybrid Maize Breeding and Seed Production*. FAO, Rome.

Julen, G. (1956). Practical aspects on tetraploid clover. *Seventh International Grassland Congress* (II), 1-18.

*Keeble, F. and Pellew, C. (1910). The mode of inheritance of stature and of time of flowering in peas (*Pisum sativum*). *Jour. Genetics*, 1, 47-56.

Kempthorne, O. (1955). The correlation between relatives in inbred populations. *Genetics*, 40, 681-691.

Kempthorne, O. (1956). The theory of diallel crosses. *Genetics*, 41, 451-459.

Kempthorne, O. (1957). *An Introduction to Genetic Statistics*, John Wiley, New York.

Kempthorne, O. and Curnow, R.W. (1961). The partial diallel cross. *Biometrics*, 17, 229-250.

Kihara, H. (1951). Triploid watermelons. *Proc. Amer. Soc. Hort. Sci.*, 58, 217-230.

Knott, D.R. (1965). The transfer of a gene for rust resistance from wheat chromosome 2 D to a chromosome in the A or B genome. *Canadian J. Genet. & Cytology*, 7, 354-355.

Knott, D.R. and Kumar, J. (1975). Comparison of early generation yield testing and single-seed descent procedure in wheat breeding. *Crop Sci.*, 15, 295-299.

Knott, D.R. and Sunshen, I. (1961). The inheritance of rust resistance to races 15 B and 56 of stem rust in eleven wheat varieties of diverse origin. *Can. J. Plant Sci.*, 41, 587-601.

Knox, R.B., Willing, R.R. and Asford, A.E. (1972). Role of Pollen-wall proteins as recognition substance in interspecific incompatibility in poplars. *Nature*, 237, 381-383.

*Knox, R.B., Willing, R.R. and Pryor, L.D. (1972). Interspecific hybridization in poplars using recognition pollen. *Silvae Genet.*, 21, 65-69.

Kosuge, T. and Nester, E.W. (eds.) (1984). *Plant-Microbe Interactions. Molecular and Genetic Perspectives, Vol. 1.* Macmillan Publishing Company, New York.

*Leclercq, P. (1966). *Une stérilité mâle Utilisable pour la production d' hybrides simples de tournesol.* Annalés de l' Amélioration des plantes, 16, 135-144.

Leppik, E.E. (1969). The life and work of N.I. Vavilov. *Economic Botany*, 23(2), 128-132.

Levitt, J. (1980). *Responses of Plants to Environmental Stresses*. Academic Press, New York.

Lewis, D. (1942). The evolution of sex in flowering plants. *Biol. Rev.*, 17, 46-47.

Lewis, D. (1949). Incompatibility in flowering plants. *Biol. Rev.*, 24, 472-496.

Lewis, D. (1954). Comparative incompatibility in angiosperms and fungi. *Advances in Genetics*, 6, 235-285.

Lewis, D. and Crowe, L.K. (1954). Structure of the incompatibility gene IV. Types of mutations in *Prunus avium* L., *Heredity*, 8, 357-363.

Lewis, D. and Crowe, L.K. (1958). Unilateral incompatibility in flowering plants. *Heredity*, 12, 233-256.

*Linskens, H.F. (1955). *Physiologische Untersuchungen der pollen schlauch Hemmung Selbsteriler Petunien. Z. Bot.*, 43, 1-44.

Linskens, H.F. (1961). Biochemical aspect of incompatibility, *Rec. Adv. Bot.*, 2, 1500-1503.

Li, C.C. (1974). *Population Genetics.* Boxwood Press, Pacific Grove, California, USA.

Lius, S., Manshardt, R.M., Fitch, M.M.M., Slightom, J.L., Sanford, J.C. and Gonsalves, D. (1997). Pathogen derived resistance provides papaya with effective protection against papaya ring spot virus. *Molecular Breeding*, 3, 161-168.

*Logemann, J., Schell, J. and Willmitzer, L. (1987). Improvement method for isolation of RNA from plant tissues. *Anal. Biochem.*, 163, 16-20.

Lonnquist, J.H. (1964). Modification of the ear-to-row procedure for the improvement of maize populations. *Crop Sci.*, 4, 227-228.

Lonnquist, J.H. and Gardner, C.O. (1961). Heterosis in intervarietal crosses of maize and its implications in breeding procedures. *Crop Sci.*, 1, 179-183.

Lonnquist, J.H. and Lindsey, M.F. (1964). Top cross versus S_1 line performance in corn (*Zea mays* L.). *Crop Sci.*, 8, 50-53.

Lower, R.L. and Edwards, M.D. (1986). Cucumber breeding. In: *Breeding Vegetable Crops*, pp. 173-207. M.J. Bassett (ed.). AVI Publishing Company, Inc., Westport, Connecticut, USA.

Markarian, D. and Andersen, R.L. (1966). The inheritance of winter hardiness in *Pisum* I. *Euphytica*, 15, 102-110.

Marshall, D.R. and Brown, A.H.D. (1973). Stability of performance of mixtures and multilines. *Euphytica*, 22, 405-412.

Mather, K. (1944). Genetic control of incompatibility in angiosperms and fungi. *Nature*, 153, 392-394.

Mather, K. (1949). *Biometrical Genetics.* Dover Publications, New York.

Mather, K. (1973). *Genetical Structure of Populations.* Chapman & Hall, London.

Mather, K. and Jinks, J.L. (1971). *Biometrical Genetics*, Chapman and Hall, London.

Matzinger, D.F. (1953). Comparison of three types of testers for the evaluation of inbred lines of corn. *Agron. J.*, 45, 493-495.

Matzinger, D.F., Sprague, G.F. and Cockerham, C.C. (1959). Diallel crosses of maize in experiments repeated over locations and years. *Agron. Jour.*, 51, 346-350.

Mayo, O. (1980). *The Theory of Plant Breeding.* Clarendon Press, Oxford.

Meshbah, M., Scholten, O.E., de Bock, T.S.M. and Lange, W. (1997). Chromosome localisation of genes for resistance to *Heterodera schachtii, Cercospora beticola* and *Polymyxa betae* using sets of *Beta procumbens* and *B. patellaris* derived monosomic additions in *B. vulgaris. Euphytica*, 97, 117-127.

Metz, P.L. J., Jacobsen E. and Stiekema, W.J. (1997). Occasional loss of expression of phosphinothricin tolerance in sexual offspring of transgenic oilseed rape (*Brassica napus* L.). *Euphytica*, 98, 189-196.

Miedema, P. (1982). The effects of low temperature on *Zea mays. Advances in Agronomy*, 35, 93-128.

Mohr, H.C. (1986). Watermelon breeding. pp. 37-66. In: *Breeding vegetable crops.* M.J. Bassett (ed.). AVI Publishing Co, Inc., Westport, Connecticut.

Muehlbauer, F.J., Slinkard, A.E. and Wilson, V.E. (1980). Lentil, pp. 417-426. In: *Hybridization of Crop Plants*, W.R. Fehr and H.H. Hadley (eds.), ASA and CSSA, Madison, Wisc., USA.

Murashige, T. and Skoog, F. (1962). A revised medium for rapid growth and bio-assays with tobacco tissue cultures, *Physiol. Plant.*, 15, 473-497.

Nelson, R.R. (1975). Horizontal resistance in plants: Concepts, controversies and applications, pp. 1-20. In: *Proc. of the Seminar on horizontal resistance to blast disease of rice*, CIAT Publ. Ser. C.E. -9, Cali, Colombia.

*Nelson, R.S., McCormick, S.M., Delannay, X., Dube, P., Layton, J., Anderson, E.J., Kaniewska, M., Proksch, R.K., Horsch, R.B., Rogers, S.G., Fraley, R.T. and Beachy, R.N. (1988). Virus tolerance, plant growth and field performance of transgenic tomato plants expressing coat protein from tobacco mosaic virus. *Bio/Technology*, 6, 403-409.

Norden, A.J. (1980). Peanut. In: *Hybridization of Crop Plants*, pp. 443-456. W.R. Fehr and H.H. Hadley (eds.). ASA & CSSA, Madison, Wisc., USA.

Old, R.W. and Primrose, S.B. (1989). Principles of Gene Manipulation. *An Introduction to Genetic Engineering.* Blackwell Scientific Publications, Oxford. (4th ed.).

Owen, F.V. (1945). Cytoplasmically inherited male sterility in sugar-beets. *J. Agr. Res.*, 71, 423-440.

Painter, R.H. (1951). Insect resistance in crop plants. Macmillan Co., New York.

Pandey, K.K. (1958). Time of the *S*-allele action. *Nature,* 181, 1220-1221.

Parker, J. (1968). Drought-resistance mechanisms. pp.195-234. In: *Water Deficits and Plant Growth* 1. T.F. Kozlowski (ed.), Academic Press, New York and London.

Parlevliet, J.E. and Zadoks, J.C. (1977). The integrated concept of disease resistance. A new view including horizontal and vertical resistance in plants. *Euphytica*, 26, 55-21.

Parthasarthy, N. and Rajan, S.S. (1953). Studies in the fertility of autotetraploids of *Brassica campestris* var. *toria, Euphytica*, 2, 25-36.

*Paterniani, E. and Vencovsky, R. (1977). Reciprocal recurrent selection in maize (*Zea mays* L.) based on test crosses of half-sib families. *Maydica*, 22, 141-152.

*Paterniani, E. (1967). Interpopulation improvement: Reciprocal recurrent selection variations. Maize. CIMMYT., Mexico.

Pedigo, L.P. (1991). *Entomology and Pest Management.* Macmillan Publishing Company, New York.

Pelletier, G. (1993). Somatic hybridization. In: *Plant Breeding. Principles and Prospects*, M.D. Hayward, N.O. Bosemark and I. Romagosa (eds.). Chapman & Hall, London.

*Pelletier, G., Primard, C., Vedel, F., Chetrit, P., Remy, R., Rousselle, P. and Renard, M. (1983). Intergeneric cytoplasmic hybridization in Cruciferae by protoplast fusion. *Mol. Gen. Genet.*, 191, 244-250.

Pesek, J. and Baker, R.J. (1969). Desired improvement in relation to selection indices. *Canadian Jour. Plant Sci.*, 49, 803-804.

Peterson, C.E. and Simon, P.W. (1986). Carrot breeding. In: *Breeding Vegetable Crops*, pp. 322-356. M.J. Bassett, (ed.). AVI Publishing Company, Inc., Westport, Connecticut, USA.

Plaisted, R.L. (1980). Potato. In: *Hybridization of Crop Plants*, pp. 483-494. W.R. Fehr and H.H. Hadley (eds.). ASA & CSSA, Madison, Wisc., USA.

Poehlman, J.M. and Borthakur, D.N. (1969). *Breeding Asian Field Crops.* Oxford and IBH Publishing Co., New Delhi.

Ponnuswamy, K.N., Das, M.N. and Handoo, M.I. (1974). Combining ability type analysis for triallel crosses in maize (*Zea mays* L.). *Theo. & App. Genet.*, 45, 170-175.

Powers, L. (1951). Gene analysis by partitioning method when interaction of genes are involved. *Bot. Gaz.*, 113, 1-23.

Potrykus, I. (1993). Gene transfer to plants: Approaches and available techniques. In: *Plant Breeding: Principles and Prospects*, pp. 126-137. M.D. Hayward, N.O. Bosemark and I. Romagosa (eds.). Chapman & Hall, London.

Rajan, S.S. and Ahuja, Y.R. (1956). Seed development in colchicine-induced autotetraploids of toria, *Ind. J. Genet & Pl. Breed.*, 16, 63-76.

Rajan, S.S. (1955). The effectiveness of mass-pedigree system of selection in the improvement of seed setting in autotetraploids of toria. *Ind. Jour. Genet. & Pl. Breed.*, 15, 47-49.

Ramachandran, M. and Prasad, M.V.R. (1995). Hybrid castor research and development in India. In: *Hybrid Research and Development*, M. Rai and S. Mauriya (eds.). Indian Society of Seed Technology, New Delhi, pp. 93-104.

Ramage, R.T. (1965). Balanced tertiary trisomics for use in hybrid seed production, *Crop Sci.*, 5, 177-178.

Ramage, R.T. and Tullen, N.A. (1964). Balanced tertiary trisomics in barley serve as a pollen source homogeneous for a recessive lethal gene. *Crop Sci.*, 4, 81-82.

Rawlings, J.O. and Thompson, D.L. (1962a). Performance level as criterion for the choice of maize testers. *Crop Sci.*, 2, 217-220.

Rawlings, J.O. and Cockerham, C.C. (1962). Triallel analysis. *Crop Sci.*, 2, 228-231.

Reinert, J. and Bajaj, Y.P.S. (eds.) (1977). *Applied and Fundamental Aspects of Plant Cell, Tissue and Organ Culture.* Springer-Verlag, Berlin.

Reynolds, J.E., Vrolijk, L.A., Hyche, R., Bias, W., Kwok, P.Y., Nickerson, D.A. and Boyce-Jacino, M.T. (1995). Single nucleotide polymorphisms and human identity testing in diverse ethnic populations. *Am. J. Hum. Genet.*, 57, Suppl. 4, A41.

Rhijn, P.V. and Vanderleyden, J. (1995). The *Rhizobium*—Plant Symbiosis. *Microbiological Reviews*, Mar. 1995, 124-142

Rhoades, M.M. (1950). Meiosis in maize. *Jour. Hered.*, 41, 59-67.

Rick, C.M. (1980). Tomato. In: *Hybridization of Crop Plants*, pp. 669-680, W.R. Fehr and H.H. Hadley (eds.). ASA and CSSA, Wisc., Madison, USA.

*Riley, R., Chapman, V. and Johnson, R. (1968). The incorporation of alien disease-resistance in wheat by genetic interference with the regulation of meiotic chromosome synapsis. *Genet. Res. Camb.*, 12, 199-219.

Riley, R. (1978). Plant Breeding—an integrating technology. In: *Technology for Increasing Food Production*, pp. 267-272. J.C. Holmes and W.M. Tahir (eds.). FAO, Rome.

Robinson, H.F., Comstock, P.E. and Harvey, P.A. (1955). Genetic variances in open-pollinated varieties of corn. *Genetics*, 40, 45-60.

Robinson, R.W., Whitaker, T.W. and Bohn, G.W. (1970). Promotion of pistillate flowering in *Cucurbita* by 2-Chloroethylphosphonic acid. *Euphytica*, 19, 180-182.

Romero, G.E. and Frey, K.G. (1966). Mass selection for plant height in oat populations. *Crop Sci.*, 6, 283-287.

Roy, N.N. (1976). Inter-genotypic plant competition in wheat under single-seed descent breeding. *Euphytica*, 25, 219-223.

Rudich, J., Kedar, N. and Halevy, A.H. (1970) Changed sex expression and possibilities for F_1 hybrid seed production in some cucurbits by application of Ethrel and Alar (B-995). *Euphytica*, 19, 47-53.

Russell, W.A. and Hallauer, A.R. (1980). Corn. In: *Hybridization of Crop Plants*, pp. 299-312. W.R. Fehr and H.H. Hadley (eds.). ASA and CSSA, Madison, Wisc., USA.

Russell, W.A. and Eberhart, S.A. (1975). Hybrid performance of selected maize lines from reciprocal recurrent and test cross selection programs. *Crop Sci.*, 15, 1-4.

Sakai, K. and Utiyamada, H. (1957). Studies on competition in plants, VIII. Chromosome number, hybridity and competitive ability in *Oryza sativa* L. *J. Genet.*, 55, 235-240.

Salamini, F. and Motto, M. (1993). The role of gene technology in plant breeding. In: *Plant Breeding: Principles and Prospects*, pp. 138-159. M.D. Hayward, N.O. Bosemark and I. Romagosa (eds.). Chapman and Hall, London.

*Sautter, C., Waldner A., Neuhausurl, G., Galli, A., Neuhaus, G. and Potrykus, I. (1991). Microtargeting; high efficiency gene transfer using a novel approach for the acceleration of microprojectiles. *Bio/Technology*, 9, 1080-1085.

Savidan, Y.H. (1982). Embryological analysis of facultative apomixis in *Panicum maximum* Jacq. *Crop Sci.*, 22, 467-469.

Schertz, K.F. and Dalton, L.G. (1980). Sorghum. In: *Hybridization of Crop Plants*, pp. 577-588. W.R. Fehr and H.H. Hadley (eds.). ASA and CSSA, Madison, Wisc., USA.

*Schwanitz, F. (1951). Untersuchungen an polyploidien pflanzen XI. Zur chlorophyllgehalt diploden and polyploiden Pflanzen. Züchter, 21, 30-36.

*Schwanitz, F. (1951). Untersuchungen an polyploiden pflanzen XII. Der gigas character der Kultur Pflanzen und siene Bedeutung fur die polyploiden züchtung. Züchter, 21, 65-75.

Sears, E.R. (1956). The transfer of leaf-rust resistance from *Aegilops umbellulata* to wheat. *Brookhaven Symposia in Biology*, 9, 1-22.

*Sears, E.R. (1977). Analysis of wheat-*Agropyron* recombinant chromosomes, pp. 63-72. In: *Interspecific Hybridization in Plant Breeding*, E. Sanchez Monge and F. Garcia Olmedo (eds.), Proc. 8th Eucarpia Congress Madrid.

Shah, D.P., Horsch, R.B., Klee, H.I., Kishore, G.M., Winter, J.A., Tumer, N.E., Hironaka C.M., Sanders, P.R., Gassers, C.S., Aykent, S., Siegel, N.R., Rogers, S.G. and Fraley, R.T. (1986). Engineering herbicide tolerance in transgenic plants. *Science* 233, 478-481.

Sharma, D. and Knott D.R. (1966). The transfer of leaf-rust resistance from *Agropyron* to *triticum* by irradiation. *Can. J. Genet. Cytol.*, 8, 137-143.

Shaw, C.H. (ed.) 1968). *Plant Molecular Biology—A Practical Approach.* IRL Press, Oxford.

Shifriss, O. (1960). Conventional and unconventional systems controlling sex variations in *Ricinus*. *J. Genet.*, 57, 361-388.

*Shull, G.H. (1914). Duplicate genes for Capsule form in *Bursa bursa pastoris*. *Ztschr. F. Induktive Abstam. U. Vererbung slehre*, 12, 97-149.

*Sidorov, V.A. Menczel, L., Nagy, F. and Maliga, P. (1981). Chloroplast transfer in *Nicotiana* based on metabolic complementation between irradiated and iodoacetate-treated protoplasts. *Planta*, 152, 341-345.

Simmonds, N.W. (1979). *Principles of Crop Improvement.* Longman, London.

Singh, D.P. (1986). *Breeding for Resistance to Diseases and Insect Pests.* Springer-Verlag, Heidelberg.

Smith, C.E. (1969). From Vavilov to the present—a review. *Economic Botany*, (23)1, 2-19.

Smith, D.C. (1966). Plant Breeding—Development and success, pp. 3-54. In: *Plant Breeding*, K.J. Frey (ed.), Iowa State Univ. Press, Ames, Ia.

Sneep, J. (1966). Some facts about plant breeding before the discovery of Mendelism. *Euphytica*, 15, 135-140.

Sneep, J., Murty, B.R. and Utz, H.F. (1979). Current breeding methods. In: *Plant Breeding Perspectives*, pp. 104-223. J. Sneep and A.J.T. Hendriksen (eds.). Centre for Agricultural Publishing and Documentation, Wageningen.

Sprague, G.F. (1966). Quantitative genetics in plant improvement, .pp. 315-354. In Frey, K.J. (ed.). Plant Breeding, Iowa State Univ. Press, Ames, Ia.

Sprague, G.F. and Eberhart, S.A. (1977). *Corn breeding.* pp. 305-362. In: Sprague, G.F. (ed.). *Corn and Corn Improvement.* American Society of Agronomy, Wisc., USA.

Stadler, L.J. (1944). Gamete selection in corn breeding. *Jour. Am. Soc. Agron.*, 36, 988-89.

Staples, R.C. and Toennissen, G.H. (eds.) (1984). *Salinity Tolerance in Plants—Strategies for Crop Improvement.* John Wiley, New York.

Starling, T.M. (1980). Barley, pp. 189-202, In: *Hybridization of Crop Plants*, W.R., Fehr and H.H. Hadley (eds.), ASA and CSSA, Madison, Wisc., USA.

Stebbins, G.L. (1950). *Variation and Evolution in Plants.* Colombia University Press, New York.

Stebbins, G.L., Jr. (1956). Artificial polyploidy as a tool in plant breeding. *Brookhaven Symposia in Biology*, 9, 37-52.

384

Steponkus, P.L. (1978). Cold hardiness and freezing injury of agronomic crops. *Advances in Agronomy*, 30, 51-98.

*Straub J. (1946). *Zur Entwicklungs Physiologie des selbsterilitat von petunia. Z. Natur forschg*, 1, 287.

*Straub, J. (1947). *Zur Ent wicklungs physiologie der Selbsterilitat von petunia II. Das Prinzipde Hemm—mechanisms. Z. Naturforschg*, 2, 433.

Stuber, C.W., Williams, W.P. and Moll, R.H. (1973). Epistasis in maize (*Zea mays* L.) III: Significance in predictions of hybrid performances. *Crop Sci.*, 13, 195-200.

Suneson, C.A. (1956). An evolutionary plant breeding method. *Agron. J.*, 48, 188-191.

Suneson, C.A. and Stevens, H. (1953). Studies with bulked hybrid populations of barley. *USDA Tech. Bull. 1067.* 14 pp.

Swaminatham, M.S. (1981). Plant breeding and agricultural transformation. *Seeds & Farms*, 7:3, 3-9. (Published by N.S.C., New Delhi).

Swaminathan, M.S. (1955). Overcoming cross-incompatibility among some Mexican diploid species of *Solanum. Nature*, 176, 887-888.

Swanson, C.P. (1957). *Cytology and Cytogenetics.* Prentice-Hall, New Jersey.

Taliaferro, C.M. and Bashaw, E.S. (1960). Inheritance and control of obligate apomixis in breeding buffel grass. *Crop Sci.*, 6, 473-476.

Tallio, G.M. (1962). A selection index for optimum genotype. *Biometrics*, 18, 120-122.

Tanno-Suenaga, L., Ichikawa, H. and Imammura, J. (1988). Transfer of the CMS trait in *Daucus carota* L. by donor-recipient protoplast fusion. *Theor. App. Genet.*, 76, 855-860.

Tee, T.S. and Quaslet, C.O. (1975). Bulk populations in wheat breeding. Comparison of single-seed-descent and random bulk methods. *Euphytica*, 24, 393-405.

Thoday, J.M. (1972). Disruptive selection. *Proc. Royal Soc. Lond.*, B, 182, 109-143.

Thomas, H. (1993). Chromosome manipulation and polyploidy, pp. 79-92. In: *Plant Breeding - Principles and Prospects.* M.D. Hayward, N.O. Bosemark and I. Romagosa (eds), Chapman & Hall, London.

Thore, D. (1963). The function of self-incompatibility alleles in red clover (*Trifolium pratense* L.). IV. Résumé. *Hereditas,* 49, 330-334.

Tigchelaar, E.C. (1986). Tomato breeding. In: *Breeding Vegetable Crops*, pp. 135-171. M.J. Bassett (ed.). AVI Publishing Company, Inc., Westport, Connecticut, USA.

Tsunoda, S. (1959*a*). A developmental analysis of yielding ability in varieties of field crops I. Leaf area per plant and leaf area ratio. *Jap. J. Breed.*, 9, 161-168.

Tsunoda, S. (1959*b*). A developmental analysis of yielding ability in varieties of field crops II. The assimilation-system of plants as affected by the form, direction and arrangement of single leaves. *Jap. J. Breed.*, 9, 237-244.

Tsunoda, S. (1960). A developmental analysis of yielding ability in varieties of field crops III. The depth of green colour and nitrogen content of leaves. *Jap. J. Breed.*, 10, 107-111.

Tsunoda, S. (1962). A developmental analysis of yielding ability in varieties of field crops IV. Quantitative and spatial development of the stem-system. *Jap. J. Breed.*, 12, 49-55.

*Tumer, N.E., O'Connell, K.M., Nelson, R.S., Saunders, P.R., Beachy, R.N., Fraley, R.T. and Shah, D.M. (1987). Expression of alfalfa mosaic virus coat protein gene confers cross-protection in transgenic tobacco and tomato plants. EMBOJ. 6, 1181-1188.

Tunistra, M.R., Grote, E.M., Goldsbrough, P.B. and Ejeta, G. (1997). Genetic analysis of post-flowering drought tolerance and components of grain development in *Sorghum bicolor* (L.) Moench. *Molecular Breeding*, 3, 439-448.

Tysdal, H.M. and Crandall, B.H. (1948). The polycross progeny performance as an index of the combining ability of alfalfa clones. *Jour. Amer. Soc. Agron.*, 40, 293-306.

Upreti., G.C. (1977). Development of virus tested nucleus stock. In: *Recent Technology in Potato Improvement and Production*. CPRI, Shimla, India.

UPOV (1996). International convention for the protection of new varieties of plants as of December 2, 1961: revised at Geneva on November 10, 1972; on October 23, 1978 and on March 19, 1991.

Vaeck, M., Reynaerts, A., Höfte, H., Jansens, S., De Bluckelers M., Dean, C., Zabeau, M., Van Montagu, M. and Leemans, J. (1987). Transgenic plants protected from insect attack, *Nature*, 328, 33-37.

Vanderplank, J.E. (1963). *Plant Diseases: Epidemics and Control*, 349 pp., Academic Press, New York & London.

Vanderplank, J.E. (1968). *Disease Resistance in Plants*, Academic Press, New York.

Vanderplank, J.E. (1975). *Principles of Plant Infection*, 216 pp., Academic Press, New York, San Francisco & London.

Vanderplank, J.E. (1982). *Host-pathogen Interactions in Plant Disease*. Academic Press, New York & London.

Vavilov, N.I. (1951). *The Origin, Variation, Immunity and Breeding of Cultivated Plants* (Translated from Russian: K.S. Chester). The Ronald Press Company, New York.

Walejko, R.N. and Russell, W.A. (1977), Evaluation of recurrent selection for specific combining ability in two open pollinated maize cultivars. *Crop Sci.*, 17, 647-651.

Walker, J.C. (1953). Disease resistance in vegetable crops. *Bot. Rev.*, 19, 606-643.

*Watrud, L.S., Perlak, F.J., Tran, M.T., Kusano, K., Mayer, E.J., Miller-Wideman, M.A., Obukowicz, M.G., Nelson, D.R., Kreitinger, J.P. and Kaufman, R.J. (1985). Cloning of *Bacillus thuringiensis* sub sp. *Kurstaki* delta endotoxin gene into *Pseudomonas fluorescens*: Molecular biology and ecology of an engineered microbial pesticide, pp. 40-46. In: *Engineered Organisms in the Environment*, H.O. Halvarson, D. Pramer and M. Rogul (eds.). American Society for Microbiology, Washington.

Wenzel, G., Hoffman, A. and Thomas, E. (1979). Comparison of single cell culture derived from *Solanum tuberosum* plant and a model for their application in breeding programmes. *Theor. & App. Genet.*, 55, 49-55.

Wery, J. Turc, O. and Lecoeur, J. (1993). Mechanisms of resistance to cold, heat and drought in cold season food legumes, with special reference to chickpea and pea. In: *Breeding for Stress Tolerance in Cool-season Food Legumes*, pp. 271-291. K.B. Singh and M.C. Saxena (eds.). John Wiley, Chichester, U.K.

Whitaker, T.W. and Robinson, R.W. (1986). Squash breeding. In: *Breeding Vegetable Crops*, pp. 210-242. M.J. Bassett (ed.). AVI Publishing Company, Inc., Westport, Connecticut, USA

Whyte, R.O. (1958). *Plant Exploration, Collection and Introduction*. FAO Agricultural Studies 41. Rome.

*Wienhues, A. (1965). *Cytogenetische untersuchungen uber die Chromosomale Grundlage der Rostresistenz der weizenroste Waique*. Züchter, 35, 352-354.

Williams, J.S. (1962). The evaluation of selection index. *Biometrics*, 18, 375-393.

Williams, J.C., Penny, C.H. and Sprague, G.F. (1965). Full-sib and half-sib estimate of genetic variance in an open pollinated variety of corn. *Crop Sci.*, 5, 125-129.

Xoconostle-Cazares, B., Lozoya-Gloria, E. and Herrera-Estrella, L. (1993). Gene cloning and identification. In: *Plant Breeding—Principles and Prospects*, pp. 107-125. M.D. Hayward, N.O. Bosemark and T. Romagosa (eds.). Chapman and Hall, London.

Yang, Z.Q., Shikanai, T., Mori, K. and Yamada, Y. (1989). Plant regeneration from Cytoplasmic hybrids of rice (*Oryza sativa* L.). *Theor. & Appl. Genet.*, 77, 305-310.

Yang, Z.Q., Shikanai. T. and Yamada, Y. (1988). Assymetric hybridization between cytoplasmic male sterile (CMS) and fertile rice (*Oryza sativa* L.) protoplasts. *Theor. & Appl. Genet.*, 76, 801-808.

*Zelcer, A., Aviv, D. and Galun, E. (1978). Interspecific transfer of cytoplasmic male sterility by fusion between protoplasts of normal *Nicotiana sylvestris* and X-ray irradiated protoplasts of male sterile *N. tabacum*. *Z. Pflanzen physiol.*, 90, 397-407.

*Zenktler, M. and Slusarkiewicz-Jarina, A. (1986). Sexual reproduction in plants by applying the method of test-tube fertilization of ovules. pp. 415-423. In: *Genetic Manipulation in Plant Breeding*, W. Horn, C.J. Jensen, W. Odenbach and O. Schieder (eds.). Walter de Gruyter, Berlin.

Zhang, Z.H. (1989). The practicability of anther culture breeding in rice. pp. 31-42. In: *Review of Advances in Plant Biotechnology*, 1985-88, A. Mujeeb-Kazi and L.A. Stich (eds.), CIMMYT & IRRI.

Zhukovsky, P.M. (1968). World genofund of plants for breeding. Macrocentres and microcentres. In: *The Life and Work of N.I. Vavilov*. E.E. Leppik (Author) (1969). *Economic Botany*, 23 (2), 128-132.

Zohary, D. (1973). Gene-pools for plant breeding. In: *Agricultural Genetics*, 177-183. R. Moav (ed.) John Wiley and Sons, New York.

---

* Original not seen.

Index